The Standard Model

The Standard Model

FROM FUNDAMENTAL SYMMETRIES
TO EXPERIMENTAL TESTS

YUVAL GROSSMAN AND
YOSSI NIR

PRINCETON UNIVERSITY PRESS
PRINCETON AND OXFORD

Published by Princeton University Press
41 William Street, Princeton, New Jersey 08540
99 Banbury Road, Oxford OX2 6JX

press.princeton.edu

Library of Congress Cataloging-in-Publication Data

Names: Grossman, Yuval, 1964– author. | Nir, Yosef, 1954– author.
Title: The standard model : from fundamental symmetries to experimental tests / Yuval Grossman and Yossi Nir.
Description: Princeton : Princeton University Press, [2023] | Includes bibliographical references and index.
Identifiers: LCCN 2022060789 (print) | LCCN 2022060790 (ebook) | ISBN 9780691239101 (hardback) | ISBN 9780691239118 (ebook)
Subjects: LCSH: Standard model (Nuclear physics) | BISAC: SCIENCE / Physics / General | SCIENCE / Physics / Nuclear
Classification: LCC QC794.6.S75 N57 2023 (print) | LCC QC794.6.S75 (ebook) | DDC 539.7/2—dc23/eng/20230609
LC record available at https://lccn.loc.gov/2022060789
LC ebook record available at https://lccn.loc.gov/2022060790

British Library Cataloging-in-Publication Data is available

Editorial: Abigail Johnson
Jacket: Wanda España
Production: Lauren Reese
Publicity: William Pagdatoon
Copyeditor: Susan McClung

Jacket Credit: Noa Geler-David, Orit Golan | Design, Photography & printing branch | Weizmann Institute of Science

This book has been composed in Arno Pro

10 9 8 7 6 5 4 3 2 1

CONTENTS

PREFACE

This book aims at teaching modern particle physics. The goal of particle physics is to understand what are the fundamental laws of nature. These are grand, yet true, words.

Particle physics is also known as high-energy physics. Particle physics experiments have probed energy scales as high as 10 TeV (which is 10,000 times the mass of the proton) and distance scales as small as 10^{-20} m. We find it amazing that by now, we have achieved a deep understanding of how nature works down to such distances. The framework that we use to describe the phenomena at this distance scale is based on quantum field theory (QFT), which is different from quantum mechanics (QM), which we use to deal with atomic physics; and which in turn is different from classical mechanics, which we apply to explain most of the macroscopic world. While these frameworks are very different from each other, the underlying principles of physics are surprisingly similar across all these scales. Specifically, in all of them, we use the principle of minimal action and symmetry arguments to construct the theory.

It is often stated that the aim of particle physics is to describe the elementary particles and the fundamental interactions among them. While this is true, particle physicists aim higher. In some sense, the particles and their interactions constitute a tool for us to obtain insights into deeper principles that describe nature. Our current understanding of the basic laws of nature is based on very elegant symmetry principles. Once we know the symmetries of the universe and how the fundamental fields respect them, much of nature is explained. In a way, the deeper we understand nature, the simpler and more abstract are the basic principles that we use to formulate this understanding.

The currently accepted theory of particle physics is called the Standard Model. The basic ingredients of the Standard Model were conceived in the late 1960s and early 1970s by Sheldon Lee Glashow, Abdus Salam, and Steven Weinberg. At that time, many of the particles that now constitute part of it were yet to be discovered. By 2012, the full list of the Standard Model particles have been directly produced and detected, and the full list of the Standard Model parameters have been measured with impressive accuracy. The Standard Model has been tested by numerous experimental measurements, and it has passed almost all of them with flying colors. The very few failures constitute the starting point for the road to an even deeper level of understanding nature.

It is customary to say that there are four forces in nature: gravitational, electromagnetic, weak, and strong forces. The Standard Model does not deal with gravity, but it does include the three other forces. As you read the book, it becomes clear that while these three forces seem very different from each other, they all arise at the fundamental level from a QFT incorporating gauge symmetries. We can think of this as a generalization of Quantum Electro Dynamics (QED) that leads to very different phenomenology. It is this unification of the underlying principle—gauge symmetries—that makes the theory so elegant.

Significant breakthroughs in physics were often achieved when realizing that phenomena that seem very different are, in fact, connected. One example is the understanding that the movement of the planets around the sun and the falling of an apple from a tree are both explained by the same law of gravity. Another example is the realization that electricity and magnetism are two manifestations of the unified electromagnetic theory. Particle physics keeps in the same path and provides a unified description of phenomena that seem very different from each other.

The way that the Standard Model has been developed was based on data that became available during the process. It took many years to reach the present status of the theory because it took time to develop new experiments and collect all the data. In this book, however, we do not follow a historical approach; rather, we describe things as we understand them today. We only briefly mention historical facts as we go. Given that all the particles of the Standard Model have been discovered and all the parameters measured, we can explain the Standard Model in a comprehensive and pedagogical way that captures the main points.

Most of the data that are relevant for particle physics came from collider experiments. In this book, we focus on the theory side and do not elaborate on the way that experiments are carried out. The various colliders differ in energy and luminosity, and in the particles that collide. It is the synergy of all of them that led to the construction of the Standard Model. At the time of writing this book, the highest energy accelerator is the Large Hadron Collider (LHC), which collides protons at a center-of-mass energy of 13 TeV. It is the energy of the LHC that sets the upper bound on the mass scale of particles that can be probed directly by being produced on-shell. High-luminosity machines help us search for quantum effects (i.e., loop effects that probe indirectly physics at even higher scales).

The data that are relevant to particle physics is collected by the Particle Data Group (PDG) in the Review of Particle Physics [1] (we loosely refer to this collection of data as the "PDG"). When we quote experimental data, they come from the PDG unless we explicitly state otherwise. For the untrained eye, the PDG may look like an (old-fashioned) phone book. There are numerous tables with data on decay rates and other properties of particles. Browsing through this data, you see large variations: some particles are stable, others have long lifetimes, and yet others have very short lifetimes before they decay. Some particles have masses that are almost a million times heavier than the electron, and others are more than a million times lighter. The decay products of various particles are very different, as are the corresponding branching ratios.

The task of particle physicists is to identify the patterns in these raw data and to organize them in such a way that the principles that explain this variety emerge. The language that we use to do this is QFT, and the main tool within this framework is the Lagrangian (which depends on fundamental fields and their couplings). Once organized in this way, we can extract the symmetry principles that dictate the Lagrangian and explain the rich phenomena observed.

In writing this book, we intended that it would serve as a textbook for a one-semester course. Thus, we omit many details and focus mainly on the phenomenological aspects. This means that we will be skipping a few interesting and relevant topics. We hope that our book will motivate you to explore these on your own.

The book, although very much self-contained, is written assuming preknowledge of basic QFT. It is aimed at students who have already taken a one-semester course in QFT and have an understanding of the concept, but may be unfamiliar with advanced topics. To gauge this statement, students familiar with the first part of Peskin and Schroeder [2] have the preknowledge necessary to follow our book. Some other books that we find useful and complementary to ours are Georgi [3], Quigg [4], Peskin [5], Burgess and Moore [6], Langacker [7], Ramond [8], Cottingham and Greenwood [9], Donoghue, Golowich, and Holstein [10], Goldberg [11], and Buras [12].

At the end of a course that follows this book, the student should gain knowledge and understanding in two areas:

- The Standard Model: the symmetry principles that define it, the fundamental interactions and elementary particles that it describes, the ways in which it has been tested, its many successes, and its very few failures.
- The principles of model building in particle physics: the tools that are used to interpret new experimental results and, in particular, to extend the Standard Model if future measurements cannot be explained by it.

Actually, in our minds, there is a third goal for this book as well. We think that the Standard Model is a scientific masterpiece, beautiful and elegant, and we hope to convey this sense of appreciation and intellectual joy to our readers.

1

Lagrangians

In this chapter, we review the basic tools that we will use in this book. In particular, we introduce the Lagrangian and present some simple Lagrangians involving scalar and fermion fields.

1.1 Introduction

Modern physics encodes the basic laws of nature in the action, S, and postulates the principle of minimal action in its quantum interpretation. In quantum field theory (QFT), the action is an integral over spacetime of the *Lagrangian density* or Lagrangian, $\mathcal{L}$, for short. For most of our purposes, it is enough to consider the Lagrangian rather than the action. In this chapter, we explain how Lagrangians are constructed. Later in the book, we discuss how the numerical values of the parameters that appear in the Lagrangian are determined and how to test if a Lagrangian provides a viable description of nature.

The QFT equivalent of the generalized coordinates of classical mechanics are *fields*. The action is given by

$$S = \int d^4x \, \mathcal{L}, \tag{1.1}$$

where $d^4x = dx^0 dx^1 dx^2 dx^3$ is the integration measure in four-dimensional Minkowski space. In general, we require the following properties for the Lagrangian:

1. It is a function of the fields and their derivatives only.
2. It depends on the fields taken at one spacetime point x^μ only, leading to a local field theory.
3. It is real, so the total probability is conserved.
4. It is invariant under the Poincaré group, which consists of spacetime translations and Lorentz transformations.
5. It is an analytic function in the fields. This is not a general requirement, but it is common to all field theories that are solved via perturbation theory. In these cases, we expand around a minimum, which means that we consider a Lagrangian that is a polynomial in the fields.

6. It is invariant under certain internal symmetry groups. The invariance of S (or $\mathcal{L}$) is in correspondence with conserved quantities and reflects basic symmetries of the physical system.
7. Every term in the Lagrangian that is not forbidden by a symmetry should appear.

We often impose an additional requirement as well:

8. Renormalizability. A renormalizable Lagrangian contains only terms that have a dimension less than or equal to four in the fields and their derivatives.

The renormalizability requirement ensures that the Lagrangian contains at most two ∂_μ operations, and it leads to classical equations of motion that are no higher than second-order derivatives. If the full theory of nature is described by a QFT, its Lagrangian should indeed be renormalizable. The theories that we consider, however, and, in particular, the Standard Model, are only low-energy-effective theories, that are valid up to some energy scale Λ. Therefore, we also must include nonrenormalizable terms, which have coefficients with inverse mass dimensions, $1/\Lambda^n$, $n=1,2,\ldots$ For most purposes, however, renormalizable terms constitute the leading terms in an expansion in E/Λ, where E is the energy scale of the physical processes under study. Therefore, the renormalizable part of the Lagrangian is a good starting point for our study. Thus, in chapters 1–10, we consider only renormalizable Lagrangians unless otherwise explicitly stated. In chapters 11–15, where we describe searches for physics beyond the Standard Model, we also consider nonrenormalizable Lagrangians.

Properties 1–5 are not the subject of this book. You should be familiar with them from your QFT course work. We do, however, deal intensively with the other requirements. Actually, the most important message that we would like to convey is the following: *(Almost) all experimental data for elementary particles and their interactions can be explained by the Standard Model of a spontaneously broken $SU(3) \times SU(2) \times U(1)$ gauge symmetry.*[1]

Writing down a specific Lagrangian is the end point of the process known as *model building*, and the starting point for a phenomenological interpretation and experimental testing. In this book, we explain both aspects of this modern way of understanding high-energy physics.

1.2 Examples of Simple Lagrangians

We next present a few examples of simple Lagrangians of scalar and fermion fields. They are simple in the sense that we are not yet imposing any internal symmetry. We use $\phi(x)$ for a scalar field and $\psi(x)$ for a fermion field. When we consider vector fields, as first done in section 2.2 of chapter 2, we use $A(x)$ for a vector field. We do not consider higher spin fields, as it is not simple to construct a QFT with them.

1. Actually, the great hope of the high-energy physics community is to prove this statement wrong and find an even more fundamental theory.

Two comments are in order:

- All fields that we consider here are functions of the spacetime coordinates $\phi(x)$, $\psi(x)$, and $A(x)$. We leave this spacetime dependence implicit except in cases where it is relevant.
- We use the notations ϕ, ψ, and A in the discussion of generic cases. When we refer to specific cases, we use different notation. For example, for the electron field, we use e instead of the generic ψ.

1.2.1 Scalars

The most general renormalizable Lagrangian for a single real scalar field ϕ is given by

$$\mathcal{L}_S = \frac{1}{2}(\partial_\mu \phi)(\partial^\mu \phi) - \frac{m^2}{2}\phi^2 - \frac{\eta}{2\sqrt{2}}\phi^3 - \frac{\lambda}{4}\phi^4. \tag{1.2}$$

We emphasize the following points:

- The term with derivatives is called the *kinetic term*. It is necessary if we want ϕ to be a dynamical field (namely, to be able to describe propagation in spacetime).
- The terms without derivatives are collectively denoted by $-V_\phi$. We then write $\mathcal{L}_S = \frac{1}{2}(\partial_\mu \phi)(\partial^\mu \phi) - V_\phi$, and V_ϕ is called the *scalar potential.*
- We work in the canonically normalized basis where the coefficient of the kinetic term is $1/2$. (This is true for a real scalar field. For a complex scalar field, the canonically normalized coefficient of the kinetic term is 1.)
- From here on, throughout the book, when we say "the most general Lagrangian," we are referring to a Lagrangian where the kinetic terms are canonically normalized, but the other terms are written in a general basis. (Question 2.8 in chapter 2 shows that there is no loss of generality in working in the canonically normalized basis.)
- We do not write a constant term since it does not enter the equation of motion for ϕ.
- We do not write a linear term in ϕ because when expanding around a minimum, the linear term vanishes.
- The quadratic term (ϕ^2) is a mass-squared term. (From here on we call it simply a *mass term.*)
- The trilinear (ϕ^3) and quartic (ϕ^4) terms describe interactions.
- Terms with five or more scalar fields (ϕ^n, $n \geq 5$) are nonrenormalizable.

1.2.2 Fermions

The basic fermion fields are two-component Weyl fermions, ψ_L and ψ_R, where L and R denote left-handed and right-handed chirality, respectively. Each of ψ_L and ψ_R has

2 degrees of freedom (DoF) and is a complex field. ψ_L and ψ_R are related to the four-component Dirac field ψ via

$$\psi_R = P_R\psi \equiv \frac{1+\gamma_5}{2}\psi, \qquad \psi_L = P_L\psi \equiv \frac{1-\gamma_5}{2}\psi. \tag{1.3}$$

It is useful to define the related left-handed Weyl fermion ψ_R^c and right-handed Weyl fermion ψ_L^c via

$$\psi_R^c = C\overline{\psi_R}^T, \qquad \psi_L^c = C\overline{\psi_L}^T, \tag{1.4}$$

where C is the *charge conjugation matrix.* (The reason for this name becomes clear once we define *charge* in chapter 2.)

The most general renormalizable Lagrangian for a single left-handed fermion field ψ_L and a single right-handed fermion field ψ_R is given by

$$\mathcal{L}_F = i\overline{\psi_L}\partial\!\!\!/\psi_L + i\overline{\psi_R}\partial\!\!\!/\psi_R - \left(\frac{m_{MR}}{2}\overline{\psi_R^c}\,\psi_R + \frac{m_{ML}}{2}\overline{\psi_L^c}\,\psi_L + m_D\overline{\psi_L}\psi_R + \text{h.c.}\right). \tag{1.5}$$

We emphasize the following points:

- The derivative terms are kinetic terms, and they are necessary if we want the field $\psi_{L,R}$ to be dynamical.
- We work in the canonically normalized basis, where the coefficient of the kinetic term is 1.
- Terms with an odd number of fermion fields violate Lorentz symmetry, and so they are forbidden.
- The quadratic terms are mass terms. The m_M terms are called *Majorana masses,* and the m_D terms are called *Dirac masses.*
- The relative factor of 1/2 between Majorana and Dirac mass terms is the analog of the similar factor between the mass terms for real and complex scalar fields.
- Terms with four or more fermion fields are nonrenormalizable.
- Given the fact that Majorana mass terms are made of a pair of identical fields, we often write

$$\frac{m_{MR}}{2}\overline{\psi_R^c}\,\psi_R \to \frac{m_{MR}}{2}\psi_R^T\,\psi_R. \tag{1.6}$$

If the Majorana masses vanish, $m_{ML} = m_{MR} = 0$, $\mathcal{L}_F$ can be written in terms of the Dirac fermion field ψ:

$$\mathcal{L}_F(m_M = 0) = i\overline{\psi}\partial\!\!\!/\psi - m_D\overline{\psi}\psi. \tag{1.7}$$

Since ψ_L and ψ_R are different fields, there are 4 DoF with the same mass, m_D. In contrast, if the Majorana masses do not vanish, there are generally only 2 DoF that have the same mass. In section 2.1.5 in chapter 2, we discuss these issues in more detail and explain why often Majorana masses vanish.

1.2.3 Fermions and Scalars

Consider the case of a single left-handed fermion ψ_L, a single right-handed fermion ψ_R, and a single real scalar field ϕ. The Lagrangian includes, in addition to terms that involve only the scalar (equation (1.2)), and terms that involve only the fermions (equation (1.5)), terms that involve both the scalar and the fermions. They can be obtained by replacing the mass parameters for the fermions with a coupling multiplied by the scalar field:

$$-\mathcal{L}_{\text{Yuk}} = \frac{Y}{\sqrt{2}}\phi\overline{\psi_L}\psi_R + \frac{Y_{MR}}{2}\phi\overline{\psi_R^c}\,\psi_R + \frac{Y_{ML}}{2}\phi\overline{\psi_L^c}\,\psi_L + \text{h.c.} \tag{1.8}$$

These terms are called *Yukawa interactions*. The Y parameters are dimensionless and are called *Yukawa couplings*. Note that in equation (1.8), we use $-\mathcal{L}$, which is a common practice when we do not write the kinetic terms.

1.3 Symmetries

We always seek deeper reasons for the laws of nature that have been discovered. These reasons are often closely related to symmetries. The term *symmetry* refers to an invariance of the equations that describe a physical system. The fact that symmetry and invariance are related concepts is obvious enough—a smooth ball has a spherical symmetry, and its appearance is invariant under rotation.

Symmetries are built into physics as invariance properties of the Lagrangian. If we construct our theories to encode various empirical facts (and, in particular, the observed conservation laws), then the equations turn out to exhibit certain invariance properties. For example, if we want to incorporate energy conservation into the theory, then the Lagrangian must be invariant under time translations (and therefore cannot depend explicitly on time). From this point of view, the conservation law is the input and the symmetry is the output.

Conversely, if we take the symmetries to be the fundamental rules, then various observed features of particles and their interactions are a necessary consequence of the symmetry principle. In this sense, symmetries provide an explanation of these features. In modern particle physics (and in particular in this book), we often take the latter point of view, in which symmetries are the input and conservation laws are the output.

In the following, we discuss the consequences of *imposing* symmetry on a Lagrangian. This is the starting point of model building in particle physics: one defines the basic symmetries and the field content and then obtains the predictions that follow from these imposed symmetries.

There are symmetries that are not imposed, however, which are called *accidental symmetries*. They are outputs of the theory rather than external constraints. Accidental symmetries arise because we truncate our Lagrangian. In particular, the renormalizable terms in the Lagrangian often have accidental symmetries that are broken by non-renormalizable terms. Since we study mostly renormalizable Lagrangians, we will often encounter accidental symmetries.

There are various types of symmetries. First, we distinguish between *spacetime* and *internal* symmetries. Spacetime symmetries include the Poincaré group of translations, rotations, and boosts. They give the energy-momentum and angular momentum conservation laws. As mentioned previously, we always impose this symmetry. The list of possible spacetime symmetries includes, in addition, space inversion (also called *parity*) P, time-reversal T, and charge conjugation C. (While C is not truly a spacetime symmetry, the way that it acts on fermions and the CPT theorem make it simpler to include C in the same class of operators.) The discrete spacetime symmetries are usually covered in QFT courses, but for completeness, we discuss them briefly in Appendix 1.A.

Internal symmetries act on the fields, not directly on spacetime. In other words, they act in internal spaces that are mathematical spaces generated by the fields. These are the kind of symmetries that we discuss in detail. In chapter 2, we introduce Abelian symmetries; in chapter 4, we introduce non-Abelian symmetries.

1.4 Model Building

As stated already, writing a Lagrangian is the end point of model building. Our procedure of constructing Lagrangians goes as follows. We start by defining the following inputs:

1. The symmetry.
2. The transformation properties of the various scalar and fermion fields under the symmetry operation.

Then we write the most general Lagrangian that depends on the fields and is invariant under the symmetry.

A renormalizable Lagrangian (or a nonrenormalizable one truncated at a certain order) has a finite number of parameters. For a theory with N parameters, we need to perform N appropriate measurements such that additional measurements, from the $(N+1)$'th on, test the theory. In principle, we do not really need to determine the values of the parameters, we can just use experimental inputs to make predictions. In practice, however, it is usually convenient to use the N measurements to determine the values of the Lagrangian parameters and then use these parameters to make further predictions. It is important to remember that the values of the parameters are not inputs to model building.

At this point, this procedure may seem abstract, but it becomes clear and concrete as we work on examples. Throughout this book, we repeat the process of model building several times. We see how Quantum ElectroDynamics (QED), the theory of electromagnetic interactions, Quantum ChromoDynamics (QCD), the theory of strong interactions, the Leptonic Standard Model (LSM), the theory of electroweak interactions among leptons, and the Standard Model itself can be understood in this way of thinking, starting from a postulate of symmetry principles.

Appendix

1.A Discrete Spacetime Symmetries: *C*, *P*, and *T*

The discrete spacetime symmetries, *C*, *P*, and *T*, play an important role in our understanding of nature. Each of these three symmetries has been experimentally shown to be violated in nature, as discussed in detail next. The *CPT* combination seems, however, to be an exact symmetry of nature. On the experimental side, no sign of *CPT* violation has been observed. On the theoretical side, *CPT* must be conserved for any Lorentz-invariant local field theory. Since we only consider such theories, we assume that *CPT* holds. In this case, *CP* and *T* are equivalent. Thus, we usually refer to *CP*.

1.A.1 *C* and *P*

We consider *C* and *P* only in theories that involve fermions. Under *C*, particles and antiparticles are interchanged by conjugating all internal quantum numbers (e.g., reversing the sign of the electromagnetic charge, $Q \to -Q$). Under *P*, the handedness of space is reversed ($\vec{x} \to -\vec{x}$), and the chirality of fermion fields is reversed ($\psi_L \leftrightarrow \psi_R$). For example, a left-handed (LH) electron e_L^- transforms under *C* into an left-handed positron e_L^+, and under *P* into a right-handed (RH) electron e_R^-.

1.A.2 *CP* Violation and Complex Couplings

The *CP* transformation combines charge conjugation *C* with parity *P*. For example, a left-handed electron e_L^- transforms under *CP* into a right-handed positron, e_R^+. *CP* is a good symmetry if there is a basis where all the parameters of the Lagrangian are real. We do not prove it here, but we do provide a simple, intuitive explanation of this statement.

Consider a theory with a single complex scalar, ϕ, and two sets of *N* fermions, ψ_L^i and ψ_R^i $(i = 1, 2, \ldots, N)$ (we define a complex scalar in chapter 2). The Yukawa interactions are given by

$$-\mathcal{L}_{\text{Yuk}} = Y_{ij}\overline{\psi_{Li}}\phi\psi_{Rj} + Y_{ij}^*\overline{\psi_{Rj}}\phi^\dagger\psi_{Li}, \tag{1.9}$$

where we write the two Hermitian conjugate terms explicitly. The *CP* transformation of the fields is defined as follows:

$$\phi \to \phi^\dagger, \qquad \psi_{Li} \to \overline{\psi_{Li}}, \qquad \psi_{Ri} \to \overline{\psi_{Ri}}. \tag{1.10}$$

Therefore, a *CP* transformation exchanges the operators

$$\overline{\psi_{Li}}\phi\psi_{Rj} \longleftrightarrow \overline{\psi_{Rj}}\phi^\dagger\psi_{Li}, \tag{1.11}$$

but leaves their coefficients, Y_{ij} and Y_{ij}^*, unchanged. This means that *CP* is a symmetry of $\mathcal{L}$ if $Y_{ij} = Y_{ij}^*$.

In practice, things are more subtle since one can define the *CP* transformation in a more general way than equation (1.10), as follows

$$\phi \to e^{i\theta}\phi^\dagger, \qquad \psi_L^i \to e^{i\theta_{Li}}\overline{\psi_L^i}, \qquad \psi_R^i \to e^{i\theta_{Ri}}\overline{\psi_R^i}, \tag{1.12}$$

with $\theta, \theta_{Li}, \theta_{Ri}$ convention-dependent phases. Then, there can be complex couplings, and yet *CP* would be an exact symmetry. The correct statement is that *CP* is violated if, using the freedom to redefine the phases of the fields, one cannot find any basis where all couplings are real.

For Further Reading

There are many books that discuss in detail the QFT-related aspects relevant to this book. For example, some of the standard textbooks are by Peskin and Schroeder [2], Zee [13], Srednicki [14], and Schwartz [15]. Other textbooks that explain many of the relevant issues include Ramond [16], Dine [17], Nagashima [18, 19], and Petrov and Blechman [20].

With regard to some specific points, we mention the following sources:

- For a formal discussion of *C* and *P*, see section 3.6 of Peskin and Schroeder [2], or sections 11.4–11.6 of Schwartz [15].
- For a discussion of the issues about quantizing theories with higher-spin fields, see Peskin [21].
- For a discussion of Majorana fermions, see section 11.3 of Schwartz [15].
- For the *CPT* theorem, see Streater and Wightman [22].

Problems

Question 1.1: Algebra

1. Draw the Feynman diagrams for the interaction terms in the Lagrangian of equation (1.2).
2. Starting from equation (1.5) and using equation (1.3), derive equation (1.7).
3. Draw the Feynman diagrams for the Yukawa interaction terms in the Lagrangian of equation (1.8).

Question 1.2: Using natural units

In high-energy physics, since relativity and quantum mechanics are essential, it is convenient to use units where

$$\hbar \approx 6.58 \times 10^{-22}\,\text{MeV s} = 1, \qquad c \approx 3 \times 10^{8}\,\text{m/s}^{-1} = 1,$$
$$\hbar c \approx 2 \times 10^{-13}\,\text{MeV m} = 1\,. \tag{1.13}$$

One can think of this convention as a choice of a unit system where the basis is $\{\hbar, c, \text{eV}\}$ instead of, for example, the {cm, g, sec} of the cgs system. In addition, it is common to make the factors of $\hbar$ and c implicit and measure everything in powers of eV. We reinstate the factors of $\hbar$ and c only when converting to a different unit system. The aim of this exercise is that you gain some practice in using these natural units.

1. The width of a particle is defined as the inverse of its lifetime. The mean lifetime for the B^+ meson is $\tau \approx 1.64 \times 10^{-12}$ s. What is its width in eV?
2. Consider a particle with a width of $\Gamma = 2.3$ eV. Recall that in the lab frame, $t = \gamma\tau$. What is the average distance that such a particle travels with $\gamma = 100$ before decaying (since $\gamma \gg 1$, you can use $\beta \approx 1$)?
3. Quantum gravity effects cannot be neglected at very short distances. This happens when the energy scale is of the order of the Planck mass:

$$M_{\text{Pl}} \equiv \sqrt{\frac{\hbar c}{G_N}}, \tag{1.14}$$

   where G_N is the Newtonian gravitational constant. (The Planck scale constitutes an upper bound on the cutoff scale of all QFTs relevant to nature.) Express M_{Pl} in GeV, and the Planck length, $L_{\text{Pl}} \equiv M_{\text{Pl}}^{-1}$, in centimeters (cm).
4. In oscillation experiments for neutrinos, it is important to know the oscillation length, $L_{\text{osc}} = 4\pi E/\Delta m^2$, where Δm^2 is the mass difference between the two neutrino states. For an experiment conducted with neutrinos of $E = 1.3$ GeV, find the value of Δm^2 in units of eV^2 that corresponds to $L_{\text{osc}} = 140$ meters.

Question 1.3: Dimensions of terms

It is useful to understand what we refer to as the *dimension of operators* or the *dimension of Lagrangian terms*. The action has dimensions of angular momentum. Therefore, in the natural unit system, the action is dimensionless and the Lagrangian has a mass dimension of four (or, more generally, of the number of spacetime dimensions).

1. Based on the Lagrangians of equations (1.2) and (1.5), show that canonical scalar fields have dimension $d = 1$, and canonical fermion fields have dimension $d = 3/2$.
2. Find the dimensions of the m^2 parameter in equation (1.2) and of the m_{MR}, m_{ML}, and m_D parameters in equation (1.5).
3. What are the dimensions of η and λ in equation (1.2) and of Y in equation (1.8)?

Question 1.4: Accidental symmetries

In this question, we study a classical system to show examples of accidental symmetries. Consider a classical one-dimensional pendulum of length ℓ. The 1 DoF can be designated as θ, the angle of the pendulum. Then the Lagrangian is given by

$$L = \frac{m\ell^2\dot{\theta}^2}{2} - mg\ell(1 - \cos\theta). \tag{1.15}$$

Assuming small oscillations ($\theta \ll 1$), we can expand the potential. Keeping only terms up to the second order, we get

$$L = \frac{m\ell^2\dot{\theta}^2}{2} - \frac{mg\ell\theta^2}{2}, \tag{1.16}$$

which is the Lagrangian of a simple harmonic oscillator. It is well known that the frequency of a simple harmonic oscillator does not depend on its amplitude. Next, we aim to understand how this result is related to accidental symmetries.

1. Show that the equation of motion (EoM) derived from the Lagrangian of equation (1.16) is invariant under dilation, $\theta \to \lambda\theta$, for any finite λ. (We are then saying that L of equation (1.16) has dilation symmetry, despite the fact that it is only the EoM that is invariant.)
2. Does the Lagrangian of equation (1.15) also have dilation symmetry?
3. Expand the Lagrangian of equation (1.15) up to $O(\theta^4)$. Show explicitly that the θ^4 term breaks the dilation invariance. Explain why this implies that this symmetry is accidental.
4. Without a formal proof, argue that dilation symmetry implies that the frequency cannot depend on the amplitude.

What we have shown here is that the dilation symmetry is accidental and is broken by higher-order terms.

2

Abelian Symmetries

In section 1.3 of chapter 1, we explained the importance of symmetries in physics and presented the various types of symmetries that we encounter. In this chapter, we introduce various concepts and definitions concerning internal symmetries. In particular, we distinguish global from local symmetries, discrete from continuous symmetries, and chiral from vectorial symmetries. We further introduce the notion of charge and its relation to symmetries. In this chapter, we only discuss Abelian symmetries (i.e., symmetries that correspond to commuting symmetry groups). Non-Abelian symmetries are discussed in chapter 4.

2.1 Global Symmetries

The term *global symmetries* refers to symmetries under transformations that are constant in spacetime. This is in distinction to *local symmetries,* which are introduced in section 2.2, where the symmetry transformation is x_μ-dependent.

2.1.1 Global Discrete Symmetries

We start with a simple example of imposing an internal global discrete symmetry. Consider a real scalar field ϕ. The most general Lagrangian is given in equation (1.2) in chapter 1, which we rewrite here:

$$\mathcal{L}_S = \frac{1}{2}(\partial^\mu \phi)(\partial_\mu \phi) - \frac{m^2}{2}\phi^2 - \frac{\eta}{2\sqrt{2}}\phi^3 - \frac{\lambda}{4}\phi^4 . \tag{2.1}$$

We now impose a symmetry: we demand that $\mathcal{L}$ is invariant under $\phi \to -\phi$, namely,

$$\mathcal{L}(\phi) = \mathcal{L}(-\phi) . \tag{2.2}$$

$\mathcal{L}$ is invariant under this symmetry if $\eta = 0$. Thus, by imposing the symmetry, we force $\eta = 0$. The most general $\mathcal{L}(\phi)$ that is invariant under $\phi \to -\phi$, then, is

$$\mathcal{L} = \frac{1}{2}(\partial^\mu \phi)(\partial_\mu \phi) - \frac{m^2}{2}\phi^2 - \frac{\lambda}{4}\phi^4 . \tag{2.3}$$

What conservation law corresponds to this symmetry? We note that the number of ϕ-particles in a system that is described by the Lagrangian of equation (2.3) can change, but only by an even number. Therefore, if we define ϕ-parity as $(-1)^n$, where n is the number of ϕ-particles in the system, this ϕ-parity is conserved. If we do not impose the symmetry and $\eta \neq 0$, then the number of particles can change by any integer and ϕ-parity is not conserved.

It is a useful exercise to describe the symmetry in terms of group theory. Here, the relevant group is Z_2. It has two elements that we call *even* (+) and *odd* (−). The multiplication table is very simple:

$$\begin{array}{c|cc} Z_2 & (+) & (-) \\ \hline (+) & (+) & (-) \\ (-) & (-) & (+) \end{array} . \tag{2.4}$$

When we say that we impose a Z_2 symmetry on $\mathcal{L}$, with the Z_2 transformation law $\phi \to -\phi$, what we mean is that $\mathcal{L}$ belongs to the even representation of Z_2 and ϕ belongs to the odd representation. Clearly, for n being an integer, ϕ^{2n} belongs to the even representation and ϕ^{2n+1} belongs to the odd representation. For $\mathcal{L}$ to be Z_2-even, each term in $\mathcal{L}$ must be Z_2-even as well, and we must omit all terms with odd powers of ϕ. Then we can construct the most general $\mathcal{L}$, given by equation (2.3).

At this point, using the language of group theory may seem cumbersome and unnecessarily complicated. Later, however, we use this vocabulary to deal with more elaborate situations, where it has proved to be very useful.

2.1.2 Global Continuous Symmetries

We now extend our discussion to global continuous symmetries. The idea is that we demand that $\mathcal{L}$ is invariant under rotation in an internal space. While some of the fields are not invariant under rotation in that space, the combinations that appear in the Lagrangian are.

The fact that we can rotate between fields should not come as a surprise. This is nothing but the idea of generalized coordinates in action. Any linear combinations of the fields can be used as our coordinates, not just the one that we choose to start with.

Consider a complex scalar field, ϕ. A complex scalar field has 2 degrees of freedom (DoF). There are two useful ways to write the DoF explicitly. First, we can use a Cartesian form:

$$\phi \equiv \frac{1}{\sqrt{2}} (\phi_R + i\phi_I) , \tag{2.5}$$

with ϕ_R and ϕ_I as real scalar fields. The most general renormalizable $\mathcal{L}(\phi_R, \phi_I)$ is given by

$$\mathcal{L}(\phi_R, \phi_I) = \frac{1}{2}\delta_{ij} \left(\partial^\mu \phi_i\right)\left(\partial_\mu \phi_j\right) - \frac{m_{ij}^2}{2}\phi_i\phi_j - \frac{\eta_{ijk}}{6}\phi_i\phi_j\phi_k - \frac{\lambda_{ijk\ell}}{24}\phi_i\phi_j\phi_k\phi_\ell ,$$

$$i, j, k, \ell = R, I. \tag{2.6}$$

We now consider rotations in the complex plane:

$$\begin{pmatrix} \phi_R \\ \phi_I \end{pmatrix} \to O \begin{pmatrix} \phi_R \\ \phi_I \end{pmatrix}, \qquad O = \begin{pmatrix} \cos\theta & \sin\theta \\ -\sin\theta & \cos\theta \end{pmatrix}. \tag{2.7}$$

When we say that we impose a *global* symmetry, we mean that θ is a number that does not depend on x_μ. Imposing that $\mathcal{L}(\phi_R, \phi_I)$ is invariant under the transformation of equation (2.7) forbids many terms and relates others, as follows:

$$\begin{aligned} \mathcal{L}(\phi_R, \phi_I) = {} & \frac{1}{2}\left(\partial^\mu \phi_R \partial_\mu \phi_R\right) + \frac{1}{2}\left(\partial^\mu \phi_I \partial_\mu \phi_I\right) - \frac{m^2}{2}\left(\phi_R\phi_R + \phi_I\phi_I\right) \\ & - \frac{\lambda}{4}\left(\phi_R^4 + \phi_I^4 + 2\phi_I^2\phi_R^2\right). \end{aligned} \tag{2.8}$$

In the language of group theory, the imposed symmetry—rotations in a two-dimensional real plane—is called $SO(2)$.

Second, we can formulate the transformation law directly in terms of the complex field ϕ:

$$\phi \to e^{i\theta}\phi, \qquad \phi^\dagger \to e^{-i\theta}\phi^\dagger. \tag{2.9}$$

Imposing that $\mathcal{L}(\phi, \phi^\dagger)$ is invariant under the transformation of equation (2.9) leads to

$$\mathcal{L}(\phi, \phi^\dagger) = \left(\partial^\mu \phi^\dagger\right)\left(\partial_\mu \phi\right) - m^2 \phi^\dagger \phi - \lambda(\phi^\dagger\phi)^2. \tag{2.10}$$

In the language of group theory, the imposed symmetry—rotations in a one-dimensional complex plane—is called $U(1)$. (Mathematically, $SO(2)$ and $U(1)$ are equivalent. The different names represent the way that we think about the underlying space.) It is easy to check that the Lagrangians in equations (2.8) and (2.10) are equivalent. Equation (2.10), however, is more compact. We emphasize the following points regarding equation (2.10):

- The three terms that appear in this equation, and in particular the mass term, do not violate any internal symmetry. Thus, there is no way to forbid them by imposing an internal symmetry.
- We would obtain the same result if we scale θ by any nonzero number. Explicitly, we would obtain the same Lagrangian with a transformation law,

$$\phi \to e^{iq\theta}\phi, \tag{2.11}$$

  for any finite q. Since q is arbitrary, we can choose the value to be 1, as we did. The situation is different when we have more than one complex field, as we discuss next.

In terms of group theory, the situation is explained as follows: we impose a $U(1)$ symmetry and assign the ϕ field to the $q = 1$ representation of that $U(1)$.

2.1.3 Charge

We are now ready to define charge. We are accustomed to the notion of charge from the specific case of electromagnetism. In electromagnetism, charge has two aspects: (1) It sets the strength of the interaction of the fermions with the photon; and (2) it is a conserved quantity. In this section, we deal with the latter point, while the aspect of interaction strength will emerge when we generalize our discussion to local symmetries in section 2.2. Charge conservation is related to symmetry. The general relation between internal global continuous symmetries and conserved charges is expressed by Noether's theorem. Here, we provide a simple example.

Consider a theory with two complex scalar fields, ϕ_1 and ϕ_2. To each field ϕ_i we assign a real number q_i. We impose a symmetry under the simultaneous phase rotation of both fields:

$$\phi_1 \to e^{iq_1\theta}\phi_1, \qquad \phi_2 \to e^{iq_2\theta}\phi_2. \tag{2.12}$$

The conjugate fields transform as follows:

$$\phi_1^\dagger \to e^{-iq_1\theta}\phi_1^\dagger, \qquad \phi_2^\dagger \to e^{-iq_2\theta}\phi_2^\dagger. \tag{2.13}$$

We say that q_i is the charge of the field ϕ_i. The charge of the conjugate field $\phi_i^\dagger$ is $-q_i$. While we can always set one of the charges, q_1 or q_2, to 1, we cannot do it for both. The ratio of charges, q_2/q_1, is a physical quantity.

The charge q_i is an input to model building: we assign charges to the fields and write the Lagrangian that is invariant under the rotations of equations (2.12) and (2.13). As a concrete example, consider a model with two complex scalar fields of charges: $q_1 = 1$ and $q_2 = 3$. Then the most general renormalizable Lagrangian that is invariant under equation (2.12) is

$$\begin{aligned}\mathcal{L} = &\left(\partial^\mu\phi_1^\dagger\right)\left(\partial_\mu\phi_1\right) + \left(\partial^\mu\phi_2^\dagger\right)\left(\partial_\mu\phi_2\right) - m_1^2\phi_1^\dagger\phi_1 - m_2^2\phi_2^\dagger\phi_2 \\ &- \lambda_{11}(\phi_1^\dagger\phi_1)^2 - \lambda_{22}(\phi_2^\dagger\phi_2)^2 - \lambda_{12}(\phi_1^\dagger\phi_1)(\phi_2^\dagger\phi_2) - (\eta\phi_1^3\phi_2^\dagger + \text{h.c.}).\end{aligned} \tag{2.14}$$

A few comments are in order here:

- For a term to be allowed, the sum of the charges of the fields in this term must be zero.
- All the interactions that are allowed by the symmetry conserve the charge. This can be seen formally (and most generally) by Noether's theorem. It can also be seen for our specific example. Each term in the Lagrangian of equation (2.14) carries an overall charge of zero. Therefore, it corresponds to creation and annihilation of particles such that the sum of charges of the initial particles equals the sum of charges of the final particles.

- Since charge is related to the phase shift of a field, it can be assigned only to complex fields. That is, real fields have $q = 0$.
- There are two terms that are often used in the physics jargon instead of *charges*: *quantum numbers* and *representations*. The use of the term quantum number might be confusing, as in quantum mechanics, charge is assigned to a particle while quantum number is assigned to a state. In quantum field theory (QFT), however, we use quantum number to describe the charges that are assigned to a field. The use of the term *representation* becomes clear when we discuss non-Abelian symmetries.

2.1.4 Product Groups and Accidental Symmetries

In the previous discussion, we introduced the simplest continuous Abelian symmetry, $U(1)$. We can generalize this idea and impose a larger symmetry, $[U(1)]^N$. This gives us another tool for model building. Consider, for example, a model with two complex scalar fields, where we require invariance of the Lagrangian under two independent rotations by phases θ_a and θ_b. Such a symmetry is called $U(1)_a \times U(1)_b$. The two fields transform as follows:

$$\phi_1 \to \exp[i(q_1^a\theta_a + q_1^b\theta_b)]\phi_1, \qquad \phi_2 \to \exp[i(q_2^a\theta_a + q_2^b\theta_b)]\phi_2. \tag{2.15}$$

The rotation of the conjugate fields is done with the replacement $q_i^{a,b} \to -q_i^{a,b}$.

Consider the assignment $(q_1^a, q_1^b) = (1,0)$ and $(q_2^a, q_2^b) = (0,1)$. In such a case, the η term of the Lagrangian of equation (2.14) is forbidden, and we get the following (repeated indexes are summed over):

$$\mathcal{L} = \partial^\mu \phi_i^\dagger \partial_\mu \phi_i - m_1^2 \phi_1^\dagger \phi_1 - m_2^2 \phi_2^\dagger \phi_2 - \lambda_{ij}(\phi_i^\dagger \phi_i)(\phi_j^\dagger \phi_j). \tag{2.16}$$

We see again how imposing a symmetry can be used to forbid terms in the Lagrangian.

The $U(1)_a \times U(1)_b$ symmetry can be defined in various ways. Instead of having θ_a and θ_b as the independent rotation angles, we can use linear combinations of them. For example, we can use $\theta_\pm \equiv \theta_a \pm \theta_b$ as the two independent angles. The corresponding charges are $q_i^\pm = q_i^a \pm q_i^b$, and the symmetry is denoted as $U(1)_+ \times U(1)_-$. In question 2.9 and in section 3.2 in chapter 3, you encounter cases that demonstrate the usefulness of such a redefinition.

It is illuminating to ask if there is a way to obtain the Lagrangian of equation (2.16) by imposing a single $U(1)$ symmetry, as in equation (2.12). The answer is in the affirmative. For example, we can assign the charges $q_1 = 1$ and $q_2 = 4$. Note that this choice allows a term of the form $\phi_1^4 \phi_2^\dagger$. The reason we do not write it is that it is a dimension-five term and therefore nonrenormalizable.

This is in fact our first example of an accidental symmetry. The symmetry that we impose is a $U(1)$ symmetry with certain charge assignments, and the resulting Lagrangian has a $U(1)\times U(1)$ symmetry. The extra $U(1)$ symmetry is accidental. In fact, the Lagrangian of equation (2.16) would be the same for any $U(1)$ symmetry with q_1 and q_2 positive integers, coprime to each other, and $q_1+q_2\geq 5$. To each such assignment corresponds a different, nonrenormalizable term, $\phi_1^{q_2}\phi_2^{\dagger q_1}$, that breaks the $[U(1)]^2$ symmetry down to the $U(1)$ that we imposed.

2.1.5 Symmetries and Fermion Masses

We can define a $U(1)$ phase transformation of Weyl fermions:

$$\psi_L \to \exp(iq_L\theta)\psi_L, \qquad \psi_R \to \exp(iq_R\theta)\psi_R. \tag{2.17}$$

The conjugate fields transform as

$$\overline{\psi_L} \to \exp(-iq_L\theta)\overline{\psi_L}, \qquad \overline{\psi_R} \to \exp(-iq_R\theta)\overline{\psi_R}. \tag{2.18}$$

Note that ψ_R and $\overline{\psi_R^c}$ transform in the same way under all symmetries. In particular, if $\psi_R \to \exp(iq_R\theta)\psi_R$, then $\overline{\psi_R^c} \to \exp(iq_R\theta)\overline{\psi_R^c}$. This is the reason why we often write $\psi_R^T\psi_R$ instead of $\overline{\psi_R^c}\psi_R$: The charge in both expressions is the same. Similarly, we often write $\psi_L^T\psi_L$ instead of $\overline{\psi_L^c}\psi_L$.

To understand the consequences of imposing such a $U(1)$ symmetry on the Lagrangian, recall that the combination of renormalizability and Lorentz invariance allows only terms with two fermion fields (or no fermion fields at all); see equation (1.5). We focus here on the fermion mass terms. There is a classification of symmetries applied to fermions that is relevant for our discussion.

Consider a theory with a single left-handed fermion field, a single right-handed fermion field, and an imposed $U(1)$ symmetry:

- The symmetry is *chiral* if $q_L \neq q_R$.
- The symmetry is *vectorial* if $q_L = q_R$.
- More generally, if there are several fermion fields, the symmetry is vectorial if all the left-handed fields and all the right-handed fields can be put into pairs with the same charge, $q_{Li}=q_{Ri}$ for each i, and chiral otherwise.
- If we include any non-Abelian group, to be discussed later in this book, the generalization of these points is that we replace the word "charge" with the word "representation."

A result that is important for our purposes is that C and P are violated in any chiral theory. Note that imposing a chiral symmetry is sufficient but not necessary for a theory to violate C and P: Vectorial theories can conserve or violate C, P, or both. (For an explicit example, see question 3.3 in chapter 3.)

Consider a theory with one or more $U(1)$ symmetries, a single left-handed field, and a single right-handed field:

- To allow a Dirac mass, the charges of $\overline{\psi_L}$ and ψ_R under these symmetries must be opposite, which is the case when $q(\psi_L) = q(\psi_R)$. (If we include any non-Abelian group, to be discussed in chapter 4, a fermion field can have a Dirac mass only if it is in a vectorial representation of the symmetry group.)
- Since the transformations of $\overline{\psi_R^c}$ and ψ_R are the same, a fermion field can have a Majorana mass only if it is neutral under all $U(1)$ symmetries. In particular, as we discuss next, fields that carry electric charges cannot acquire Majorana masses. (If we include any non-Abelian group, a fermion field can have a Majorana mass only if it is in a real representation of the symmetry group.)

Consider a theory with m left-handed fields and n right-handed fields:

- If all of these $(m+n)$ fields carry the same charges, $q \neq 0$, the Majorana mass terms vanish, while the Dirac mass terms form an $m \times n$ general complex matrix m_D:

$$(m_D)_{ij}(\overline{\psi_L})_i(\psi_R)_j + \text{h.c.} \tag{2.19}$$

  The mass eigenstates are, for $m \leq n$, m Dirac fermions and $(n-m)$ massless, right-handed fermions or, for $m \geq n$, n Dirac fermions and $(m-n)$ massless, left-handed fermions. In the Standard Model, as we discuss later, in chapter 8, charged fermion fields are present in three copies ($m = n = 3$) with the same quantum numbers, and the Dirac mass matrices are 3×3.
- If all of these $(m+n)$ fields are neutral, $q = 0$, the mass terms form an $(m+n) \times (m+n)$ symmetric complex matrix, M_ψ. This matrix consists of an $m \times m$ block of Majorana mass terms for the left-handed fields, an $n \times n$ block of Majorana mass terms for the right-handed fields, and an $m \times n$ block of Dirac mass terms:

$$\begin{pmatrix} \overline{\psi_L} & \overline{\psi_R^c} \end{pmatrix} \begin{pmatrix} m_{ML}^{(m\times m)} & m_D^{(m\times n)} \\ m_D^{T(n\times m)} & m_{MR}^{(n\times n)} \end{pmatrix} \begin{pmatrix} \psi_L^c \\ \psi_R \end{pmatrix}. \tag{2.20}$$

  The $(m+n)$ mass eigenstates are Majorana fermions. In the Standard Model, neutrinos are the only neutral fermions with $m = 3$ and $n = 0$. If they have Majorana masses, then their mass matrix is 3×3.

We summarize the differences between Dirac and Majorana masses in table 2.1.

An important lesson that we can draw from these observations is the following: Charged fermions in a chiral representation are massless. In other words, if we encounter massless fermions in nature, there is a way to explain their masslessness using symmetry principles.

Table 2.1. Dirac and Majorana masses. The question of whether neutrinos have Dirac or Majorana masses is not yet experimentally settled, as explained in chapter 14.

	Dirac	Majorana
Number of DoF	4	2
Representation	Vectorial	Neutral
Mass matrix	$m \times n$, general	$(m+n) \times (m+n)$, symmetric
SM fermions	Quarks, charged leptons	Neutrinos (?)

2.2 Local Symmetries

2.2.1 *Introducing Local Symmetries*

So far, we have discussed global symmetries (i.e., symmetries where θ is independent of spacetime). In this section, we discuss local symmetries (also called *gauge symmetries*), which are continuous symmetries in which the transformation can be different at different spacetime points $\theta(x)$, where x stands for x_μ, the spacetime coordinates. (We discuss only continuous local symmetries, not discrete local symmetries.) Local symmetries have far-reaching consequences. All the internal symmetries that are imposed in defining the Standard Model are local.

We generalize equation (2.9) and define a local transformation of a complex scalar field:

$$\phi(x) \to e^{iq\theta(x)}\phi(x), \qquad \phi^\dagger(x) \to e^{-iq\theta(x)}\phi^\dagger(x). \tag{2.21}$$

(From here on, we do not write explicitly the transformation of the conjugate field.) All terms in the Lagrangian that do not involve derivatives of fields and that are invariant under a global symmetry are also invariant under the corresponding local symmetry. This is, however, not the case for derivative terms. The local transformation of $\partial^\mu\phi$ is given by

$$\partial^\mu\phi(x) \to \partial^\mu[e^{iq\theta(x)}\phi(x)] = e^{iq\theta(x)}\partial^\mu\phi(x) + iqe^{iq\theta(x)}[\partial^\mu\theta(x)]\phi(x)\,. \tag{2.22}$$

Consequently, the kinetic term of a scalar is not invariant under the local symmetry:

$$\begin{aligned}\partial^\mu\phi^\dagger(x)\,\partial_\mu\phi(x) &\to \Big[\partial^\mu\phi^\dagger(x) - iq[\partial^\mu\theta(x)]\phi^\dagger(x)\Big]\Big[\partial_\mu\phi(x) + iq[\partial_\mu\theta(x)]\phi(x)\Big] \\ &\neq \partial^\mu\phi^\dagger(x)\partial_\mu\phi(x).\end{aligned} \tag{2.23}$$

The kinetic term also violates the local symmetry in the case of fermion fields (see question 2.1). We learn that in a theory that includes only scalars and fermions, a local symmetry acting on these scalar and fermion fields forbids the kinetic terms. A field without a kinetic term is not dynamical and cannot describe the particles observed in nature.

Can we have a theory of dynamical scalars and fermions that is invariant under a local symmetry? The answer is in the affirmative. To do that, we have to correct for the extra terms that arise in the transformation of the kinetic terms while keeping in the kinetic terms themselves. These two requirements give us an idea of how to solve the situation for the local case. We should replace[1] $\partial^\mu\phi$ with a so-called covariant derivative $D^\mu\phi$, with D^μ transforming in such a way that the transformation law for $D^\mu\phi$ under the local symmetry is similar to the transformation law of $\partial^\mu\phi$ under the global symmetry, that is,

$$D^\mu\phi \to e^{iq\theta} D^\mu\phi. \tag{2.24}$$

If we have a globally invariant $\mathcal{L}(\partial^\mu\phi, \phi)$, and the transformation law (equation (2.24)) applies, then $\mathcal{L}(D^\mu\phi, \phi)$ is guaranteed to be invariant under the corresponding local symmetry.

How do we find what D^μ is? Given the transformation law (equation (2.21)), and the desired transformation law (equation (2.24)), let us try

$$D^\mu = \partial^\mu + igqA^\mu, \tag{2.25}$$

where g is a dimensionless constant, which we take to be positive, and $A^\mu(x)$ is a vector field with the following transformation law:

$$A_\mu \to A_\mu - \frac{1}{g}\partial_\mu\theta. \tag{2.26}$$

We leave it for the reader (in question 2.1 item 5) to check whether equations (2.25) and (2.26) indeed lead to equation (2.24). We thus achieved our goal of obtaining a locally invariant Lagrangian with dynamical scalar fields, fermion fields, or both. We do so by taking a Lagrangian $\mathcal{L}$ that is invariant under the global symmetry and replacing ∂^μ with D^μ. The price to pay for this is that we must add vector fields to the model.

The transformation of equation (2.26) is familiar from classical electromagnetism, where A_μ is just the vector potential. In that context, the invariance of the vector potential under equation (2.26) is often stated as the fact that only E and B are physical, while A_μ is redundant, and thus we have an extra freedom, which we call the *gauge freedom*. Here, we reverse the logic: we impose gauge symmetry, and the result is that the photon is massless and only E and B are physical.

The field A^μ is called a *gauge field*. The constant g is called the *coupling constant* (for reasons that will become clear shortly). Following our principle that we include all the terms that are allowed, we add a kinetic term for A^μ. To do that, we define *field strength* $F^{\mu\nu}$ via the following relation:

$$[D^\mu, D^\nu] = igqF^{\mu\nu}, \tag{2.27}$$

1. From this point on, we leave the x-dependence implicit. We add it only when we introduce new x-dependent entities.

which leads to the familiar definition:

$$F^{\mu\nu} = \partial^\mu A^\nu - \partial^\nu A^\mu. \tag{2.28}$$

The Lorentz invariant kinetic term for the gauge field is given by

$$\mathcal{L} = -\frac{1}{4} F^{\mu\nu} F_{\mu\nu}. \tag{2.29}$$

The Lagrangian of equation (2.29) is the most general renormalizable one that involves only the vector field and is invariant under the local symmetry.

Note the following points:

- Given the kinetic term, A^μ is a dynamical field and its excitations are physical particles. For example, as we show in chapter 3, the photon is associated with such excitations.
- We work in the canonically normalized basis, where the coefficient of the kinetic term is $1/4$.
- While a kinetic term is invariant under the local phase transformation, a mass term, $\frac{1}{2}m^2 A^\mu A_\mu$, is not. You will prove it by doing question 2.5 at the end of this chapter. Here, we just emphasize the result: *local invariance implies massless gauge fields.* Thus, these gauge bosons have only 2 DoF.
- One may wonder why we do not include a term of the form $F_{\mu\nu}\tilde{F}_{\mu\nu}$ (with $\tilde{F}_{\mu\nu} \equiv \varepsilon_{\mu\nu\rho\sigma} F^{\rho\sigma}$). The reason is that such a term is a total derivative which, for Abelian symmetries, is unphysical.
- A gauge field related to an Abelian symmetry does not couple to itself (i.e., there is no term in the renormalizable Lagrangian that involves more than two gauge fields).
- You may be puzzled by the fact that the transformation law for the gauge field is additive, not multiplicative, as is the case for the scalar and fermion fields. You should not be entirely surprised, though. When we write the transformation law for a scalar field as $\phi \to e^{iq\theta}\phi$, the phase of the field is shifted: $\arg(\phi) \to \arg(\phi) + iq\theta$. The A^μ field transforms like a phase.
- Consider Lagrangian terms involving only scalar fields, fermion fields, or both, and no derivatives of fields. If such a term is invariant under the global symmetry, then it is also invariant under the corresponding local symmetry. This is not the case for the terms involving the gauge fields. Indeed, a mass term and a quartic interaction term are invariant under the global $U(1)$, but not under the local $U(1)$. This is related to the fact that the $U(1)$ transformation law for scalars and fermions involves a multiplicative phase factor, while the one for gauge fields involves an additive shift.
- If a local symmetry decomposes into several independent factors, each factor requires its own corresponding gauge field and has its own independent coupling constant. For example, if the symmetry is a local $U(1)_a \times U(1)_b$, we must include

two massless gauge fields, A^a_μ and A^b_μ, and their transformation laws involve two independent coupling constants, g_a and g_b. (Note that the subindexes a and b on the symmetry groups are just labels that do not change the mathematical properties. In the example here, these are arbitrary labels, but later, similar subindexes are more meaningful. For example, we use $U(1)_{\rm EM}$ to denote the symmetry group that corresponds to the electromagnetic (EM) interaction.)

2.2.2 *Charge*

The notion of charge under a global symmetry was introduced in equation (2.12). The global symmetry implies charge conservation. In this subsection, we show that in the case of local symmetry, there is an additional implication to the charge: it sets the strength of the interaction with the gauge boson.

Consider a local $U(1)$ symmetry. The covariant derivative of a field of charge q is given by

$$D^\mu = \partial^\mu + igqA^\mu \,. \tag{2.30}$$

Take, for example, a fermion field with charge q under the local $U(1)$. Its kinetic term is

$$i\overline{\psi}\not{D}\psi = i\overline{\psi}\not{\partial}\,\psi - gq\overline{\psi}\not{A}\psi. \tag{2.31}$$

The second term is an interaction term between the fermion and vector fields. The strength of the interaction is given by the gauge-coupling constant g times charge q. In particular, the larger the charge, the stronger the coupling to the gauge boson.

Two comments are in order with regard to equations (2.30) and (2.31):

1. Unlike the derivative ∂^μ, the covariant derivative D^μ depends on the charge q of the field on which it acts. It is different for fields with different charges.[2]
2. What appears in a Lagrangian is the combination gq, not g and q separately. Indeed, for the Abelian case, one can rescale the coupling constant g and the charge q such that gq remains the same and the physics of the model is unaffected. When we have several fields with different charges, one should rescale all of them in the same way. The ratio between different charges (e.g., q_2/q_1), is physical and sets the relative strength of the interactions of ϕ_2 and ϕ_1 with the gauge field A_μ. As we discuss later in this book, the situation is different in the non-Abelian case.

We can summarize this section as follows. To have a Lagrangian of dynamical scalar and/or fermion fields that is invariant under a local $U(1)$ symmetry requires the introduction of a vector boson. The vector boson necessarily interacts with all scalars and fermions

2. Perhaps it would have been helpful to denote a $U(1)$-related covariant derivative by D^μ_q, emphasizing the q-dependence. However, it became customary to keep the q-dependence implicit and write just D^μ. We are choosing to comply with this norm of the community.

that are charged under the symmetry. Conversely, the gauge boson does not couple to fields that are neutral ($q=0$) under the symmetry. The global symmetry implies charge conservation. In addition, the local symmetry implies that the charge sets the strength of the interaction with the corresponding gauge field.

2.3 Summary

We described the first steps in the process of model building for scalars and fermions. We need to provide as input the following two ingredients:

1. The symmetry
2. The charges of the fermions and scalars

Then we write the most general renormalizable Lagrangian that depends on the scalar and fermion fields and is invariant under the symmetry. If the imposed symmetry is local, corresponding vector fields must be added.

The most general renormalizable Lagrangian with scalar, fermion, and gauge fields can be decomposed into

$$\mathcal{L}=\mathcal{L}_{\rm kin}+\mathcal{L}_{\psi}+\mathcal{L}_{\phi}+\mathcal{L}_{\rm Yuk}. \tag{2.32}$$

Here, $\mathcal{L}_{\rm kin}$ describes the free propagation in spacetime of all dynamical fields, as well as the gauge interactions; $\mathcal{L}_{\psi}$ gives the fermion mass terms; $\mathcal{L}_{\rm Yuk}$ describes the Yukawa interactions; and $V_{\phi}\equiv-\mathcal{L}_{\phi}$ is the scalar potential. In all the examples we will study in this book, our task will be to find the specific form of each of these four parts of the renormalizable Lagrangian.

In the Standard Model, only local symmetries are imposed. Similarly, in most of the extensions of the Standard Model, only local symmetries and global discrete symmetries are imposed. While it is possible, in principle, to impose global continuous symmetries as well, this is rarely done in current model building. There are two reasons for that. First, there are arguments suggesting that continuous global symmetries are always broken by gravitational effects and thus can only arise as accidental rather than imposed symmetries. Second, there is no obvious phenomenological motivation to impose such symmetries. Thus, in all the models presented in this book that aim to describe nature, global continuous symmetries are not imposed.

In chapter 3, we show how Quantum ElectroDynamics (QED), the theory of electromagnetic interactions can be derived by starting with these principles.

For Further Reading

A discussion of Noether's theorem can be found, for example, in section 3.3 of Schwartz [15] and chapter 3 of Goldberg [11].

Problems

Question 2.1: Algebra

1. Show that the Lagrangians in equations (2.8) and (2.10) are equivalent.
2. Using equations (2.25) and (2.27), derive equation (2.28).
3. Consider a kinetic term of a fermion field, $i\overline{\psi}\not{\partial}\,\psi$, and show that it is not invariant under a local transformation: $\psi \to e^{i\theta(x)}\psi$.
4. Explain why, for terms that involve scalar and/or fermion fields but do not involve derivatives, the $\theta \to \theta(x)$ substitution has no effect on the symmetry properties.
5. Show that the covariant derivative of the field, defined in equation (2.25), transforms in the same way as the field itself:

$$D_\mu\phi \to e^{iq\theta(x)}D_\mu\phi. \tag{2.33}$$

6. Show that equation (2.33) can be written equivalently as

$$D_\mu \to e^{iq\theta(x)}D_\mu e^{-iq\theta(x)}. \tag{2.34}$$

7. Show that a mass term for A_μ (i.e., $m^2A_\mu A^\mu$) is not invariant under the transformation law of equation (2.26).
8. Show that $F^{\mu\nu}$, defined in equation (2.28), is gauge invariant.
9. Show that $F^{\mu\nu}\tilde{F}_{\mu\nu}$ (with $\tilde{F}_{\mu\nu} \equiv \varepsilon_{\mu\nu\rho\sigma}F^{\rho\sigma}$) is a total derivative.

Question 2.2: Charges

Consider a system with three complex fields, ϕ_1, ϕ_3, and ϕ_4, where ϕ_q carries charge q under $U(1)$ symmetry. Write all the allowed interaction terms with three and four fields.

Question 2.3: Imposing a symmetry

The most general mass terms for a theory with two real scalar fields, ϕ_1 and ϕ_2, are

$$-\mathcal{L}_m = \frac{1}{2}\left(m_{11}^2\phi_1^2 + m_{22}^2\phi_2^2 + 2m_{12}^2\phi_1\phi_2\right). \tag{2.35}$$

1. Show that one can always rotate to a basis where $m_{12}^2 = 0$.
2. Find a symmetry that you can impose to get a universal mass term; that is,

$$-\mathcal{L}_m = \frac{m^2}{2}\left(\phi_1^2 + \phi_2^2\right). \tag{2.36}$$

   In other words, the symmetry forces $m_{11}^2 = m_{22}^2$ and $m_{12}^2 = 0$.
3. Define a complex field:

$$\phi \equiv \frac{\phi_1 + i\phi_2}{\sqrt{2}}, \tag{2.37}$$

   and write the Lagrangian of equation (2.36) in terms of this field.

4. Argue, based on comparing equation (2.36) and the Lagrangian that you found in the previous item, that a simple way to talk about fields that transform nontrivially under an Abelian symmetry is in terms of complex fields.

Question 2.4: Gauge boson interactions

1. Explain why a gauge boson associated with a local symmetry necessarily interacts with all fermions and scalars that are charged under the symmetry.
2. Explain why the gauge boson necessarily does not couple to fields that are neutral under the symmetry ($q=0$).

Question 2.5: Gauging a symmetry

In the process of gauging a symmetry, we take a theory that is invariant under a global symmetry and modify it to make it invariant under the corresponding local symmetry. This is done by making the symmetry transformation parameter x-dependent, $\theta=\theta(x)$, and adding a gauge field.

In this question, we consider a simple theory with a single complex scalar field ϕ that carries charge $q=1$ under $U(1)$ symmetry.

1. Write the most general Lagrangian, taking the $U(1)$ symmetry to be global.
2. Write the most general Lagrangian, taking the $U(1)$ symmetry to be local. In other words, gauge the theory by adding a gauge field, denoted by A_μ, and replacing the derivative with a covariant derivative.
3. Writing the kinetic term with the A_μ field appearing explicitly, obtain all the interaction terms involving A_μ.

Question 2.6: Chiral symmetry

The Lagrangian for a single, massless, free Dirac fermion field is given by

$$\mathcal{L}=i\overline{\psi}\,\partial\!\!\!/\,\psi\,. \tag{2.38}$$

Consider the following two transformations:

$$\psi\to e^{i\theta}\,\psi, \qquad \psi\to e^{i\theta\gamma_5}\,\psi\,. \tag{2.39}$$

1. Show that under these two transformations, the conjugate field transforms, respectively, as

$$\overline{\psi}\to\overline{\psi}\,e^{-i\theta}, \qquad \overline{\psi}\to\overline{\psi}\,e^{i\theta\gamma_5}\,. \tag{2.40}$$

   Hint: For the second transformation, use a Taylor expansion of $e^{-i\theta\gamma_5}$.

2. The $\psi \to e^{i\theta}\, \psi$ transformation is clearly vectorial; that is, ψ_L and ψ_R transform in the same way:

$$\psi \to e^{i\theta}\, \psi = e^{i\theta}\, \psi_L + e^{i\theta}\, \psi_R. \qquad (2.41)$$

The $\psi \to e^{i\theta\gamma_5}\, \psi$ transformation, on the other hand, is chiral. That is, ψ_L and ψ_R transform differently under it:

$$\psi \to e^{i\theta\gamma_5}\, \psi = e^{-i\theta}\, \psi_L + e^{+i\theta}\, \psi_R. \qquad (2.42)$$

Prove equation (2.42).

3. Show that the Lagrangian (2.38) is invariant under both transformations of equation (2.39).
4. Add a Dirac mass term to the Lagrangian, $m\overline{\psi}\psi$, and show that it breaks the chiral symmetry, $\psi \to e^{i\theta\gamma_5}\, \psi$, and conserves the vectorial one, $\psi \to e^{i\theta}\, \psi$.

This result is the source of the statement that we can use chiral symmetries to forbid mass terms for fermions.

Question 2.7: Vectorial and chiral symmetries

Consider a system with N Dirac fermion fields, $\psi_1, \psi_2, \ldots, \psi_N$, and one real scalar field, ϕ. The most general part of the Lagrangian that involves the fermions is

$$\mathcal{L} = \overline{\psi}_i[i\partial\!\!\!/\delta_{ij} - m_{ij} - \lambda_{ij}\phi]\psi_j. \qquad (2.43)$$

Generally, the symmetry of equation (2.43) is $U(1)$, under which all the fermions carry the same charge and the scalar carries a charge of zero.

1. Show that if $\lambda_{ij} \propto m_{ij}$, the symmetry is larger, $[U(1)]^N$.
2. Find a symmetry to impose on the Lagrangian in equation (2.43) that sets $m_{ij} = 0$ but allows $\lambda_{ij} \neq 0$. Explain why this symmetry must be chiral.

Question 2.8: Canonical normalization

According to the rules of writing Lagrangians, listed in section 1.1 of chapter 1, we should be writing the most general Lagrangian, consistent with the imposed symmetry and the field content. Yet we always use canonically normalized fields rather than the most general kinetic terms. Consider, for example, a theory with N real scalar fields:

$$\mathcal{L} = \mathcal{L}_{\rm kin} - V_\phi. \qquad (2.44)$$

The most general scalar potential is given by

$$V_\phi = \frac{m_{ij}^2}{2}\phi_i\phi_j + \frac{\eta_{ijk}}{6}\phi_i\phi_j\phi_k + \frac{\lambda_{ijk\ell}}{24}\phi_i\phi_j\phi_k\phi_\ell. \qquad (2.45)$$

For the kinetic terms, we use the canonically normalized form:

$$\mathcal{L}_{\text{kin}} = \frac{1}{2}\delta_{ij}\partial^{\mu}\phi_i\partial_{\mu}\phi_j. \tag{2.46}$$

Yet the most general $\mathcal{L}_{\text{kin}}$ is given by

$$\mathcal{L}_{\text{kin}} = Z_{ij}\partial^{\mu}\phi_i\partial_{\mu}\phi_j, \tag{2.47}$$

where Z is a general, symmetric and real matrix.

1. Show explicitly how to perform a change of basis by which $\mathcal{L}_{\text{kin}}$ of equation (2.47) assumes the form of equation (2.46).
2. Taking the scalar potential of equation (2.45) to apply in the basis of equation (2.47), obtain the scalar potential in the basis of equation (2.46).
3. Based on your answers to the previous two items, argue that there is no loss of generality when choosing to work with canonically normalized fields.

Question 2.9: Accidental symmetries

Consider a variation of the model discussed in subsection (2.1.4). We impose $U(1)_{\text{imp}}$ symmetry. The model has two complex scalar fields, ϕ_1 and ϕ_2, that carry charges $q_1^{\text{imp}} = 1$ and $q_2^{\text{imp}} = 5$. The most general Lagrangian for this model is given by equation (2.16) that we rewrite here:

$$\mathcal{L} = \partial^{\mu}\phi_i^{\dagger}\partial_{\mu}\phi_i - m_i^2\phi_i^{\dagger}\phi_i - \lambda_{ij}(\phi_i^{\dagger}\phi_i)(\phi_j^{\dagger}\phi_j). \tag{2.48}$$

1. Show that $\mathcal{L}$ has a $U(1)_a \times U(1)_b$ symmetry, where the charges of the fields are

$$q_1 = (1, 0), \qquad q_2 = (0, 1). \tag{2.49}$$

   Here, the first (second) number in the parentheses refers to the charge under $U(1)_a$ $[U(1)_b]$.

The $U(1)_a \times U(1)_b$ symmetry is partially accidental; that is, we imposed a single $U(1)$ and ended with a $[U(1)]^2$ symmetry. We can rewrite the symmetry as a product of $U(1)_{\text{imp}} \times U(1)_{\text{acc}}$, such that $U(1)_{\text{imp}}$ is the imposed symmetry (with $q_1^{\text{imp}} = 1$ and $q_2^{\text{imp}} = 5$) and $U(1)_{\text{acc}}$ is an accidental one. While there are many possible choices for $U(1)_{\text{acc}}$, the one that we consider next is particularly useful.

2. We choose $U(1)_{\text{acc}}$ such that the charges under $U(1)_{\text{imp}}$ and $U(1)_{\text{acc}}$ are orthonormal:

$$q_1^{\text{imp}}q_1^{\text{acc}} + q_2^{\text{imp}}q_2^{\text{acc}} = 0, \qquad \left(q_1^{\text{imp}}\right)^2 + \left(q_2^{\text{imp}}\right)^2 = \left(q_1^{\text{acc}}\right)^2 + \left(q_2^{\text{acc}}\right)^2. \tag{2.50}$$

   Find one possibility for q_1^{acc} and q_2^{acc}.

3. A generic $[U(1)]^2$ transformation can be written in the two bases as

$$\phi \to e^{i(q_a\theta_a + q_b\theta_b)}\phi = e^{i(q^{\rm imp}\theta_{\rm imp} + q^{\rm acc}\theta_{\rm acc})}\phi. \tag{2.51}$$

   The angles, $\theta_{\rm imp}$ and $\theta_{\rm acc}$, can be expressed in terms of θ_a and θ_b. Find that relation.
4. Write a dimension-six operator that breaks the $U(1)_a \times U(1)_b$ symmetry into $U(1)_{\rm imp}$.

3

QED

You are familiar with Quantum ElectroDynamics (QED) from quantum field theory (QFT) courses, where the QED Lagrangian is given and its implications are studied. In this chapter, we introduce QED using tools of model building: We postulate a symmetry principle and the transformation properties of the fields, and then derive the Lagrangian.

This is also our first example of a model that describes nature. QED is an effective theory describing electromagnetic interactions at low energy. It encompasses, however, very important and well-known phenomena, including the Coulomb potential, the anomalous magnetic moment $(g-2)$ of the leptons, and Compton scattering.

3.1 QED with One Fermion

3.1.1 *Defining QED*

The simplest version of QED is defined as follows:

1. The symmetry is a local

$$U(1)_{\rm EM}. \tag{3.1}$$

2. There are two fermion fields:

$$e_L(-1), \qquad e_R(-1). \tag{3.2}$$

   The subindexes L, R denote the chirality: left-handed and right-handed, respectively. The number in parentheses is the $U(1)_{\rm EM}$ charge.
3. There are no scalars.

3.1.2 *The Lagrangian*

As discussed in section 2.3 in chapter 2, the most general renormalizable Lagrangian with scalar, fermion, and gauge fields can be decomposed into

$$\mathcal{L} = \mathcal{L}_{\rm kin} + \mathcal{L}_\psi + \mathcal{L}_{\rm Yuk} + \mathcal{L}_\phi. \tag{3.3}$$

It is now our task to find the specific form of the Lagrangian made of the fermion fields e_L and e_R subject to the $U(1)_{\rm EM}$ gauge symmetry.

The imposed $U(1)_{\rm EM}$ gauge symmetry requires that we include a single gauge boson, A^μ (of charge $q=0$) that we call the *photon field*. The corresponding field strength is given by

$$F^{\mu\nu} = \partial^\mu A^\nu - \partial^\nu A^\mu. \tag{3.4}$$

The covariant derivative is

$$D^\mu = \partial^\mu + ieqA^\mu, \tag{3.5}$$

where e is the coupling constant and, specifically for the $e_{L,R}$ fields, $q=-1$.

$\mathcal{L}_{\rm kin}$ includes the kinetic terms of all the fields:

$$\mathcal{L}_{\rm kin} = i\overline{e_L}\not{D}e_L + i\overline{e_R}\not{D}e_R - \frac{1}{4}F^{\mu\nu}F_{\mu\nu}. \tag{3.6}$$

$\mathcal{L}_\psi$ includes a Dirac mass term for the electron fields:

$$-\mathcal{L}_\psi = m_e\overline{e_L}e_R + \text{h.c.} \tag{3.7}$$

Finally, since there are no scalar fields in the model,

$$\mathcal{L}_{\rm Yuk} = 0, \qquad \mathcal{L}_\phi = 0. \tag{3.8}$$

3.1.3 The Spectrum

The spectrum of the model defined here consists of a massive Dirac fermion of mass m_e and a massless gauge boson—the photon. We call the Dirac fermion the *electron* and denote it by e. (It is somewhat unfortunate that the QED coupling constant and the electron field are both denoted by e. However, which of the two options is meant should be clear from the context.)

We emphasize the following points in this discussion:

- The electron is a Dirac field, which has 4 degrees of freedom (DoF).
- The reason that we can write a Dirac mass term for the electron is that the e_L and e_R fields are assigned the same charge (i.e., QED is vectorial).
- The reason that we cannot write Majorana mass terms for the $e_{L,R}$ fields is that they carry charge $q \neq 0$.
- The masslessness of the photon is a consequence of the gauge symmetry.

3.1.4 The Interactions

Expanding D^μ, we obtain the photon–fermion interaction term:

$$\mathcal{L}_{\rm int} = e\,\bar{e}\not{A}e\,, \tag{3.9}$$

where the coupling constant e is positive, and we used $q=-1$ for the electrons. When the charge of the electron is set by convention to $q=-1$ (as we do in our definition of the model), the coupling constant e is related to the fine structure constant as follows:

$$\alpha = \frac{e^2}{4\pi}. \tag{3.10}$$

We emphasize the following points here:

- The photon is not charged under QED ($q_A=0$), and consequently, photons do not interact with photons at tree level.
- The QED interaction is *vector-like*: the photon couples equally to e_L and e_R.
- QED conserves C, P, and CP. That is, C, P, and CP are accidental symmetries of the theory.

3.1.5 Parameter Counting

The QED Lagrangian has two free parameters that can be chosen to be the electron mass, m_e, and the coupling constant e or, equivalently, α. Both have been measured with impressive accuracy:

$$m_e = 0.51099895000(15)\ \text{MeV}, \qquad \alpha^{-1} = 137.035999084(21). \tag{3.11}$$

This model has no internal accidental symmetry.

3.2 QED with More Fermions

3.2.1 Two Dirac Fermions

We study a generalization of QED in which we add a second fermion to the theory. We obtain interesting lessons regarding symmetries. Our definition of the model is as follows:

1. The symmetry is a local

$$U(1)_{\rm EM}. \tag{3.12}$$

2. There are four fermion fields:

$$\ell_L^i(-1), \qquad \ell_R^i(-1), \qquad (i=1,2). \tag{3.13}$$

3. There are no scalars.

The Lagrangian is a simple extension of the previous model. Since the model has no scalars, we still have $\mathcal{L}_{\rm Yuk} = \mathcal{L}_\phi = 0$. The canonically normalized kinetic terms for the fermions read as

$$\mathcal{L}_{\rm kin} = i\overline{\ell_L^1}\not{D}\ell_L^1 + i\overline{\ell_L^2}\not{D}\ell_L^2 + i\overline{\ell_R^1}\not{D}\ell_R^1 + i\overline{\ell_R^2}\not{D}\ell_R^2. \tag{3.14}$$

The fermion mass terms read as

$$-\mathcal{L}_\psi = \overline{\ell_L^i}m_{ij}\ell_R^j + \text{h.c.} = (\overline{\ell_L^1}\,\overline{\ell_L^2}) \begin{pmatrix} m_{11} & m_{12} \\ m_{21} & m_{22} \end{pmatrix} \begin{pmatrix} \ell_R^1 \\ \ell_R^2 \end{pmatrix} + \text{h.c.} \tag{3.15}$$

We can transform to a different basis for the fermion fields, however, such that in the new basis the mass matrix is diagonal (see question 3.5 at the end of this chapter). Consider two unitary matrices, V_L and V_R, which diagonalize the mass matrix:

$$V_L \begin{pmatrix} m_{11} & m_{12} \\ m_{21} & m_{22} \end{pmatrix} V_R^\dagger = \begin{pmatrix} m_e & \\ & m_\mu \end{pmatrix}. \tag{3.16}$$

Then, inserting the unit matrix twice and rewriting $\mathcal{L}_\psi$ of equation (3.15) as

$$-\mathcal{L}_\psi = \overline{\ell_L^i}V_L^\dagger V_L m_{ij} V_R^\dagger V_R \ell_R^j, \tag{3.17}$$

it is clear that these two matrices are the rotation matrices from the original basis to the new one:

$$\begin{pmatrix} \ell_L^1 \\ \ell_L^2 \end{pmatrix} \to \begin{pmatrix} e_L \\ \mu_L \end{pmatrix} = V_L \begin{pmatrix} \ell_L^1 \\ \ell_L^2 \end{pmatrix}, \quad \begin{pmatrix} \ell_R^1 \\ \ell_R^2 \end{pmatrix} \to \begin{pmatrix} e_R \\ \mu_R \end{pmatrix} = V_R \begin{pmatrix} \ell_R^1 \\ \ell_R^2 \end{pmatrix}. \tag{3.18}$$

In terms of these new basis states, the Lagrangian reads as

$$-\mathcal{L}_\psi = m_e\overline{e}e + m_\mu\overline{\mu}\mu. \tag{3.19}$$

Since we work with canonical kinetic terms, the rotation of equation (3.17) keeps the kinetic terms invariant. The basis where the mass matrix is diagonal is called the *mass basis.* The fermion states in this basis are called *mass eigenstates.* The two mass eigenstates are the electron, e, and the muon, μ. By definition, the muon is heavier than the electron, $m_\mu > m_e$.

In the language of the e and μ Dirac fields, the full QED Lagrangian reads as

$$\mathcal{L}_{\rm QED}^{e,\mu} = i\overline{e}\not{D}e + i\overline{\mu}\not{D}\mu - \frac{1}{4}F^{\mu\nu}F_{\mu\nu} - m_e\overline{e}e - m_\mu\overline{\mu}\mu. \tag{3.20}$$

The QED model with two Dirac fermions thus has three parameters: α and m_e, whose experimental values are given in equation (3.11), and m_μ:

$$m_\mu = 105.6583755 \pm 0.0000023 \text{ MeV}. \tag{3.21}$$

3.2.2 Accidental Symmetries

An interesting feature of this model is that it exhibits an internal accidental symmetry. The imposed symmetry is a local $U(1)_{\text{EM}}$ symmetry. The Lagrangian, however, is symmetric under a global $[U(1)]^2$ symmetry, which we choose to denote as the $U(1)_e \times U(1)_\mu$ symmetry. The conventional charge assignment under this symmetry is

$$e(+1, 0), \qquad \mu(0, +1). \tag{3.22}$$

The symmetry is manifest in equation (3.20), where there is no term that involves both e and μ fields. Thus, an independent phase rotation of each of them represents a symmetry. However, the global $U(1)_{\text{EM}}$ (which is automatically imposed when imposing local $U(1)_{\text{EM}}$) is part of $U(1)_e \times U(1)_\mu$.

We can rephrase our findings in the following way. Equivalently to $U(1)_e \times U(1)_\mu$, we can use

$$U(1)_{\text{EM}} \times U(1)_{e-\mu}, \tag{3.23}$$

with the charges

$$q_e^{\text{EM}} = q_\mu^{\text{EM}} = -1, \qquad q_e^{e-\mu} = -q_\mu^{e-\mu} = +1. \tag{3.24}$$

$U(1)_{\text{EM}}$ is the imposed symmetry, while $U(1)_{e-\mu}$ is an accidental symmetry.

The experimental implication of the accidental symmetry is that QED conserves the muon and electron numbers. In any QED process, the number of electrons minus the number of positrons is conserved, and the number of muons minus the number of antimuons is also conserved. The symmetry that we impose—electric charge conservation—is the sum of these two conservation laws.

The accidental symmetry is broken by higher-dimension operators. For example, the dimension-six term,

$$G_{\mu eee} \overline{e_L} e_R \overline{e_L} \mu_R, \tag{3.25}$$

where $G_{\mu eee}$ is a dimensionful constant, breaks the accidental symmetry (it carries $U(1)_{e-\mu}$ charge of -2), but it is still invariant under the imposed local symmetry (its $U(1)_{\text{EM}}$ charge is zero). This term allows processes that are forbidden by the accidental symmetry. For example, the decay process $\mu^- \to e^- e^+ e^-$, which is allowed by the imposed $U(1)_{\text{EM}}$ but violates the accidental $U(1)_{e-\mu}$ symmetry, occurs if $G_{\mu eee} \neq 0$. In question 3.2, you are asked to work out the details.

So far, no decay process that violates the $U(1)_{e-\mu}$ symmetry has been observed. In neutrino oscillation experiments, however, such a violation has been observed. We discuss this issue in more detail in chapter 14.

3.2.3 Even More Fields

We can add even more fields to QED. So long as we add pairs of left-handed and right-handed fields with the same charge, the situation is similar to the model of two Dirac fermions as described previously. There is always a basis where the mass terms are diagonal, and thus each pair of left-handed and right-handed fields can be combined into a Dirac field that does not interact with the other fields. In this basis,

$$\mathcal{L}_{\text{QED}} = -\frac{1}{4}F^{\mu\nu}F_{\mu\nu} + i\sum_j \overline{\psi^j}\not{D}\psi^j - \sum_j m_j\overline{\psi^j}\psi^j, \tag{3.26}$$

where $j = 1, \ldots, N$ runs over all the Dirac fields. This model has a $[U(1)]^N$ symmetry, which in this form is manifest.

Note that not all fields necessarily carry the same $U(1)_{\text{EM}}$ charge. (The charge of each field is implicit in the covariant derivative.) Indeed, the charged elementary fermions known to us come in three charges: the charged leptons (electron, muon, and tau-lepton) with $q_\ell = -1$, the up-type quarks (up, charm, and top) of charge $q_u = +2/3$, and the down-type quarks (down, strange, and bottom) of charge $q_d = -1/3$. There are three Dirac fermions of each of these three types. The QED Lagrangian for these nine Dirac fields has a global $[U(1)]^9$ symmetry. In other words, the number of fermions minus the number of anti-fermions of a given charge and a given mass is conserved.

3.3 Experimental Tests of QED

QED had been tested to a very high degree of precision by numerous experiments. QED passed these tests successfully (i.e., the QED calculations agree very well with the experimental results). In fact, the tests go well beyond tree-level calculations and often require a very high loop-level calculation. Moreover, some of the experiments are precise enough that tiny contributions from beyond QED need to be incorporated into the corresponding calculations.

The masslessness and charge-neutrality of the photon are predictions of QED, independent of the values of its parameters. The experimental upper bounds are

$$m_A < 1 \times 10^{-18}\ \text{eV}, \qquad |q_A| < 1 \times 10^{-35}\ e. \tag{3.27}$$

The massless photon field should generate a long-range potential of the form e^2q/r. This is indeed the form of the Coulomb potential.

Next, we mention a few of the processes that were used to form the theory of QED and further test it. Some of the well-known tests of QED include the Lamb shift, $g - 2$

of the muon, positronium decays, Compton scattering ($\gamma e^- \to \gamma e^-$), Moeller scattering ($e^-e^- \to e^-e^-$), Bhabha scattering ($e^+e^- \to e^+e^-$), muon pair production ($e^+e^- \to \mu^+\mu^-$), and photon pair production ($e^+e^- \to \gamma\gamma$). You should have encountered most of these processes in your QFT course, but if you did not, you are encouraged to study them and carry out some of the calculations yourself.

For Further Reading

The phenomenology of QED is discussed in many textbooks and reviews. In particular, see chapter 5 of Peskin and Schroeder [2] and chapter 13 of Schwartz [15].

Problems

Question 3.1: Algebra

1. Draw the Feynman diagrams for the gauge interaction of equation (3.14).

Question 3.2: Nonrenormalizable terms in QED

In this question, we discuss aspects of nonrenormalizable terms in QED. Consider the model defined in section 3.2.1, where we impose C and P symmetries and add the following nonrenormalizable term:

$$-\mathcal{L}_{\rm NR} = G\,\overline{e}e\overline{e}\mu. \tag{3.28}$$

Here, G is a coupling of mass dimension $d = -2$, and we use Dirac field notations for the electron and muon fields.

1. Draw the tree-level diagram that contributes to the decay $\mu^+ \to e^+e^-e^+$ and obtain the resulting amplitude. Denote the amplitude as $\mathcal{A}_{\rm NR}$.

We now study a renormalizable model that generates the term of equation (3.28). Add to the model defined in section 3.2.1 a scalar field S with charge $q = +2$.

2. Write the covariant derivative $D_\mu S$, as well as the scalar potential $\mathcal{L}_S$. Denote the scalar mass term by m_S^2 and assume that it is positive.
3. Write the Yukawa couplings in this model.
4. Draw all the tree-level diagrams that contribute to the decay $\mu^+ \to e^+e^-e^+$, and obtain the resulting amplitude. Denote this amplitude as $\mathcal{A}_{\rm UV}$.
5. Consider the case where $m_S \gg m_\mu$. Identify G in terms of the model parameters such that $\mathcal{A}_{\rm UV} = \mathcal{A}_{\rm NR}$ to leading order in m_μ^2/m_S^2.

This model provides an example of how nonrenormalizable terms, made of light DoF, can be generated from a full theory that includes also much heavier DoF.

Question 3.3: A mirror world

Consider the following model:

(*i*) The symmetry is a local $U(1)_{\rm EM} \times U(1)_{\rm D}$. We denote the corresponding gauge bosons by A_μ and C_μ, and their field strengths by $F_{\mu\nu}$ and $C_{\mu\nu}$, respectively.

(*ii*) There are four Weyl fermion fields:

$$e_L(-1,0), \qquad e_R(-1,0), \qquad d_L(0,-1), \qquad d_R(0,-1), \tag{3.29}$$

where the first number in the parenthesis is the charge under $U(1)_{\rm EM}$ and the second is the charge under $U(1)_{\rm D}$.

(*iii*) There are no scalars.

1. Write the covariant derivative D_μ for a generic field with charges $(q_{\rm EM}, q_{\rm D})$, and then write it specifically for the four fermion fields. Use a normalization such that the coupling constants of the two groups are the same (i.e., $g_{\rm EM} = g_{\rm D} = e$).
2. Draw the Feynman diagrams for this theory (i.e., draw all vertices from the interaction terms in the Lagrangian and write their corresponding rules). Be sure to label the fields and take care of particle flow.
3. Find $\mathcal{L}_\psi$ and state what are the masses of the fermions and how many DoF each has.
4. Consider the term $C_{\mu\nu}F^{\mu\nu}$. (A term of this form is called a "kinetic mixing" term.) Argue that this term is gauge invariant and Lorentz invariant and has mass dimension $d = 4$.

Even though the kinetic mixing term is allowed, we do not write it since we use canonical normalization. (The reason is similar for why we do not write a term of the form $\bar{e}\gamma_u D^\mu \mu$.) Thus, we can write $\mathcal{L} = \mathcal{L}_{\rm EM} + \mathcal{L}_{\rm D}$, and the two sectors are completely decoupled. In particular, the process $e\bar{e} \to d\bar{d}$ is not allowed, as it connects the two sectors.

5. We now add a scalar field $S(-1,+1)$ to the theory. Write explicitly the covariant derivative $D_\mu S$.
6. Write the most general coupling of S to the fermions (up to dimension-four terms). Write the hermitian conjugate terms explicitly.
7. Explain why the theory violates parity. What are the consequences of imposing parity?

From now on, assume that parity is imposed and we use Dirac fields.

8. Draw the Feynman diagrams of the gauge and Yukawa interactions involving S.
9. With the addition of the field S, the process $e\bar{e} \to d\bar{d}$ is allowed. Draw a tree-level Feynman diagram for this process.
10. We assume that the mass of the scalar, M_S, is much larger than E, the center of mass energy of the incoming electron (i.e., $M_S \gg E$). In this limit, how does $\sigma(e\bar{e} \to d\bar{d})$ scale with M_S?

11. We now add another Dirac fermion field, $b(0, -2)$. Write its couplings to the gauge bosons, the scalar, and other fermions, and draw the corresponding Feynman diagrams (again, up to dimension-four).
12. The process $e\bar{e} \to b\bar{b}$ does not occur at tree level, but it does at the one-loop level. Draw one such loop diagram.

The model that we considered in this question is representative of a class of models where, as well as a sector with the particles and interactions known to us, one adds a sector that is either completely decoupled from the first sector or coupled to it only via very heavy DoF. The (almost) decoupled sector is often called a "dark sector." Thus, C_μ would be called a "dark photon," and its interactions can be termed as "dark QED." We return to this model in question 6.4 in chapter 6.

Question 3.4: Light-by-light scattering

One of the well-known properties of waves is the superposition principle that states that two electromagnetic waves do not interact with each other. The superposition principle is only a classical approximation, however, and it does not hold at the quantum level. To understand this statement, we discuss the cross section for the scattering process $\gamma\gamma \to \gamma\gamma$. We only consider photons that have a center of mass energy of less than $2m_e$.

1. Explain why at tree level (also referred to as the *classical level*),

$$\sigma(\gamma\gamma \to \gamma\gamma) = 0. \tag{3.30}$$

2. Within the framework of QED with only an electron field, draw a one-loop diagram that contributes to this process.
3. The cross section was calculated to leading order in E/m_e (where E is the photon energy) in the 1930s by Euler and Kockel. In the center of mass frame, the differential cross section is given by

$$\frac{d\sigma}{d\Omega} = \frac{139\alpha^4}{(180\pi E)^2}\frac{E^8}{m_e^8}\left(3+\cos^2\theta\right)^2, \tag{3.31}$$

where Ω is the solid angle, and θ is the (smaller) angle between an incoming photon and an outgoing photon. Without explicit calculation, explain why the cross section depends on α^4 and on $1/m_e^8$.
4. Calculate the total cross section—that is, integrate over Ω and show that

$$\sigma = \frac{973}{10125\pi}\frac{E^6\alpha^4}{m_e^8}. \tag{3.32}$$

Remember to include a factor of $1/2$ due to having identical particles at the final state.

5. The results in equations (3.31) and (3.32) take into account only an intermediate electron. Estimate the effect of including the muon. Consider only the case where $E \ll m_e$.
6. The cross section is numerically very small for visible light. Calculate the probability to have at least one light-by-light interaction when two laser beams of visible light (of 500 nm) cross each other for 1 second. Assume that the beam area is 1 mm^2 and the flux of photons, F, is 10^{21} photons per second per m^2. Recall that the event rate is given by

$$\frac{dN}{dt} = L\sigma = F^2 AT\sigma, \tag{3.33}$$

where A is the beam area and T is the overlap time. For colliding beams, luminosity L is given by $L = F^2 AT$.
7. At what photon energies do you expect the effect to become significant?

Question 3.5: Matrix diagonalization

Consider a 2×2 Hermitian matrix:

$$M = \begin{pmatrix} a & c \\ c^* & b \end{pmatrix}, \tag{3.34}$$

with a and b being real.

1. Show that the eigenvalues are

$$\lambda_{1,2} = \frac{\mathrm{Tr}M \pm \sqrt{(\mathrm{Tr}M)^2 - 4\det M}}{2}. \tag{3.35}$$

2. Prove that

$$(\mathrm{Tr}M)^2 \geq 4\det M. \tag{3.36}$$

This is needed to ensure that the eigenvalues are real.
3. Assume that c is real. In this case, the matrix M can be diagonalized by an orthogonal matrix O:

$$O = \begin{pmatrix} \cos\theta & \sin\theta \\ -\sin\theta & \cos\theta \end{pmatrix}. \tag{3.37}$$

We call θ the *mixing angle*. Show that

$$\tan 2\theta = \frac{2c}{b-a}. \tag{3.38}$$

4. Consider a general matrix:

$$M = \begin{pmatrix} a & b \\ c & d \end{pmatrix}. \tag{3.39}$$

In general, it can be diagonalized by a biunitary transformation, as we did in equation (3.16). That is,

$$M_{\text{diag}} = V_L M V_R^\dagger . \tag{3.40}$$

To find V_L and V_R, first show that $MM^\dagger$ and $M^\dagger M$ are Hermitian and they are diagonalized by V_L and V_R, respectively. Then use this formalism to find the diagonalization angles in the case where M is real.

4

Non-Abelian Symmetries

In previous chapters, we have discussed Abelian symmetries, such as $U(1)$. For this type of symmetries, also known as *commutative symmetries*, the result of applying two symmetry transformations does not depend on the order in which they are applied. In this chapter, we discuss non-Abelian symmetries, such as $SU(2)$ and $SU(3)$. For this type of symmetries, also known as *noncommutative symmetries*, the result of applying two symmetry transformations might depend on the order in which they are applied. Within the Standard Model, and in models that extend it, both Abelian and non-Abelian symmetries play an important role.

4.1 Introduction

To discuss non-Abelian symmetries, we use the language of Lie groups (we provide a short review of the topic in appendix A). For the sake of concreteness, we focus on $SU(N)$ groups, but most of the results are generic and apply to any Lie group.

We are interested in internal symmetries where the symmetry transformation is unitary:

$$\phi \to U\phi, \qquad UU^\dagger = U^\dagger U = \mathbf{1}. \tag{4.1}$$

We now emphasize some of the differences between the transformation laws for Abelian and non-Abelian symmetries. Specifically, we list differences between $U(1)$ and $SU(N)$ symmetries.

Consider a $U(1)$ symmetry and a field ϕ that carries charge $q \neq 0$. We note the following features:

- The field ϕ is necessarily complex.
- The transformation law for this field is

$$\phi \to e^{iq\theta}\phi. \tag{4.2}$$

- The charge, q, is a real number.
- Thus, for an Abelian symmetry, the transformation law is defined for each single complex field separately. A $U(1)$ symmetry operation changes the phase of the field proportionally to the charge.

- For a Lagrangian to be invariant under a $U(1)$ symmetry, each term must consist of a product of fields such that the sum of their charges is zero.

Consider an $SU(N)$ symmetry and a field ϕ in a representation R of dimension $M > 1$. Here, ϕ is a vector with M components, ϕ_i with $i = 1, \ldots, M$. We note the following features:

- If R is a complex representation, ϕ is complex. If R is a real representation, ϕ can be real.
- The transformation law for this field is

$$\phi_i \to \left(e^{iT_a\theta_a}\right)_{ij} \phi_j. \tag{4.3}$$

Here, $i, j = 1, \ldots, M$ and $a = 1, \ldots, (N^2 - 1)$. The T_as are the generators of the $SU(N)$ algebra:

$$[T_a, T_b] = if_{abc}T_c. \tag{4.4}$$

For field ϕ in the M-dimensional representation R, the T_as are represented by $M \times M$ matrices.
- Thus, for a non-Abelian symmetry, the transformation law is defined for each multiplet of fields separately. The $SU(N)$ symmetry operations consist of rotations among the various components within each multiplet.
- For a Lagrangian to be invariant under a non-Abelian symmetry, each term must consist of a product of fields such that the various representations are contracted into a singlet of the symmetry group.

If the symmetry group is not simple, we can consider an independent rotation within each simple subgroup. Then, it is convenient to represent the field as a vector under each of the simple Lie subgroups. When a field undergoes a transformation under one simple subgroup, it is not affected by the other subgroups. For example, consider an $SU(3) \times SU(2)$ group and a field that is a triplet under $SU(3)$ and a doublet under $SU(2)$. We can denote the field by $\phi_{\alpha i}$, where $\alpha = 1, 2, 3$ is the $SU(3)$-triplet index and $i = 1, 2$ is the $SU(2)$-doublet index. We can write separately the transformation laws under an $SU(3)$ symmetry transformation and an $SU(2)$ symmetry transformation:

$$\phi_{\alpha i} \to \left(e^{(i/2)\lambda_a\theta_a}\right)_{\alpha\beta} \phi_{\beta i}, \qquad \phi_{\alpha i} \to \left(e^{(i/2)\tau_b\theta_b}\right)_{ij} \phi_{\alpha j}. \tag{4.5}$$

Here, λ_a are the eight 3×3 Gell-Mann matrices, such that $\lambda_a/2$ are the $SU(3)$ generators of the triplet representation, and τ_b are the three 2×2 Pauli matrices, such that $\tau_b/2$ are the $SU(2)$ generators of the doublet representation.

The notation that we use is best explained by examples. For the case of an $SU(3) \times SU(2)$ symmetry, we write the representation of a field ϕ that is an octet under $SU(3)$

and a singlet under $SU(2)$ as $\phi(8,1)$: The first number in parentheses is the representation under $SU(3)$, and the second one is under $SU(2)$. When we consider product groups that include both non-Abelian and Abelian factors, the charges under $U(1)$ symmetries are written as subindexes. For example, if the symmetry is $SU(3) \times U(1)$, a field that is a triplet under $SU(3)$ and has charge $+1$ under $U(1)$ is written as $\phi(3)_{+1}$. As a final example, we take the symmetry to be $SU(3) \times SU(2) \times U(1)$, with a field ϕ that is an $SU(3)$-triplet, an $SU(2)$-doublet, and carries a $U(1)$ charge of $+1/6$: $\phi(3,2)_{+1/6}$.

4.2 Global Symmetries

As described in subsection 4.1, in each Lagrangian term, the product of the various representations must be contracted into a singlet of the imposed symmetry. The way that we combine representations is presented in Appendix A.8. In most of the upcoming sections, we only make sure that the product of representations contains a singlet, and we keep the group indexes of the various fields implicit. (We can do so because, in most cases discussed in this book, there is a single way to combine the product of representations into a singlet. In other cases, where there is more than one way to do so, it is important to write the contraction explicitly.)

In this section, we present three models:

1. A model that demonstrates that mass terms for scalars cannot be forbidden
2. A model that demonstrates that fermions in vectorial representations have Dirac masses
3. A model that demonstrates that fermions in chiral and complex representations are massless

4.2.1 *Scalars and SO(N)*

Consider the following model:

(*i*) The symmetry is a global

$$SO(N), \tag{4.6}$$

with $N \geq 2$.

(*ii*) There is a single scalar field in the fundamental representation,

$$\phi(N). \tag{4.7}$$

(*iii*) There are no fermions.

In this model, there are N scalar degrees of freedom (DoF), but we combine them into an N representation and thus we refer to it as a single scalar field.

The most general renormalizable Lagrangian is given by

$$\mathcal{L} = \frac{1}{2}\partial_\mu\phi^\dagger\partial^\mu\phi - \frac{1}{2}m^2\phi^\dagger\phi - \frac{1}{4}\lambda(\phi^\dagger\phi)^2 . \tag{4.8}$$

What we mean by $\phi^\dagger\phi$ is the contraction of $(\overline{N}) \times (N)$ into a singlet of $SO(N)$. Explicitly, the contraction of the $SO(N)$ indexes is given by $\phi_i^*\phi_i$.

We note the following points:

- There is no internal symmetry that one can impose to set $m^2 = 0$ or $\lambda = 0$.
- In the case of $N = 2$, the symmetry is Abelian, and the model is the same as the one discussed in section 2.1.2 in chapter 2.

4.2.2 Vectorial Fermions and $U(N)$

Consider the following model of vectorial fermions:

(*i*) The symmetry is a global

$$U(N) = SU(N) \times U(1), \tag{4.9}$$

with $N \geq 2$.

(*ii*) There are two fermion fields in the fundamental representation:

$$\psi_L(N)_{+1}, \qquad \psi_R(N)_{+1}. \tag{4.10}$$

(*iii*) There are no scalars.

Here, similar to the model of section 4.2.1, many DoF are combined into an irreducible representation. Given that each Weyl fermion has 2 DoF, the model has a total of $4N$ DoF. The most general renormalizable Lagrangian is given by

$$\mathcal{L} = i\overline{\psi_L}\,\partial\!\!\!/\,\psi_L + i\overline{\psi_R}\,\partial\!\!\!/\,\psi_R - (m\overline{\psi_L}\psi_R + \text{h.c.}), \tag{4.11}$$

where the group indexes are implicit.

We note the following points:

- We cannot write Majorana mass terms since the fermions are charged under $U(1)$.
- Since this model is vectorial (i.e., the left-handed ψ_L and right-handed ψ_R transform in the same way), we can combine them into a Dirac fermion ψ and rewrite equation (4.11) as follows:

$$\mathcal{L} = i\overline{\psi}\,\partial\!\!\!/\,\psi - m\overline{\psi}\psi. \tag{4.12}$$

4.2.3 Chiral Fermions and $U(N) \times U(N)$

Consider the following model of chiral fermions:

(*i*) The symmetry is a global

$$U(N)_L \times U(N)_R = SU(N)_L \times SU(N)_R \times U(1)_L \times U(1)_R, \tag{4.13}$$

with $N \geq 2$.

(*ii*) There are two fermion fields:

$$\psi_L(N,1)_{1,0}, \qquad \psi_R(1,N)_{0,1}. \tag{4.14}$$

(*iii*) There are no scalars.

The most general renormalizable Lagrangian has only kinetic terms:

$$\mathcal{L} = i\overline{\psi_L}\,\partial\!\!\!/\,\psi_L + i\overline{\psi_R}\,\partial\!\!\!/\,\psi_R. \tag{4.15}$$

A few comments are in order:

- As discussed in section 2.1.5 in chapter 2 and demonstrated here, fermion mass terms can be forbidden by a symmetry. This stands in contrast to scalar mass terms, which cannot be forbidden by a symmetry, as demonstrated in the model of section 4.2.1.
- The reason that the mass term vanishes in the Lagrangian of equation (4.15) is that the fermion fields of the model are in a chiral representation of the symmetry. This stands in contrast to the model of section 4.2.2, where the fermion fields are in a vectorial representation of the symmetry, and therefore the Lagrangian of equation (4.11) includes a Dirac mass term.
- The $U(N)$ symmetry of equation (4.9) is a subgroup of the $U(N)_L \times U(N)_R$ symmetry of equation (4.13). It is often called the "vectorial subgroup."

4.3 Local Symmetries

Abelian local symmetries were discussed in section 2.2 in chapter 2. Here, we discuss non-Abelian local symmetries, commonly called "Yang-Mills theories." The following properties apply for both Abelian and non-Abelian local symmetries:

- Terms that depend on scalar and/or fermion fields, but not on their derivatives, and which are invariant under a global symmetry, are also invariant under the corresponding local symmetry.
- The kinetic terms are not invariant under the local symmetry.

From here on, we consider specifically the case of $SU(N)$, but the main points apply to all Lie groups. To achieve invariance under a local non-Abelian symmetry, we must add gauge fields, which we denote by G_a^μ, and replace the derivative $\partial^\mu\phi$ with a covariant derivative $D^\mu\phi$, such that $D^\mu\phi$ and ϕ transform in the same way:

$$\phi \to e^{iT_a\theta_a(x)}\phi, \qquad D^\mu\phi \to e^{iT_a\theta_a(x)}D^\mu\phi. \tag{4.16}$$

The T_as are the N^2-1 generators of the $SU(N)$ algebra and, for ϕ in an M-dimensional representation, they are represented by $M\times M$ matrices.

The gauge fields that we introduce must restore the local symmetry for N^2-1 independent rotations. Therefore, they themselves must carry an index $a=1,2,\ldots,N^2-1$, in contrast to the single gauge field needed for an Abelian local symmetry. Furthermore, the fields in an $SU(N)$-symmetric theory must transform in a well-defined way under the symmetry. It is suggestive, then, that the gauge fields G_a^μ are in the adjoint representation of $SU(N)$. You are asked to prove that this is indeed the case in question 4.6 at the end of this chapter. We conclude that the covariant derivative is given by

$$D^\mu = \partial^\mu + igT_aG_a^\mu, \tag{4.17}$$

where g is the dimensionless, positive coupling constant. The transformation law for G_a^μ is given by (see question 4.6)

$$G_a^\mu \to G_a^\mu - f_{abc}\theta_b G_c^\mu - \frac{1}{g}\partial^\mu\theta_a. \tag{4.18}$$

The fact that the non-Abelian gauge field is in the adjoint representation of the gauge group—and, in particular, that, unlike in the Abelian case, it is not a singlet—has significant consequences: It leads to self-interactions of the gauge fields, as we discuss next.

To promote G_a^μ to a dynamical field, we must introduce a kinetic term. Similar to our treatment of the $U(1)$ case, as in equation (2.27) in chapter 2, we now define the field strength $G_a^{\mu\nu}$ as

$$[D^\mu, D^\nu] = igT_aG_a^{\mu\nu}. \tag{4.19}$$

Inserting equation (4.17) into equation (4.19), and using the $SU(N)$ algebra, we obtain

$$T_aG_a^{\mu\nu} = T_a(\partial^\mu G_a^\nu - \partial^\nu G_a^\mu - gf_{abc}G_b^\mu G_c^\nu), \tag{4.20}$$

Further using $\mathrm{tr}(T_aT_b) \propto \delta_{ab}$, we can rewrite equation (4.20) as follows:

$$G_a^{\mu\nu} = \partial^\mu G_a^\nu - \partial^\nu G_a^\mu - gf_{abc}G_b^\mu G_c^\nu. \tag{4.21}$$

Compared to the Abelian case (equation (2.28)), the non-Abelian case (equation (4.21)), has an extra term. This term is the source of self-interactions of the gauge fields.

The kinetic term of non-Abelian gauge fields is given by

$$\mathcal{L}_V = -\frac{1}{4}G_a^{\mu\nu}G_{a\mu\nu}. \tag{4.22}$$

Note the following points with regard to equation (4.22):

- Given the kinetic term, G_a^μ is a dynamical field, and its excitations are physical particles. For example, as we show in chapter 5, the gluon is associated with such excitations.
- We work in the canonically normalized basis, where the coefficient of the kinetic term is 1/4.
- While a kinetic term is invariant under the gauge transformation, a mass term—$\frac{1}{2}m^2 G_a^\mu G_{a\mu}$—is not. Local invariance under non-Abelian symmetry implies massless gauge fields in the adjoint representation.
- A total derivative term of the form $G_a^{\mu\nu}\tilde{G}_{a\mu\nu}$ (with $\tilde{G}_{a\mu\nu} \equiv \varepsilon_{\mu\nu\rho\sigma}G_a^{\rho\sigma}$) is Lorentz invariant, gauge invariant, and dimension-four. For the Abelian case, we did not include such a term in equation (2.29) in chapter 2 because it is not physical. For the non-Abelian case, we still do not include it in equation (4.22) because it is not physical at the classical level. However, at the quantum level, it is physical. We briefly present the physical implications within the Standard Model of such a term for a local $SU(3)$ symmetry in section 8.A.1 in chapter 8.

The non-Abelian gauge fields couple to themselves. This can be seen by replacing $G_a^{\mu\nu}$ in equation (4.22) with the explicit expression in equation (4.21), which leads to interaction terms that are trilinear and quartic in the gauge fields:

$$\mathcal{L}_{\text{self-interactions}} = g f_{abc}(\partial_\mu G_{a\nu})G_b^\mu G_c^\nu - \frac{1}{4}g^2(f_{abc}G_b^\mu G_c^\nu)(f_{ade}G_{d\mu}G_{e\nu}). \tag{4.23}$$

The self-interactions of the non-Abelian gauge fields stand in contrast to the Abelian gauge fields which have no self-interactions. As mentioned below equation (4.21), the source of this difference is the fact that $U(1)$ gauge fields are neutral under $U(1)$, while $SU(N)$ gauge fields are in the adjoint representation of $SU(N)$, and thus are charged under their own group.

If the symmetry group decomposes into several commuting factors, each factor has its own gauge field in the corresponding adjoint representation and an independent coupling constant. For example, if the symmetry is $SU(3) \times SU(2) \times U(1)$, we must introduce three irreducible representations of gauge fields: $(8,1)_0$ representation, with coupling constant g_3, to make the Lagrangian invariant under the local $SU(3)$; $(1,3)_0$, with coupling constant g_2, to achieve invariance under the local $SU(2)$; and $(1,1)_0$, with coupling constant g_1, to achieve invariance under the local $U(1)$.

4.4 Running Coupling Constants

In quantum field theory (QFT), the coupling constants depend on the energy scale. In physics jargon, we say that the coupling constants "run." Here, we do not discuss the theoretical background for this effect and assume that the reader is familiar with it from a QFT course. We just present the relevant equations and mention several important consequences.

The running—namely, the energy dependence—of coupling g is given by the beta function:

$$\frac{\partial g}{\partial \log \mu} = \beta(g), \tag{4.24}$$

where μ is the relevant energy scale. The beta function depends on the field content of the theory. The leading order effects depend only on fields that are charged under the symmetry. The fact that an Abelian gauge boson is neutral under the corresponding Abelian symmetry while a non-Abelian gauge field is charged under the corresponding non-Abelian symmetry has important implications for the running of the respective coupling constants.

Consider a local $U(1)$ theory, with n_f Weyl fermion fields with charge $|q| = 1$. The beta function for the coupling constant g_1 is given, to leading order, by

$$\beta(g_1) = \frac{n_f g_1^3}{24\pi^2}. \tag{4.25}$$

(This result is often quoted for the case of n_f Dirac fermions, where the factor of $1/24$ is replaced by $1/12$.)

Consider a local $SU(N)$ theory, with n_f fermion fields in the fundamental representation N and n_f fermion fields in the antifundamental representation $\overline{N}$. The beta function for the coupling constant g_N is given, to leading order, by

$$\beta(g_N) = \left(\frac{2n_f - 11N}{3}\right)\frac{g_N^3}{16\pi^2}. \tag{4.26}$$

The important difference between the running effects of the Abelian case, equation (4.25), and the non-Abelian case, equation (4.26), is in the sign of the beta function. In the $U(1)$ case, the beta function is always positive. Consequently, the lower the relevant energy scale, the smaller the coupling. In the $SU(N)$ case, the beta function can assume either sign. If the number of fermions is not too large, $n_f < (11N/2)$ (and in particular in the pure gauge case, $n_f = 0$), the sign is negative, and consequently, the lower the relevant energy scale, the larger the coupling. The phenomenon that couplings become smaller at higher energies goes under the name of "asymptotic freedom."

If a coupling grows large, one loses the ability to make accurate predictions using perturbation theory, with the coupling as the small parameter. For $U(1)$ theories, we can use perturbative calculations to make accurate predictions at a sufficiently low energy (often called the "IR" [infrared]). For $SU(N)$ theories, if the number of charged fermions is small, $n_f < 11N/2$, we can use perturbative calculations to make accurate predictions only at high enough energies (the "UV" [ultraviolet]), while in the IR, perturbativity is lost. We discuss the implications of this situation in chapter 5, where we present Quantum ChromoDynamics (QCD)—the theory of strong interactions—as a specific and important example.

Table 4.1. Abelian vs non-Abelian symmetries

	Abelian	Non-Abelian
Field	Φ_q	$\Phi(n) \equiv \Phi_i, \quad i=1,\ldots,n$
Transformation	$\Phi \to e^{iq\theta}\Phi$	$\Phi_i \to \left(e^{iT_a\theta_a}\right)_{ij}\Phi_j$
	q a real number	T_a in the irrep of Φ
The gauge field	A_μ	G^a_μ
Irrep of the gauge field	$q=0$	Adjoint
Covariant derivative	$D^\mu = \partial^\mu + igqA^\mu$	$D^\mu = \partial^\mu + igT_aG^\mu_a$
Field strength tensor	$F^{\mu\nu} = \partial^\mu A^\nu - \partial^\nu A^\mu$	$G^{\mu\nu}_a = \partial^\mu G^\nu_a - \partial^\nu G^\mu_a - gf_{abc}G^\mu_b G^\nu_c$
Gauge kinetic term	$-\frac{1}{4}F_{\mu\nu}F^{\mu\nu}$	$-\frac{1}{4}G_{\mu\nu a}G^{\mu\nu}_a$
Gauge boson self-interactions	No	Yes
Sign of the beta function	Positive	Can assume either sign

4.5 Summary

We summarize the main differences between the Abelian and non-Abelian cases in table 4.1.

For Further Reading

QFT aspects of non-Abelian symmetries can be found, for example, in chapters 15 and 16 of Peskin and Schroeder [2]. More on running coupling constants can be found, for example, in chapter 23 of Schwartz [15].

Problems

Question 4.1: Algebra

1. Insert equation (4.17) into equation (4.19) and use the $SU(N)$ algebra to obtain equation (4.20).
2. Use the definition given in equation (4.21) to derive equation (4.23).
3. Derive the Feynman rules for the self-interactions of the gauge fields.

Question 4.2: Constructing invariants

Consider the following model:

(*i*) The symmetry is a global

$$SU(3) \times SU(2) \times U(1). \tag{4.27}$$

(*ii*) There are three complex scalars:

$$A(3,2)_x, \qquad B(T,3)_{+2}, \qquad C(\bar{3},2)_{+1}. \tag{4.28}$$

(*iii*) There are no fermions.

1. Find all possible values of x and T such that the term ABC is allowed.
2. Repeat the question for a term of the form $ABC^\dagger$.

Question 4.3: SU(3) and a flavor symmetry

Consider the following model:

(*i*) The symmetry is a global

$$SU(3)_G. \tag{4.29}$$

(*ii*) There are N complex scalar fields, which transform as

$$\phi_i(3), \qquad i=1,\ldots,N. \tag{4.30}$$

(*iii*) There are no fermions.

We call the index i a *flavor index*.

1. What is the number of real scalar DoF in this model?
2. Take $N=3$. Write the most general renormalizable Lagrangian. For each of the quadratic, cubic, and quartic terms, state explicitly whether it has to have a special structure (symmetric, antisymmetric, or hermitian) in flavor space.
3. Take the $N=3$ case again. Impose an additional global symmetry that we call flavor $SU(3)_F$, such that the three flavors constitute an $SU(3)_F$-triplet $\Phi = (\phi_1,\phi_2,\phi_3)$. In other words, the scalars of this model are in the (3, 3) representation of $SU(3)_G \times SU(3)_F$. What are the implications of imposing $SU(3)_F$ symmetry on the parameters of $\mathcal{L}$?

Question 4.4: Scalar Lagrangians

Consider a model with four real scalar fields, ϕ_i $(i=1,\ldots,4)$.

1. Write the kinetic terms for this model. Argue that $\mathcal{L}_{\text{kin}}$ has an $SO(4)$ symmetry, where the four real scalars form the fundamental representation of $SO(4)$.
2. Impose an $SO(4)$ symmetry, with four real scalars in the fundamental representation of $SO(4)$; that is, they form a vector in four-dimensional real space, $\Phi = (\phi_1,\ldots,\phi_4)^T$. Write the most general renormalizable $\mathcal{L}_{SO(4)}$. Explain why the mass terms and interaction terms must be proportional to the unit matrix in that space.

3. Impose a $U(1)$ symmetry (instead of the $SO(4)$ of item 2, not in addition to it). To do so, pair the four real scalar fields to create two complex scalar fields, $\phi_a = (\phi_1 + i\phi_2)/\sqrt{2}$ and $\phi_b = (\phi_3 + i\phi_4)/\sqrt{2}$. The $U(1)$ charges are $q_a = +1$ and $q_b = +4$. Write the most general renormalizable $\mathcal{L}_{U(1)}$.
4. $\mathcal{L}_{U(1)}$ has an accidental symmetry. What is it, and what are the charges of ϕ_a and ϕ_b under it?
5. Write a nonrenormalizable term that breaks the accidental symmetry.

Question 4.5: Discrete subsymmetries

Consider the following Lagrangian for two Dirac fermions:

$$\mathcal{L} = \overline{\psi}_i(i\partial\!\!\!/ - m)\delta_{ij}\psi_j, \tag{4.31}$$

where the flavor indexes i, j run from 1 to 2. This Lagrangian is manifestly invariant under some discrete symmetries in flavor space. These symmetries, however, are part of the $U(2)$ flavor symmetry, under which (ψ_1, ψ_2) form a doublet. For example, the symmetry operation $\psi_i \to -\psi_i$ can be generated by the $U(1)$ transformation: $U = e^{i\alpha}$, with $\alpha = \pi$. Find the $U(2) = SU(2) \times U(1)$ rotations for the following two symmetries:

$$(i)\ \ \psi_1 \to \psi_1, \quad \psi_2 \to -\psi_2 \tag{4.32}$$

$$(ii)\ \ \psi_1 \leftrightarrow \psi_2 .$$

Hint: You can write your answer as a series of several rotations. The following identity can be useful:

$$\exp(i\alpha\sigma_a) = I\cos\alpha + i\sigma_a \sin\alpha. \tag{4.33}$$

Question 4.6: Non-Abelian gauge bosons

In this question, you are asked to prove that the gauge bosons belong to the adjoint representation. Consider a field ϕ that transforms as an M-dimensional representation, R:

$$\phi \to U\phi, \qquad U = e^{iT_a\theta_a}, \tag{4.34}$$

where a runs from 1 to the dimension of the group (for $SU(N)$, $a = 1, 2, \ldots, N^2 - 1$) and U and T^a are $M \times M$ matrices. We take θ_a to be independent of x_μ. This may seem weird, as the whole reason to introduce the gauge fields is to let θ_a depend on x_μ. Yet once the gauge fields are introduced, they also must transform under the corresponding global symmetry, so $\theta_a = \text{const}(x_\mu)$ is a special case. The covariant derivative is

$$D_\mu = \partial_\mu + igG_\mu, \qquad G_\mu \equiv G_\mu^a T_a. \tag{4.35}$$

1. Show that the infinitesimal transformation is

$$U = 1 + iT_a\theta_a + O(\theta_a^2). \tag{4.36}$$

2. Write the infinitesimal transformation of ϕ_k explicitly with the group indexes.
3. To promote a global symmetry to a local one, $D_\mu\phi$ must transform in the same way as ϕ; that is,

$$D_\mu\phi \to UD_\mu\phi. \tag{4.37}$$

Show that equation (4.37) implies that

$$G_\mu \to UG_\mu U^\dagger - \frac{1}{g}T_a\partial_\mu\theta_a. \tag{4.38}$$

4. Show that equation (4.38), together with the algebra of the group, imply that for an infinitesimal gauge transformation,

$$G_\mu^a \to G_\mu^a + \theta^c f^{abc} G_\mu^b - \frac{1}{g}\partial_\mu\theta^a. \tag{4.39}$$

5. Using equation (4.36), argue that the irrep of the gauge field is

$$(T^c)^{ab} = -if^{cab}. \tag{4.40}$$

Using this result, argue that the gauge fields transform as the adjoint.

5

QCD

Quantum ChromoDynamics (QCD), the theory of strong interactions, is our first example of a non-Abelian theory. The theory exhibits various interesting aspects. In particular, at low energy (often called the "IR" [infrared]), the perturbative approach cannot be applied. In this chapter, we define the theory at high energy (the "UV" [ultraviolet]) in a perturbative way. We discuss IR aspects in chapter 10. QCD is a successful theory that has passed many experimental tests. We defer the discussion of these successes to chapter 10.

5.1 Defining QCD

QCD is defined as follows:

(*i*) The symmetry is a local

$$SU(3)_C. \tag{5.1}$$

(*ii*) There are six left-handed and six right-handed fermion fields,

$$Q_{Li}(3), \qquad Q_{Ri}(3), \qquad i=1,\dots,6. \tag{5.2}$$

From here on, we call fermions in the triplet representation of $SU(3)_C$ *quarks*.

(*iii*) There are no scalars.

5.2 The Lagrangian

The most general renormalizable Lagrangian is given in equation (2.32). For a model with fermion fields and no scalar fields, we have

$$\mathcal{L}=\mathcal{L}_{\text{kin}}+\mathcal{L}_{\psi}. \tag{5.3}$$

It is now our task to find the specific form of the Lagrangian made of the fermion fields Q_{Li} and Q_{Ri}, subject to $SU(3)_C$ gauge symmetry.

The imposed $SU(3)_C$ gauge symmetry requires that we include a gauge field in the adjoint representation of $SU(3)_C$, G_a^μ, which we call the *gluon field*. The corresponding field strength is given by

$$G_a^{\mu\nu} = \partial^\mu G_a^\nu - \partial^\nu G_a^\mu - g_s f_{abc} G_b^\mu G_c^\nu. \tag{5.4}$$

Here, g_s is the coupling constant and f_{abc} are the structure constant of $SU(3)$. The covariant derivative is

$$D^\mu = \partial^\mu + ig_s G_a^\mu L_a, \tag{5.5}$$

where L_a, $a = 1, \ldots, 8$, are the $SU(3)$ generators, $[L_a, L_b] = if_{abc}L_c$. In the triplet representation, $L_a = \frac{1}{2}\lambda_a$, where λ_a are the eight 3×3 Gell-Mann matrices given explicitly in equation (A.26) in the appendix. The covariant derivative is the same for all the fermions in the theory and is given by

$$D^\mu = \partial^\mu + \frac{i}{2} g_s G_a^\mu \lambda_a. \tag{5.6}$$

$\mathcal{L}_{\text{kin}}$ includes the kinetic terms of all the fields:

$$\mathcal{L}_{\text{kin}} = i\overline{Q_{Li}} \not{D} Q_{Li} + i\overline{Q_{Ri}} \not{D} Q_{Ri} - \frac{1}{4} G_a^{\mu\nu} G_{a\mu\nu}. \tag{5.7}$$

$\mathcal{L}_\psi$ includes Dirac mass terms for the quarks:

$$-\mathcal{L}_\psi = m_{ij}^Q \overline{Q_{Li}} Q_{Rj} + \text{h.c.} \tag{5.8}$$

We can always, however, make a unitary transformation of the quark fields:

$$Q_{Li} \to (V_L)_{ij} Q_{Lj}, \qquad Q_{Ri} \to (V_R)_{ij} Q_{Rj}, \tag{5.9}$$

such that in the new basis, the mass matrix is diagonal (this basis is called the *mass basis*):

$$V_L m^Q V_R^\dagger = \text{diag}(m_u, m_d, m_s, m_c, m_b, m_t). \tag{5.10}$$

The transformation (equation (5.9)) by unitary matrices preserves the canonical normalization of the kinetic terms. The six mass eigenstates are given individual names: up (u), down (d), strange (s), charm (c), bottom (b), and top (t) quarks, where by definition, $m_u < m_d < m_s < m_c < m_b < m_t$.

In the language of the quark Dirac mass eigenstate fields, $q_i \equiv (Q_L^i, Q_R^i)^T$, the full QCD Lagrangian reads as

$$\mathcal{L}_{\text{QCD}} = -\frac{1}{4} G_a^{\mu\nu} G_{a\mu\nu} + \sum_{q_i} (i\overline{q_i} \not{D} q_i - m_{q_i} \overline{q_i} q_i), \qquad q_i = u, d, s, c, b, t. \tag{5.11}$$

5.3 The Spectrum

When discussing the spectrum of QCD, the energy scale at which it is probed makes a significant difference. The reason is that the strong coupling constant g_s, unique among the Standard Model coupling constants, become stronger the lower the energy or equivalently, the larger the distance between quarks. This leads to confinement: quarks bind into $SU(3)_C$-singlet states (*color-singlets*). Here, we discuss the UV spectrum. The IR spectrum is discussed in chapter 10.

At short distances, the spectrum of QCD consists of six massive Dirac quarks q_i of masses m_{q_i} in the color-triplet representation, as well as a massless gluon in the color-octet representation. We emphasize the following points:

- The masslessness of the gluon is a consequence of the gauge symmetry.
- The reason that we can write mass terms for all six quarks is that QCD is vectorial (i.e., the six Q_L and six Q_R fields are all color-triplets).
- Since each of the six quarks is a color-triplet Dirac fermion, it has 12 degrees of freedom (DoF).
- The situation where a theory has six fermion mass eigenstates that carry the same quantum numbers and differ only in mass is described, in physics jargon, by the statement that "the model has six flavors."

5.4 The Interactions

Expanding D^μ, we obtain the quark–gluon interaction term:

$$\mathcal{L}_{\rm int} = -\frac{g_s}{2}\overline{q_i}\lambda_a \not{G}_a q_i\,. \tag{5.12}$$

Expanding $G_a^{\mu\nu}G_{a\mu\nu}$ we obtain the gluon self-interactions:

$$\mathcal{L}_{\rm self-int} = g_s f_{abc}(\partial^\mu G_a^\nu)G_b^\mu G_c^\nu - g_s^2(f_{abc}G_b^\mu G_c^\nu)(f_{ade}G_d^\mu G_e^\nu). \tag{5.13}$$

We emphasize the following points:

- The quark–gluon interaction and the gluon self-interaction depend on a single coupling constant, g_s.
- The quark–gluon interaction is vector-like, diagonal, and universal in the flavor space.
- The strong interaction conserves P, T, C, and any combination of them.

5.5 The Parameters

The QCD Lagrangian has seven free parameters: the six masses, m_{q_i}, and the coupling constant, g_s or, equivalently,

$$\alpha_s \equiv \frac{g_s^2}{4\pi}. \tag{5.14}$$

The values of the parameters depend on the scale where they are measured and the renormalization scheme that is used. We do not discuss these issues here, and just quote the values from the Particle Data Group (PDG):

$$m_u = 2.2^{+0.5}_{-0.3}\ \text{MeV}, \qquad m_d = 4.7^{+0.5}_{-0.2}\ \text{MeV}, \qquad m_s = 93^{+11}_{-5}\ \text{MeV}, \tag{5.15}$$
$$m_c = 1.27 \pm 0.02\ \text{GeV}, \quad m_b = 4.18^{+0.03}_{-0.02}\ \text{GeV}, \quad m_t = 172.9 \pm 0.4\ \text{GeV}.$$

The coupling constant is given by

$$\alpha_s(m_Z^2) = 0.1181 \pm 0.0011, \tag{5.16}$$

where $m_Z \sim 91$ GeV is the mass of the Z boson, which we will discuss in chapter 7.

5.6 Confinement

The confinement hypothesis states that quarks, which are $SU(3)_C$ triplets, must be confined within color-singlet bound states. This implies that all asymptotic states are singlets of $SU(3)_C$. This prediction is confirmed by the fact that we do not observe free quarks in nature.

While there is no formal proof of this phenomenon, there are many indications that it is true. In particular, the confinement hypothesis is consistent with the way that the strong coupling runs. The beta function for the $SU(N)$ coupling constant is given in equation (4.26). For QCD, where the gauge group is $SU(3)$ and the number of relevant quarks at $\mu > m_t$ is six, we have

$$\beta(g_s)(\mu > m_t) = -\frac{7g_s^3}{16\pi^2} < 0. \tag{5.17}$$

For lower energy scales, where the number of relevant quarks is smaller, $\beta(g_s)$ becomes even more negative, $\beta(g_s) = -(33 - 2n_q^{\text{light}})g_s^3/(48\pi^2)$, where n_q^{light} is the number of quarks with $m_q < \mu$. The fact that the beta function is negative is very important. It implies that the strong coupling constant becomes larger for lower energies. This behavior is what leads to confinement: when a quark and an antiquark are traveling away from each other, at some point it becomes energetically favorable to pop up a quark–antiquark pair from the vacuum. The result is that we do not observe free quarks at the IR. All quarks are confined to color-singlet objects.

The scale where the transition between the UV and IR descriptions of QCD occurs is called Λ_{QCD}. Roughly speaking, it is the scale where the coupling constant becomes strong, $g_s \approx 4\pi$. It occurs around $\Lambda_{\text{QCD}} \approx 300$ MeV (the value depends on the renormalization scheme). We can treat QCD as a perturbative theory when considering processes where $q^2 \gg \Lambda_{\text{QCD}}^2$.

Note that Λ_{QCD} is not a fundamental parameter of the theory, but rather an emergent one. Its value is determined by the masses of the quarks and the coupling constant (see question 5.3 at the end of this chapter).

5.7 Accidental Symmetries

The kinetic term of QCD has a large accidental global symmetry, $U(6)_L \times U(6)_R$. In general, the mass terms break it to $[U(1)]^6$:

$$U(1)_u \times U(1)_d \times U(1)_s \times U(1)_c \times U(1)_b \times U(1)_t. \tag{5.18}$$

In the mass basis, equation (5.11), this symmetry is manifest.

The following comments are in order:

- The $[U(1)]^6$ symmetry predicts that the quarks are stable. This is because under each of the $U(1)$s, only a single particle is charged, so it must be stable.
- More generally, the accidental symmetry implies that the quark flavor is conserved. For example, $u\bar{u} \to c\bar{c}$ is allowed, but $u\bar{c} \to c\bar{u}$ is not.
- Both predictions are violated in nature. In fact, the weak interaction breaks the $[U(1)]^6$ symmetry down to a single $U(1)$ called the *baryon number*. This $U(1)_B$ is one where all quarks rotate with the same phase. It is also the diagonal $U(1)$. The $[U(1)]^6$-breaking by the weak interactions allows the QCD-forbidden processes to be mediated by weak interactions. We discuss this point in chapter 8.
- At and below the scale of present experiments, the weak interactions are considerably weaker than the strong interactions. Thus, the QCD Lagrangian and its accidental $[U(1)]^6$ symmetry constitute a good approximation.
- At low energies, the QCD Lagrangian has even larger approximate accidental symmetries: isospin, $SU(3)$-flavor, and heavy quark symmetry. These are discussed in section 10.4 in chapter 10.
- Similar to quantum electro dynamics (QED), QCD conserves C, P, and CP because it is a vectorial theory with masses and gauge interaction only.

5.8 Combining QCD with QED

In this section, we combine QCD, the local $SU(3)_C$ theory of strong interactions, with QED, the local $U(1)_{EM}$ theory of electromagnetic interactions. This model describes the quark spectrum and interactions in nature, neglecting the weak and Yukawa interactions.

The model is defined as follows:

(*i*) The symmetry is a local

$$SU(3)_C \times U(1)_{EM}. \tag{5.19}$$

(*ii*) There are six left-handed and six right-handed quark fields,

$$U_{Li}(3)_{+2/3}, \quad U_{Ri}(3)_{+2/3}, \quad D_{Li}(3)_{-1/3}, \quad D_{Ri}(3)_{-1/3}, \quad i = 1, \ldots, 3. \tag{5.20}$$

(*iii*) There are no scalars.

The resulting Lagrangian combines QED with QCD. The imposed symmetry requires that we include the following gauge fields:

$$G_a^\mu(8)_0, \qquad A^\mu(1)_0. \tag{5.21}$$

The covariant derivative acting on the quark fields is given by

$$D^\mu = \partial^\mu + \frac{i}{2} g_s G_a^\mu \lambda_a + ieqA^\mu, \tag{5.22}$$

where $q = +2/3$ for the U_i fields and $q = -1/3$ for the D_i fields.

$\mathcal{L}_{\text{kin}}$ includes the kinetic terms of all the fields:

$$\mathcal{L}_{\text{kin}} = -\frac{1}{4} G_a^{\mu\nu} G_{a\mu\nu} - \frac{1}{4} F^{\mu\nu} F_{\mu\nu} + i\overline{U_{Li}} \not{D} U_{Li} + i\overline{U_{Ri}} \not{D} U_{Ri}. + i\overline{D_{Li}} \not{D} D_{Li} + i\overline{D_{Ri}} \not{D} D_{Ri}. \tag{5.23}$$

The gauge interactions are extracted from the kinetic terms. The gluon interactions are identical to the pure QCD Lagrangian, and the photon interactions are identical to the pure QED Lagrangian except that the quark charges are different from those of the electron. The mass terms are the same as in the QCD Lagrangian. In terms of the Dirac mass eigenstate quark fields, they read as

$$-\mathcal{L}_\psi = m_u \bar{u} u + m_c \bar{c} c + m_t \bar{t} t + m_d \bar{d} d + m_s \bar{s} s + m_b \bar{b} b, \tag{5.24}$$

where u, c, t are the three $q = +2/3$ fields and d, s, b are the three $q = -1/3$ fields.

The resulting theory is not much different from just putting together $\mathcal{L}_{\text{QCD}} + \mathcal{L}_{\text{QED}}$. The global symmetry of the kinetic terms is now $[U(3)]^4$ (instead of the $[U(6)]^2$ symmetry of the pure QCD case), but the mass terms break it to the same symmetry as that of QCD: $[U(1)]^6$.

We can also add the three charged leptons to this model. They are in the $(1)_{-1}$ representation of $SU(3)_C \times U(1)_{\text{EM}}$. The terms that involve them are the same as in equation (3.20) in chapter 3.

For Further Reading

Quantum field theory (QFT) aspects of QCD can be found, for example, in chapter 17 of Peskin and Schroeder [2]. The issue of quark masses is discussed in the PDG review of quark masses.

Problems

Question 5.1: Algebra

Draw the Feynman diagrams for the quark-gluon gauge interaction of equation (5.12) and the gluon self-interaction of equation (5.13) and indicate how they depend on g_s.

Question 5.2: Testing QCD with e^+e^- Scattering

Consider the following model:

(*i*) The symmetry is a local

$$SU(3)_C \times U(1)_{EM}. \tag{5.25}$$

(*ii*) There are four quark fields:

$$A_L(N_a)_{q_a}, \qquad A_R(N_a)_{q_a}, \qquad B_L(N_b)_{q_b}, \qquad B_R(N_b)_{q_b} \tag{5.26}$$

and four lepton fields (in fact, the electron and muon fields):

$$e_L(1)_{-1}, \qquad e_R(1)_{-1}, \qquad \mu_L(1)_{-1}, \qquad \mu_R(1)_{-1}. \tag{5.27}$$

(*iii*) There are no scalars.

We assume that the masses of the fermions are much smaller than the energy scale $\sqrt{s}$ of the scattering processes that we consider here.

1. Draw the diagrams of $e^+e^- \to f\bar{f}$ for $f=\mu$ and $f=A$.
2. Show that

$$\frac{\sigma(e^+e^- \to A\bar{A})}{\sigma(e^+e^- \to \mu^+\mu^-)} = N_a q_a^2. \tag{5.28}$$

3. Calculate

$$\frac{\sigma(e^+e^- \to A\bar{A}) + \sigma(e^+e^- \to B\bar{B})}{\sigma(e^+e^- \to \mu^+\mu^-)}. \tag{5.29}$$

Consider now the model defined in section 5.8.

4. We define the ratio as

$$R = \frac{\sum_q \sigma(e^+e^- \to q\bar{q})}{\sigma(e^+e^- \to \mu^+\mu^-)}, \tag{5.30}$$

where the sum runs over all quarks with $4m_q^2 < s$. Calculate R for the following two cases: $m_s^2 \ll s/4 \ll m_c^2$ and $m_c^2 \ll s/4 \ll m_b^2$.

We discuss the ratio R in more detail in section 10.5.1 in chapter 10. The measurements of R constitute an important test of QCD, which the theory passes.

Question 5.3: The β function

In section 4.4 in chapter 4, we presented the idea that coupling constants run (i.e., that higher-order corrections can be absorbed into the definition of the coupling constants). Consider, for example, the cross section for $e^+e^- \to \mu^+\mu^-$. In the center of mass (CM) frame, at tree level, and for $E \gg m_\mu$, this is given by (up to normalization factors)

$$\sigma \propto \frac{\alpha^2}{E^2}, \tag{5.31}$$

where E is the energy of the electron. Higher-order effects change this result. Most of these effects can be absorbed into the running of α; that is,

$$\sigma \propto \frac{\alpha(\mu)^2}{E^2}, \tag{5.32}$$

where $\mu \sim E$ is the energy scale in the problem and $\alpha(\mu)$ is a running coupling constant that satisfies a differential equation:

$$\frac{\partial \alpha}{\partial \log(\mu)} = \beta(\alpha). \tag{5.33}$$

The beta function can be calculated to the desired precision in perturbation theory. In QED at one loop, with only electrons in the loop, we have

$$\beta(\alpha) = B\alpha^2, \qquad B = \frac{2}{3\pi}. \tag{5.34}$$

This equation is valid for $\mu > m_e$.

1. Verify that the solution of the beta function equation is

$$\frac{1}{\alpha(\mu_1)} = \frac{1}{\alpha(\mu_2)} + B \log\left(\frac{\mu_2}{\mu_1}\right). \tag{5.35}$$

2. Use $\alpha(m_e) \approx 1/137.0$ to calculate $\alpha(m_Z)$, where $m_Z \approx 91$ GeV.

The fact that the coupling constant becomes larger for higher energy scale raises the possibility that it diverges at some scale, which would be an indication that the theory breaks down. The energy scale at which it diverges is called the *Landau pole*. That is, the Landau pole is $\mu_{\rm LP}$ where $\alpha(\mu_{\rm LP}) \to \infty$.

3. Find the Landau pole for α. How does it stand with respect to the Planck scale, which constitutes an upper bound on the cutoff scale of all QFTs?

Measurements find that $\alpha(m_Z) \approx 1/128$. The reason for the disagreement with your result to item 2 is that there are other particles in the loop besides the electron. The generalization of equation (5.35) to that case is

$$\frac{1}{\alpha(\mu_1)} = \frac{1}{\alpha(\mu_2)} + B \sum_i Q_i^2 N_C^i \log\left(\frac{\max(m_i, \mu_2)}{\mu_1}\right), \tag{5.36}$$

where $N_C^i = 1(3)$ for leptons (quarks) and Q_i is the electric charge. The sum is over all the charged particles with mass below μ_1.

4. Use the physical masses and charges of the known fermions,

$$q = -1 : m_{e,\mu,\tau} \approx (0.5, 100, 1777) \text{ MeV},$$

$$q = +2/3 : m_{u,c,t} \approx (0.3, 1.4, 174)\ \text{GeV},$$
$$q = -1/3 : m_{d,s,b} \approx (0.3, 0.4, 4.2)\ \text{GeV}, \tag{5.37}$$

to calculate $\alpha(m_Z)$. How close is your result to the measured value? (Note that the values of m_u, m_d, and m_s are much larger than the values quoted in PDG. The reason is that we use "valence quark masses" rather than "running quark masses," which is what the PDG quotes values for. We discuss this point in chapter 10.)

We now move to QCD. The beta function is given by $\beta(\alpha) = B\alpha_s^2$ with

$$B = -\left(11 - \frac{2n_f}{3}\right)\frac{1}{2\pi}, \tag{5.38}$$

where n_f is the number of quark flavors with masses below the relevant scale. In what follows, we use the input $\alpha_s(m_Z) \approx 0.12$.

5. The sign of the beta function depends on the number of flavors. How many flavors are needed to change the sign of the beta function? We denote this number as N_{Cri}.
6. Sketch the shape of the function $\alpha_s(\mu)$ for μ between 1 and 10^4 GeV for (i) a theory with $n_f < n_{Cri}$, and (ii) a theory with $n_f > n_{Cri}$. Use log scale for μ.
7. Estimate Λ_{QCD} (i.e., the scale where $\alpha_s = 4\pi$). For simplicity, you can neglect all quark masses except m_t.
8. The proton mass is roughly $m_p \approx 3\Lambda_{\text{QCD}}$. Can you tell if the mass of the proton would be lighter or heavier if we did not have the third generation, assuming the same measured value of $\alpha_s(m_Z)$?

6

Spontaneous Symmetry Breaking

Spontaneously broken symmetries play an important role in physics, and in particle physics in particular. In this chapter, we introduce the idea of spontaneously broken symmetries and discuss the consequences. The role of such symmetries in the weak interaction part of the Standard Model is discussed in chapter 7.

6.1 Introduction

The notion of broken symmetries may seem strange: in what sense is there a difference between the case that we call a *broken symmetry* and the case of not having the symmetry at all? The idea of a broken symmetry, however, is meaningful in two scenarios:

- Explicit breaking of a symmetry by a small parameter. The Lagrangian includes terms that break the symmetry, but these terms are characterized by a small parameter. The small parameter can be either a small dimensionless coupling, or a small ratio between mass scales. The symmetry is then approximate, and one can expand around the symmetry limit.
- Spontaneous symmetry breaking (SSB). The Lagrangian is symmetric, but the vacuum state is not. Even though with SSB, the symmetry is not manifest but rather hidden, the number of parameters is the same as in the case of unbroken symmetries. In this sense, the predictive power of a spontaneously broken symmetry is as strong as that of an unbroken symmetry.

SSB is based on the following ingredients: Symmetries of interactions are determined by the symmetries of the Lagrangian. The states, however, do not have to obey these symmetries. Consider, for example, the hydrogen atom. While the Lagrangian is invariant under rotation, an eigenstate does not have to be. Specifically, a state with a finite m quantum number is not invariant under rotation around the z-axis. The fact that there are eigenstates that are not invariant under the symmetry of the system is a generic feature when there are degenerate states.

In perturbative quantum field theory (QFT), we expand around the lowest energy state. This lowest state is called the *vacuum state*. When the vacuum state is degenerate, the fact that the physics would remain the same for any choice of vacuum to expand around is a consequence of the symmetry. Yet, when we expand around a specific vacuum state out of the degenerate set of vacuums, we expand around a state that does not conserve the symmetry.

The name "spontaneously broken" indicates that there is no preference as to which state is chosen. A very simple example is that of a hungry donkey. Consider a donkey that stands exactly halfway between two stacks of hay. Symmetry tells us that it costs the same amount of energy to go to either stack. Thus, we may expect that the donkey is unable to choose and would starve! Yet, in reality, the donkey would make an arbitrary choice and go to either one of the stacks to eat. We say that the donkey "spontaneously breaks" the symmetry between the two sides.

In previous chapters, we have encountered the predictive power of imposed symmetries. In this chapter, we show that spontaneously broken symmetries are no less predictive than exact ones, even though the predictions are different. While the symmetry is no longer manifest, in the sense that processes that are forbidden in the symmetry limit may become allowed if it is spontaneously broken, there are subtle relations between these forbidden processes and the allowed ones. These relations reveal that the Lagrangian does have this symmetry. This is why a spontaneously broken symmetry is also called a *hidden symmetry*.

6.2 Global Discrete Symmetries: Z_2

Consider a model with an imposed Z_2 symmetry, similar to the one discussed in section 2.1.1 in chapter 2. There is a single real scalar field ϕ, which is odd under the symmetry:

$$\phi \to -\phi. \tag{6.1}$$

Thus, the symmetry is simply ϕ-parity. The Lagrangian reads as

$$\mathcal{L} = \frac{1}{2}(\partial_\mu \phi)(\partial^\mu \phi) - \frac{\mu^2}{2}\phi^2 - \frac{\lambda}{4}\phi^4. \tag{6.2}$$

In particular, the symmetry forbids a ϕ^3 term. Hermiticity of $\mathcal{L}$ requires that μ^2 and λ are real, and we must have $\lambda > 0$. ($\lambda < 0$ leads to a *runaway* potential; i.e., one that is not bounded from below.) As for the μ^2 term, we can have either $\mu^2 > 0$ or $\mu^2 < 0$. The $\mu^2 > 0$ case is considered in section 2.1.1. It corresponds to an ordinary ϕ^4 theory, and μ is the mass of ϕ. The case of interest for our purposes is

$$\mu^2 < 0. \tag{6.3}$$

The minimum of the scalar potential should satisfy

$$0 = \frac{\partial V}{\partial \phi} = \phi(\mu^2 + \lambda\phi^2). \tag{6.4}$$

Thus, the potential has two possible minima:

$$\phi_\pm = \pm\sqrt{\frac{-\mu^2}{\lambda}} \equiv \pm v. \tag{6.5}$$

The classical solution would be either ϕ_+ or ϕ_-. We say that ϕ acquires a vacuum expectation value (VEV):

$$\langle\phi\rangle \equiv \langle 0|\phi|0\rangle \neq 0. \tag{6.6}$$

Perturbative calculations should involve expansion around the classical minimum. Since the two solutions are physically equivalent, the physics cannot depend on our choice, but we must make a choice. Let us choose—without loss of generality—to expand around ϕ_+. We define the field ϕ' with a vanishing VEV:

$$\phi' = \phi - v. \tag{6.7}$$

In terms of ϕ', the Lagrangian reads as

$$\mathcal{L} = \frac{1}{2}(\partial_\mu\phi')(\partial^\mu\phi') - \frac{1}{2}(2\lambda v^2)\phi'^2 - \lambda v\phi'^3 - \frac{\lambda}{4}\phi'^4, \tag{6.8}$$

where we used $\mu^2 = -\lambda v^2$ and discarded a constant term.

We recall that the most general Lagrangian of a scalar field is given in equation (1.2) in chapter 1, and we rewrite it here in terms of ϕ':

$$\mathcal{L}_S = \frac{1}{2}\partial_\mu\phi'\partial^\mu\phi' - \frac{m^2}{2}\phi'^2 - \frac{\eta}{2\sqrt{2}}\phi'^3 - \frac{\lambda}{4}\phi'^4. \tag{6.9}$$

Examining the Lagrangians of equations (6.2), (6.8) and (6.9) raises the following points:

- The Lagrangian (equation (6.8)) includes all possible terms for the real scalar field ϕ'. In particular, it has no ϕ'-parity symmetry. Thus, the $\phi \to -\phi$ symmetry is hidden. It is spontaneously broken by our choice of the ground state $\langle\phi\rangle = +v$.
- Yet the Lagrangian (equation (6.8)) is not the most general renormalizable Lagrangian for a scalar field. While the most general one, equation (6.9), depends on three independent parameters, equation (6.8) depends on only two. In terms of the parameters of the general Lagrangian (equation (6.9)), the relation is

  $$\eta^2 = 4\lambda m^2. \tag{6.10}$$

  This relation is the clue that the symmetry is spontaneously rather than explicitly broken.
- The two parameters can be chosen to be v and λ or μ^2 and λ. The first choice is the one that we made in writing equation (6.8). The second choice employs the same parameters of the original, manifestly symmetric $\mathcal{L}(\phi)$; see equation (6.2). It demonstrates that SSB does not introduce additional new parameters.
- The coefficients of the quadratic and trilinear terms in equation (6.8) are different from those of the quadratic and trilinear terms in equation (6.2). In contrast, the coefficients of the quartic terms are the same. This is a general result: so long as we consider only the renormalizable terms, SSB changes dimensionful parameters, but not dimensionless ones.

- While the symmetry is manifest in equation (6.2), the phenomenological interpretation of this model should start from equation (6.8). Specifically, the model describes a scalar particle of mass $2\lambda v^2 = -2\mu^2$. This particle is an excitation of the ϕ' field.
- Hypothetical particles with negative mass are called *tachyons*, and they travel faster than light. The example given here shows why tachyons do not appear in QFT. To make a physical interpretation, we have to expand around the minimum, and thus physical particles have positive mass. Fields with negative mass terms are sometimes referred to as *tachyonic fields*.
- In nonrelativistic quantum mechanics, the analogous case—a particle in a double-well potential—does not have a degenerate vacuum due to tunneling effects. Such tunneling effectively vanishes in QFT.

6.3 Global Abelian Continuous Symmetries: $U(1)$

Consider a model with an imposed $U(1)$ symmetry, similar to the one discussed in section 2.1.2 in chapter 2. There is a single complex scalar field ϕ, with $q = +1$, so the theory is required to be invariant under

$$\phi \to e^{i\theta}\phi. \tag{6.11}$$

The Lagrangian reads as

$$\mathcal{L} = (\partial_\mu \phi^\dagger)(\partial^\mu \phi) - \mu^2 \phi^\dagger \phi - \lambda(\phi^\dagger \phi)^2. \tag{6.12}$$

Equivalently, we can rewrite the Lagrangian in terms of two real scalar fields, as in equation (2.5):

$$\phi \equiv \frac{1}{\sqrt{2}}(\phi_R + i\phi_I) \tag{6.13}$$

and impose an $SO(2)$ symmetry:

$$\begin{pmatrix} \phi_R \\ \phi_I \end{pmatrix} \to \begin{pmatrix} \cos\theta & \sin\theta \\ -\sin\theta & \cos\theta \end{pmatrix} \begin{pmatrix} \phi_R \\ \phi_I \end{pmatrix}. \tag{6.14}$$

The Lagrangian reads as (equation, see (2.8))

$$\mathcal{L} = \frac{1}{2}(\partial^\mu \phi_R)(\partial_\mu \phi_R) + \frac{1}{2}(\partial^\mu \phi_I)(\partial_\mu \phi_I) - \frac{\mu^2}{2}\left(\phi_R^2 + \phi_I^2\right) - \frac{\lambda}{4}\left(\phi_R^2 + \phi_I^2\right)^2. \tag{6.15}$$

The μ^2 and λ parameters are real, and we must have $\lambda > 0$. We consider the case that $\mu^2 < 0$. (The $\mu^2 > 0$ case is considered in section 2.1.2.) We define $v^2 = -\mu^2/\lambda$. The scalar potential can be written (up to a constant term) as

$$V = \lambda\left(\phi^\dagger\phi - \frac{v^2}{2}\right)^2. \tag{6.16}$$

Thus, ϕ acquires a VEV:

$$2\left\langle\phi^\dagger\phi\right\rangle=\langle\phi_R^2+\phi_I^2\rangle=v^2=-\frac{\mu^2}{\lambda}. \tag{6.17}$$

In the (ϕ_R, ϕ_I) plane, there is a circle of radius v that corresponds to minima of the potential. We have to choose a specific vacuum to expand around. We choose the real component of ϕ to carry the VEV:

$$\langle\phi_R\rangle=v, \qquad \langle\phi_I\rangle=0. \tag{6.18}$$

We define the real scalar fields as

$$h=\phi_R-v, \qquad \xi=\phi_I \tag{6.19}$$

with vanishing VEVs:

$$\langle h\rangle=\langle\xi\rangle=0. \tag{6.20}$$

We obtain the Lagrangian in terms of h and ξ:

$$\mathcal{L}=\frac{1}{2}(\partial_\mu h)(\partial^\mu h)+\frac{1}{2}(\partial_\mu\xi)(\partial^\mu\xi)-\lambda v^2h^2-\lambda vh(h^2+\xi^2)-\frac{\lambda}{4}(h^2+\xi^2)^2. \tag{6.21}$$

Note the following points:

- The $SO(2)$ symmetry is spontaneously broken. This can be seen from the presence of the $h(h^2+\xi^2)$ term.
- Since the symmetry is spontaneously broken, the Lagrangian is not invariant under a transformation similar to the one in equation (6.14). Explicitly,

$$\begin{pmatrix} h \\ \xi \end{pmatrix} \to \begin{pmatrix} \cos\theta & \sin\theta \\ -\sin\theta & \cos\theta \end{pmatrix}\begin{pmatrix} h \\ \xi \end{pmatrix} \tag{6.22}$$

  is not a symmetry of $\mathcal{L}$.
- The Lagrangian describes one massive scalar, h, with $m^2=2\lambda v^2$, and one massless scalar, ξ. If the symmetry were not broken, it would be impossible to distinguish the two components of the complex scalar field. With the symmetry spontaneously broken, these 2 degrees of freedom (DoF) are distinguishable by their different masses.
- The Lagrangian of equation (6.21) is not the most general one for two real scalar fields. Many terms are missing, while others, which would have been independent in the general case, are related. In particular, there are only two independent parameters, as for a Lagrangian with an unbroken $SO(2)$.
- The quartic terms, with dimensionless couplings, are the same in equations (6.15) and (6.21). Only dimensionful couplings are modified.

FIGURE 6.1. The "Mexican hat" potential. The masses of the two scalar DoF correspond to the second derivative of the potential around the minimum. One direction (left) is flat, while the other (center) is not. The plot on the right shows the symmetric maximum. It is unstable and does not correspond to a particle. In the case of global symmetry, the flat direction corresponds to the massless Goldstone boson, while the nonflat direction corresponds to the massive DoF. In the case of a local symmetry, the flat direction corresponds to the longitudinal component of the vector boson, while the nonflat direction corresponds to the massive Higgs boson.

- We chose a basis by assigning the VEV to the real component of ϕ. This is an arbitrary choice. We made it since it is convenient. The physics does not depend on this choice.
- We write the VEV as $\langle\phi_R\rangle = v$, or equivalently, as $\langle\phi\rangle = v/\sqrt{2}$. The factor of $\sqrt{2}$ between the two VEVs is just the one that we encounter many times when moving between real and complex fields.

One of the most interesting features of the model presented here is the existence of a massless scalar field. This feature is not particular to our specific model, but rather the result of a general theorem called *Goldstone's theorem*: the spontaneous breaking of a global continuous symmetry is accompanied by massless scalars. Their number and quantum numbers equal those of the broken generators. The massless scalars are called *Nambu-Goldstone bosons*.

While we do not prove the theorem in this discussion, we briefly describe the intuition behind it. SSB is only possible when the vacuum is degenerate. For a continuous symmetry, the set of degenerate vacuums is also continuous. In the case of a $U(1)$ symmetry, the shape of the potential is usually called a "Mexican hat" (see figure 6.1). When expanding around any point in the so-called valley, one direction is flat. A flat direction in the potential corresponds to a massless DoF. Goldstone's theorem is a generalization of this simple picture. Figure 6.1 demonstrates the point.

6.4 Global Non-Abelian Continuous Symmetries: $SO(3)$

Consider a model with an imposed $SO(3)$ symmetry and a real scalar field, ϕ, that transforms as a triplet under the symmetry:

$$\phi \to e^{iL_a\theta_a}\phi. \tag{6.23}$$

The three L_a matrices, $(L_a)_{bc} = i\varepsilon_{abc}$, constitute the triplet representation of $SO(3)$ algebra. They are given explicitly in equation (6.49), later in this chapter. This model provides an intuitive picture of SSB. The triplet constitutes a vector in a real three-dimensional space. Once a vector is fixed in space, the symmetry under three-dimensional rotation breaks, but not completely, as the symmetry under rotations in the plane that is perpendicular to the vector remains. Here, we translate this intuitive picture into a rigorous analysis.

The Lagrangian reads as

$$\mathcal{L} = \frac{1}{2}(\partial_\mu \phi^T)(\partial^\mu \phi) - \frac{\mu^2}{2}\phi^T\phi - \frac{\lambda}{4}(\phi^T\phi)^2. \tag{6.24}$$

We take $\mu^2 < 0$ and define $v^2 = -\mu^2/\lambda$. Then ϕ acquires a VEV: $|\langle\phi\rangle| = v$. The triplet ϕ has 3 DoF. We choose the VEV to lie in the ϕ_3-direction:

$$\phi = \begin{pmatrix} \phi_1 \\ \phi_2 \\ v + \phi_3 \end{pmatrix}, \tag{6.25}$$

such that $\langle\phi_i\rangle = 0, i = 1, 2, 3$. The $SO(3)$ symmetry is only *partially* broken:

$$SO(3) \to SO(2), \tag{6.26}$$

where the $SO(2)$ symmetry refers to rotations in the (ϕ_1, ϕ_2) plane.

The Lagrangian for the ϕ_i fields can be written as

$$\mathcal{L} = \mathcal{L}_{\text{kin}} + \mathcal{L}_2 + \mathcal{L}_3 + \mathcal{L}_4, \tag{6.27}$$

where $\mathcal{L}_n$ includes terms that are to the nth power in the ϕ_i fields. Let us comment on the significance of each of these parts of the Lagrangian:

- Quadratic terms:

$$-\mathcal{L}_2 = \lambda v^2 \phi_3^2. \tag{6.28}$$

  The model has one massive scalar, ϕ_3, of mass $m_3^2 = 2\lambda v^2$, and two massless scalars, $m_1^2 = m_2^2 = 0$. This is a manifestation of Goldstone's theorem. SSB requires the appearance of massless Goldstone bosons in correspondence to the broken generators. Since $SO(3)$ has three generators and it is spontaneously broken to $SO(2)$ that has one generator, our model must have two Goldstone bosons.
- Trilinear terms:

$$-\mathcal{L}_3 = \lambda v \phi_3(\phi_1\phi_1 + \phi_2\phi_2 + \phi_3\phi_3). \tag{6.29}$$

  The fact that $\mathcal{L}_3 \neq 0$ is a manifestation of the $SO(3)$ breaking.
- Quartic terms:

$$-\mathcal{L}_4 = \frac{\lambda}{4}\left(\phi_1^2 + \phi_2^2 + \phi_3^2\right)^2. \tag{6.30}$$

  This part of the Lagrangian has dimensionless couplings, and therefore, it is unchanged from the symmetric form.

The model presented here is an example of partial breaking of the symmetry. In general, a generator corresponds to a spontaneously broken symmetry if the vacuum is not invariant under an operation of the corresponding group element. Conversely, a generator corresponds to an unbroken symmetry if the vacuum is invariant to an operation by the corresponding group element. Given the fact that the group element is the exponent of the generator, these conditions can be represented in terms of the generators as follows. We denote the the vacuum state by $\langle\phi\rangle$. A broken generator gives

$$T_a\langle\phi\rangle \neq 0. \tag{6.31}$$

An unbroken generator gives

$$T_a\langle\phi\rangle = 0. \tag{6.32}$$

More details are given in question 6.3 at the end of this chapter.

6.5 Fermion Masses

SSB can give masses to chiral fermions. We explain this statement by providing an explicit example.

Consider a model with $U(1)$ symmetry. The field content consists of a left-handed fermion ψ_L, a right-handed fermion ψ_R, and a complex scalar ϕ with the following $U(1)$ charges:

$$q(\psi_L) = +1, \qquad q(\psi_R) = +2, \qquad q(\phi) = +1. \tag{6.33}$$

The most general Lagrangian we can write is

$$\mathcal{L} = \mathcal{L}_{\text{kin}} - \mu^2\phi^\dagger\phi - \lambda(\phi^\dagger\phi)^2 - (Y\phi\,\overline{\psi_R}\psi_L + \text{h.c.}). \tag{6.34}$$

Since the fermions are charged and chiral, we cannot write mass terms for them ($\mathcal{L}_\psi = 0$).

We take $\mu^2 < 0$, such that the scalar potential is the one given in equation (6.15), leading to a VEV for ϕ: $|\langle\phi\rangle| = v/\sqrt{2} \neq 0$. As in section 6.3, we choose $\langle\phi_R\rangle = v$, $\langle\phi_I\rangle = 0$ and define the real fields h and ξ in such a way that they have vanishing VEVs:

$$\phi = \frac{h + v + i\xi}{\sqrt{2}}, \tag{6.35}$$

Expanding around the chosen vacuum, we find

$$\mathcal{L} = \mathcal{L}_{\text{kin}} - V(h,\xi) - \left[\frac{Yv}{\sqrt{2}}\overline{\psi_R}\psi_L + \frac{Y}{\sqrt{2}}(h + i\xi)\overline{\psi_R}\psi_L + \text{h.c.}\right], \tag{6.36}$$

where $V(h,\xi)$ can be read off equation (6.21). We learn that ψ_L and ψ_R combine to form a Dirac fermion with mass

$$m_\psi = \frac{Yv}{\sqrt{2}}. \tag{6.37}$$

This is possible because the symmetry under which the fermion is chiral is broken.

In a more general case, the symmetry might be only partially broken—namely, a subgroup of the original group remains unbroken. In this case, necessary conditions for generating fermion masses are the following:

- Dirac mass: The fermion representation is vector-like under the unbroken subgroup.
- Majorana mass: The fermion is neutral under the unbroken $U(1)$ groups and in a real representation of the unbroken non-Abelian subgroups.

6.6 Local Symmetries: The Higgs Mechanism

In this section, we discuss spontaneous breaking of local symmetries. We demonstrate this by studying a $U(1)$ gauge symmetry. One of the main results is that the breaking of a local symmetry generates mass terms for the gauge bosons that correspond to the broken generators. At first sight, this result might seem surprising since the spontaneous breaking of a global symmetry gives massless Nambu-Goldstone bosons. In the case of local symmetry, however, these would-be Nambu-Goldstone bosons are "eaten" by the gauge bosons and become the longitudinal components of the resulting massive vector bosons.

An explanation of the terms "eaten" and "would-be Goldstone bosons," which are commonly used in the physics jargon, is called for. The key point for the term "eaten" is that, for a model with a spontaneously broken local symmetry, the number of DoF is the same in the interaction basis and in the mass basis. Concretely, if N_H of the symmetry generators are spontaneously broken then, in the mass basis, there are N_H massive vector-boson fields, each with 3 DoF (two transverse and one longitudinal polarizations). In the interaction basis, these $3N_H$ DoF are assigned to N_H gauge fields, each with 2 DoF (the two transverse polarizations), and N_H real scalar fields. These N_H real scalar fields of the interaction basis, which become part (the longitudinal components) of the N_H massive vector bosons in the mass basis, are referred to as the "eaten" DoF. The term "would-be Goldstone bosons" refers to the fact that, if the spontaneously broken symmetry were global instead of local, these N_H real scalar fields of the interaction basis would correspond to the massless Goldstone bosons in the mass basis.

Consider a theory similar to the one discussed in section 6.3, where we have a single scalar field that is charged under $U(1)$ symmetry. The difference is that here, we impose a local $U(1)$ symmetry:

$$\phi \to e^{i\theta(x)}\phi. \tag{6.38}$$

The Lagrangian is given by

$$\mathcal{L} = (D_\mu\phi)^\dagger(D^\mu\phi) - \frac{1}{4}F_{\mu\nu}F^{\mu\nu} - \mu^2\phi^\dagger\phi - \lambda(\phi^\dagger\phi)^2, \tag{6.39}$$

where the covariant derivative is given by

$$D^{\mu}\phi = (\partial^{\mu} + igA^{\mu})\phi, \tag{6.40}$$

A^{μ} is the gauge field, $F_{\mu\nu}$ is defined in equation (2.28), and g is the coupling constant.

We consider the case of $\mu^2 < 0$, leading to SSB via a VEV of ϕ:

$$\langle\phi\rangle = \frac{v}{\sqrt{2}}, \qquad v^2 = -\frac{\mu^2}{\lambda}. \tag{6.41}$$

We choose the real component of ϕ to carry the VEV. We again write the complex scalar in terms of two real scalar fields with vanishing VEVs, $\langle h\rangle = \langle\xi\rangle = 0$, but, unlike the global case, it is convenient to write the 2 DoF as a phase, $\xi(x)$ and a magnitude, $h(x)$:

$$\phi(x) = e^{i\xi(x)/v}\frac{v + h(x)}{\sqrt{2}}. \tag{6.42}$$

Note that we normalized $\xi(x)$ such that it has a mass-dimension 1. To linear order in the fields, equation (6.42) is the same as equation (6.35). We usually refer to equation (6.42) as a *nonlinear realization* and to equation (6.35) as a *linear realization.*

When a symmetry is spontaneously broken and we write the Lagrangian in terms of the VEV-less fields, the Lagrangian is no longer manifestly invariant under the broken symmetry transformation. Instead, the transformation constitutes a change of basis, which we can use to our advantage by choosing a basis that makes the physics of the model more transparent. This is what we are doing here by choosing a specific gauge: $\theta(x) = -\xi(x)/v$. (It is fully legitimate to choose the phase to be related to a field.) This gauge is called the *unitary gauge.*

With this choice of gauge,

$$\phi \to \phi' = \frac{1}{\sqrt{2}}(h + v), \qquad A_{\mu} \to V_{\mu} = A_{\mu} + \frac{1}{gv}\partial_{\mu}\xi, \tag{6.43}$$

such that ϕ' has 1 DoF and V_{μ} has 3 DoF. The Lagrangian in terms of h and V_{μ} reads as

$$\begin{aligned}\mathcal{L} = &-\frac{1}{4}V_{\mu\nu}V^{\mu\nu} + \frac{1}{2}(\partial_{\mu}h)(\partial^{\mu}h) + \frac{1}{2}(g^2v^2)V_{\mu}V^{\mu} - \frac{1}{2}(2\lambda v^2)h^2 \\ &+ \frac{g^2}{2}V_{\mu}V^{\mu}h(2v + h) - \lambda v h^3 - \frac{\lambda}{4}h^4.\end{aligned} \tag{6.44}$$

The kinetic term of the gauge boson is independent of the gauge fixing. This can be seen from the fact that

$$\partial_{\mu}V_{\nu} - \partial_{\nu}V_{\mu} = \partial_{\mu}A_{\nu} - \partial_{\nu}A_{\mu}. \tag{6.45}$$

The spectrum of the model consists of a massive vector boson of mass $m_V^2 = (gv)^2$ and a massive scalar of mass $m_h^2 = 2\lambda v^2$. We note the following points:

- The sign of a mass term for a vector boson is opposite to that of a mass term for a scalar.

- The h scalar is called a *Higgs boson*. The related field, which acquires a VEV, in our case ϕ, is called the *Brout-Englert-Higgs* (BEH) *field* or the *Higgs field*.
- The source of the mass term for the vector boson is the kinetic term of the Higgs field.
- The propagator of a massive gauge boson depends on the gauge choice. In the unitary gauge, it is given by

$$(-i)\frac{g^{\mu\nu} - (k^{\mu}k^{\nu})/m_V^2}{k^2 - m_V^2}. \tag{6.46}$$

We do not discuss in detail here the issues of gauge fixing for massive gauge bosons.

- The ξ field is "eaten" to give mass to the gauge boson. It was a convenient choice to make the phase the eaten DoF. The total number of DoF does not change: instead of the scalar ξ, we have the longitudinal component of a massive vector boson.
- In the limit $g \to 0$, we have $m_V \to 0$. This situation describes a massless gauge boson and a massless scalar. We see that in that limit, the longitudinal component is the massless Nambu-Goldstone boson, as expected.

The interactions of the model include scalar self-interactions and interactions of the scalar with the vector boson. We note the following points:

- The hVV coupling is proportional to the mass of the vector boson.
- The dimensionless $VVhh$ and $hhhh$ couplings are unchanged from the symmetric Lagrangian.

The Lagrangian of equation (6.39) depends on three parameters. They can be taken to be g, v, and λ. The Lagrangian of equation (6.44) has two mass terms and four interaction terms, which depend on the same three parameters. Thus, the six terms, which would be independent in the absence of a symmetry, obey three relations among them. This is a sign of SSB.

In the example given here, we consider SSB of a local $U(1)$ symmetry. However, the basic ingredients are much more generic and also apply to non-Abelian symmetries and product groups. In fact, the Standard Model incorporates SSB of a local $SU(2) \times U(1)$ symmetry. The following lessons are generic to all cases of spontaneous breaking of a local symmetry:

- SSB gives masses to the gauge bosons related to the broken generators.
- Gauge bosons related to an unbroken subgroup remain massless because their masslessness is protected by the symmetry.

The following points are common to the spontaneous breaking of both local and global symmetries:

- The field that acquires a VEV (the BEH field) must be a scalar field. Otherwise, its VEV would break Lorentz invariance.
- Spontaneous breaking of a symmetry, whether global or local, can give masses to fermions as well, via Yukawa interactions.
- States with different quantum numbers under the broken symmetry, but with the same quantum numbers under the unbroken subgroup, can mix. By "mixing," we mean that a mass eigenstate can be a linear combination of such states. We do not elaborate on that point here. Chapter 7 provides an example of mixing among vector bosons, and chapter 14 provides an example of mixing among fermions.

6.7 Summary

Symmetries in QFT have a strong predictive, or explanatory, power. The main consequences of the various types of symmetries are summarized in table 6.1.

To construct a model, we provide as input the following ingredients:

1. The symmetry
2. The transformation properties of the fermions and the scalars

Then we write the most general Lagrangian that is invariant under the symmetry up to some mass dimension in the fields. Unless explicitly stated otherwise, we truncate the Lagrangian at the renormalizable level (i.e., at dimension four in the fields).

The resulting Lagrangian has a finite number of parameters, which we need to determine by experiment. In principle, for a theory with N independent parameters, we need to perform N appropriate measurements to extract the values of the parameters. Additional measurements test this theory.

The values of the parameters can have minor or major implications. In particular, variation of values of parameters can lead to different patterns of SSB, which result in very different phenomenology. Thus, when we define a theory, the various SSB branches of it are often given different names.

Our process of building a model X starts with defining the imposed symmetry and the transformation properties of fermions and scalars under this symmetry, and continues

Table 6.1. Symmetries and some of their main consequences

Type	Consequences
Spacetime	Conservation of energy, momentum, and angular momentum
Discrete	Selection rules
Global (exact)	Conserved charges
Global (spontaneously broken)	Massless scalars
Local (exact)	Interactions, massless spin-1 mediators
Local (spontaneously broken)	Interactions, massive spin-1 mediators

with writing the most general Lagrangian that is consistent with these definitions. At this stage, we can obtain the predictions of model X that are independent of the values of the model parameters. Additional predictions can be made when the values of the parameters are determined experimentally. We sometimes use the term "*a* model X" for the *class* of models that are described by the Lagrangian, with any possible values of its parameters; and "*the* model X" for the *specific* model with the values of the parameters as realized in nature (namely, as determined by experiments). In particular, we use the terminology of "a Standard Model" and "the Standard Model" in this sense later in this book.

For Further Reading

More on the formal aspects of SSB can be found, for example, in section 2.2 of Dine [17], chapters 20 and 21 of Peskin and Schroeder [2], chapter 28 of Schwartz [15], and section 4 of Peskin [21].

Problems

Question 6.1: Algebra

1. Show that, up to linear order in all fields, equation (6.42) leads to equation (6.35).
2. Use equations (6.39), (6.42), and (6.43) to derive equation (6.44).

Question 6.2: SSB with many scalars

Consider the following model. The symmetry is global $SO(N)$. There is a real scalar field Φ in the N representation, such that there are N real scalar DoF, $\Phi = (\phi_1, \phi_2, \ldots, \phi_N)^T$. The Lagrangian is given by

$$\mathcal{L} = \frac{1}{2}\partial_\mu \Phi \partial^\mu \Phi - \frac{1}{2}\mu^2 \Phi^2 - \frac{1}{4}\lambda \Phi^4, \qquad (6.47)$$

with $\mu^2 < 0$ and $\lambda > 0$. This Lagrangian is a generalization of equation (6.12).

1. Show that $\mathcal{L}$ describes a theory with a single massive scalar of mass $m^2 = -2\mu^2$ and $N-1$ massless scalars.
2. What is the unbroken symmetry group?
3. Goldstone's theorem states that the number of massless bosons is equal to the number of broken generators. Show this explicitly for this model.

Question 6.3: Broken and unbroken symmetries

In this question, we elaborate on equations (6.31) and (6.32), which state that an unbroken generator T_a annihilates the vacuum ($T_a\langle\phi\rangle = 0$), while a spontaneously broken one does not ($T_a\langle\phi\rangle \neq 0$). Consider the operation of a group element on the vacuum:

$$\langle\phi'\rangle = e^{iT_a\theta_a}\langle\phi\rangle. \tag{6.48}$$

1. Explain why the symmetry is unbroken if $\langle\phi'\rangle = \langle\phi\rangle$ for any θ_a and that it is broken if there is a θ_a such that $\langle\phi'\rangle \neq \langle\phi\rangle$.
2. Explain why this implies that $T_a\langle\phi\rangle = 0$ if T_a corresponds to an unbroken symmetry.
3. A familiar example is the case of a vector in three dimensions, which breaks the symmetry from $SO(3)$ to $SO(2)$ (i.e., from rotations in three dimensions to rotations in the plane perpendicular to the vector). Consider a case where we choose the normalized vector to be $\vec{v} = (0, 0, 1)^T$. Show that L_z is still a symmetry, while L_x and L_y are not. It is useful to recall the explicit representation of L_i for a vector in the basis that corresponds to rotations in real space:

$$L_x = -i\begin{pmatrix} 0 & 0 & 0 \\ 0 & 0 & 1 \\ 0 & -1 & 0 \end{pmatrix} \quad L_y = -i\begin{pmatrix} 0 & 0 & -1 \\ 0 & 0 & 0 \\ 1 & 0 & 0 \end{pmatrix} \quad L_z = -i\begin{pmatrix} 0 & 1 & 0 \\ -1 & 0 & 0 \\ 0 & 0 & 0 \end{pmatrix}. \tag{6.49}$$

4. Now consider a generic normalized vector, $\vec{v} = (a, b, c)$, such that $a^2 + b^2 + c^2 = 1$. Show that

$$\left[aL_x + bL_y + cL_z\right]\vec{v} = 0. \tag{6.50}$$

   This shows that there is always one generator that is not broken, so indeed the unbroken symmetry is $SO(2)$.
5. Consider $SU(2)$ transformations. Item 4 demonstrates that a vector (the 3 irrep) breaks $SU(2)$ to $U(1)$. Here, you are asked to show that a spinor (the 2 irrep) breaks $SU(2)$ completely. To show this, we need to prove that there is no combination of generators that annihilates a spinor. The $SU(2)$ generators in the spinor representation are the Pauli matrices. Consider the spinor $\vec{s} = (0, 1)$ and show that no nonzero linear combination with real coefficients of the Pauli matrices annihilates it.

Question 6.4: More on the dark photon

Here we consider a model that is an extension of the one discussed in question 3.3 in chapter 3:

(*i*) The symmetry is a local

$$U(1)_{\mathrm{EM}} \times U(1)_D. \tag{6.51}$$

We denote the gauge bosons by A_μ and C_μ, respectively.

(*ii*) There are four fermion fields:

$$e_L(-1,0), \qquad e_R(-1,0), \qquad d_L(0,-1), \qquad d_R(0,-1). \tag{6.52}$$

(*iii*) There is a single complex scalar:

$$\phi(q_{\rm EM}, q_D). \tag{6.53}$$

We assume no kinetic mixing and use a normalization such that the coupling constants of the two groups is the same (i.e., $g_{\rm EM} = g_D = e$).

1. There are five specific charge assignments that allow Yukawa interactions (i.e., couplings between ϕ and the fermions). What are these charge assignments?

From this point on, we do not consider any of the previous options (i.e., we consider only cases where all Yukawa interactions are forbidden).

2. Write the scalar potential. What is the condition for ϕ to acquire a VEV? From here on, assume that this condition is satisfied.
3. One way to make the model possibly consistent with nature is to have partial SSB such that the photon A_μ is massless but the dark photon C_μ is massive. Explain why this is the case when $q_{\rm EM} = 0$ and $q_D \neq 0$.
4. In the case where $q_{\rm EM} = 0$ and $q_D \neq 0$, write the mass of the dark photon in terms of the model parameters.
5. Take $q_{\rm EM} \neq 0$ and $q_D \neq 0$. In this case, both $U(1)_{\rm EM}$ and $U(1)_D$ are broken. Show, however, that the breaking pattern is $[U(1)]^2 \to U(1)$. We denote the massless gauge boson as A'_μ and the massive one as C'_μ.
6. Write the couplings of the fermions to A'_μ and C'_μ.
7. We now assume that $q_{\rm EM} \ll q_D$ (and $g_{\rm EM} = g_D$). In this case, we can think of A'_μ as a small deviation from A_μ, and still call it the photon. We further assume that $m_d \sim m_e$. Experimentally, a particle with a mass of order of the electron mass and with an electromagnetic (EM) charge greater than about 10^{-3} times that of the electron is ruled out. Obtain the resulting constraint on $q_{\rm EM}/q_D$.

Question 6.5: Physics of the Higgs boson

We consider the model of section 6.3 with the Lagrangian of equation (6.21).

1. Draw the tree-level diagrams for the $hh \to hh$ scattering and write the amplitude. Note that there is more than one diagram.
2. Estimate the cross section in the limit where $E \gg v$. Here, E is the center of mass energy of the collision.
3. Consider the same model, but with $\mu^2 > 0$. Estimate the $\phi\phi^* \to \phi\phi^*$ cross section in the limit where $E^2 \gg \mu^2$. Explain the similarity of this cross section to the result of the $hh \to hh$ scattering cross section obtained in the previous item.

Question 6.6: The Sigma model

A classic example of SSB with Nambu–Goldstone bosons is provided by the σ-model. It predated quantum chromodynamics (QCD), and we now think of it as an effective theory of the strong interactions at low energies. It aims to describe the effective strong interactions between the nucleons (the proton p and the neutron n) via the exchange of three scalars (the pions π^a $(a=1,2,3)$).

Consider the following model:

(*i*) The symmetry is a global

$$SU(2)_L \times SU(2)_R \times U(1)_B. \tag{6.54}$$

(*ii*) There are two fermion fields:

$$N_L(2,1)_{+1} = \begin{pmatrix} p_L \\ n_L \end{pmatrix}, \qquad N_R(1,2)_{+1} = \begin{pmatrix} p_R \\ n_R \end{pmatrix}. \tag{6.55}$$

(*iii*) There is a single scalar field:

$$\Sigma(2,2)_0. \tag{6.56}$$

The most general Lagrangian can be written as

$$\mathcal{L} = i\overline{N_L}\partial\!\!\!/ N_L + i\overline{N_R}\partial\!\!\!/ N_R + \frac{1}{4}\mathrm{Tr}\left[\partial_\mu \Sigma^\dagger \partial^\mu \Sigma\right] - \left[g(\overline{N}_L \Sigma N_R + \text{h.c.})\right] - V(\Sigma). \tag{6.57}$$

The infinitesimal symmetry transformations on the fermion fields are chiral and given by

$$\delta N_L = i\epsilon_L^a T^a N_L, \qquad \delta N_R = i\epsilon_R^a T^a N_R. \tag{6.58}$$

1. $\mathcal{L}$ is invariant under these chiral symmetries. The fermion kinetic terms can be written in terms of the Dirac field $N=(N_L\ N_R)^T$. Write the infinitesimal symmetry transformations in the form

$$\delta N = i\epsilon^a T^a N, \qquad \delta N = i\gamma_5 \epsilon_5^a T^a N, \tag{6.59}$$

and express ϵ^a and ϵ_5^a in term of ϵ_L^a and ϵ_R^a.

What you have shown is that we can write the symmetry in a different basis. Instead of $SU(2)_L \times SU(2)_R$, we can write $SU(2)_V \times SU(2)_A$. The $SU(2)_V$ group is also called the "diagonal $SU(2)$" or "isospin symmetry," while $SU(2)_A$ is usually called the "axial $SU(2)$."

2. Show that a mass term, $m\overline{N}N$, is invariant under $SU(2)_V$, but not under $SU(2)_A$. We learn that it is the axial symmetry that forbids fermion masses.

The Σ field has 4 DoF, so we can write it in terms of four real scalar fields, σ and π^a $(a=1,2,3)$:

$$\Sigma = \sigma + i\tau_a\pi^a, \qquad a=1,2,3, \tag{6.60}$$

where τ_a are the Pauli matrices. We aim to have a model that describes nature, so we need to provide masses to the proton and the neutron. This is done via spontaneous breaking of the chiral symmetry by the Σ field acquiring a VEV.

3. The scalar potential, $V(\Sigma)$, can be written as

$$V(\Sigma) = \frac{\lambda}{4}(\Sigma^\dagger \Sigma)^2 - \frac{m^2}{2}(\Sigma^\dagger \Sigma). \tag{6.61}$$

What are the conditions for $V(\Sigma)$ to be bounded from below and for Σ to acquire a VEV? Below we assume that these conditions are satisfied.

4. Show that, up to a constant term, the potential in equation (6.61) is equivalent to

$$V(\Sigma) = \frac{1}{4}\lambda \left[\sigma^2 + \vec{\pi}^2 - F_\pi^2\right]^2, \tag{6.62}$$

with λ and F_π real and positive. F_π is the so-called pion decay constant, and it is the only mass scale in the theory.

5. What are the minima of $V(\Sigma)$? Find a minimum where only σ acquires a VEV, but the π_as do not.
6. Rewrite $\mathcal{L}$ in terms of fields that do not carry a VEV (i.e., N_L, N_R, π_a, and $s \equiv \sigma - F_\pi$). What are the masses of these fields? How many DoF are massless?
7. How many generators are broken? Check your result against Goldstone's theorem—that is, check that the number of massless scalars is the same as the number of broken generators.
8. We denote the $\overline{N}N\pi$ interaction coupling by $g_{\pi NN}$. Show that the following relation between masses and couplings holds:

$$m_N = g_{\pi NN} F_\pi. \tag{6.63}$$

This relation is known as the *Goldberger-Treiman relation*. It is satisfied in nature to a good level of accuracy. Such a relation between masses and couplings is a signal of SSB, as discussed in this chapter.

9. In nature, the pions have small masses (compared to the nucleon), which reflects a small, explicit breaking of a symmetry. What is this broken symmetry: $SU(2)_V$ or $SU(2)_A$?
10. In nature, there is a very small mass splitting between the proton and the neutron, while the model predicts that they are degenerate. Thus the model provides a very good approximation of nature. What symmetry has to be broken to generate the splitting, $SU(2)_V$ or $SU(2)_A$?

We conclude that the Sigma model predicts that the proton and neutron are degenerate and the pions are massless. These two predictions are approximately fulfilled in nature. The model also predicts the existence of the s particle, which can be identified as the $f_0(500)$ resonance. We discuss low-energy QCD in more detail in chapter 10.

7

The Leptonic Standard Model

The Leptonic Standard Model (LSM) incorporates the three aspects of imposed symmetries that have been discussed in previous chapters: Abelian symmetries, non-Abelian symmetries, and spontaneous symmetry breaking (SSB). Moreover, the model is relevant to what happens in nature. It accounts for the weak, electromagnetic, and Yukawa interactions of the leptons. The LSM is part of the Standard Model, which adds quarks and strong interactions to the LSM. We discuss the Standard Model in chapter 8.

7.1 Defining the LSM

In section 6.7, we presented the ingredients that are required to define a model. For the LSM, these ingredients are defined as follows:

1. The symmetry is a local
$$SU(2)_L \times U(1)_Y. \tag{7.1}$$
2. There are three fermion generations, each consisting of two different representations:
$$L_L^i(2)_{-1/2}, \qquad E_R^i(1)_{-1}, \qquad i=1,2,3. \tag{7.2}$$
3. There is a single scalar multiplet:
$$\phi(2)_{+1/2}. \tag{7.3}$$

What we define as the LSM is the theory where SSB occurs. Given the scalar representation of equation (7.3), the pattern of SSB is

$$SU(2)_L \times U(1)_Y \to U(1)_{\text{EM}}. \tag{7.4}$$

We use the notation $(N)_Y$, such that N is the representation under $SU(2)_L$ and Y is the charge under $U(1)_Y$, which we call *hypercharge*. What we mean by equation (7.2) is that there are nine Weyl fermion degrees of freedom (DoF) that are grouped into three copies (*generations*) of the same gauge representations. The three fermionic DoF in each generation form an $SU(2)$-doublet (of hypercharge $-1/2$) and an $SU(2)$-singlet (of hypercharge -1).

7.2 The Lagrangian

As explained in section 2.3, the most general renormalizable Lagrangian with scalar, fermion, and gauge fields can be decomposed into

$$\mathcal{L} = \mathcal{L}_{\text{kin}} + \mathcal{L}_{\psi} + \mathcal{L}_{\phi} + \mathcal{L}_{\text{Yuk}}. \tag{7.5}$$

It is now our task to find the specific form of the Lagrangian made of the L_L^i and E_R^i fermion fields of equation (7.2) and the ϕ scalar field of equation (7.3), subject to the gauge symmetry of equation (7.1) and leading to the SSB of equation (7.4).

7.2.1 $\mathcal{L}_{\text{kin}}$ and the Gauge Symmetry

The gauge group was given in equation (7.1), earlier in this chapter. It has four generators: three T_as that form the $SU(2)$ algebra,

$$[T_a, T_b] = i\varepsilon_{abc} T_c, \tag{7.6}$$

where $a, b, c = 1, 2, 3$; and a single Y that corresponds to the $U(1)$ group. The $SU(2)$ generators commute with the $U(1)$ generator, as they belong to different gauge groups:

$$[T_a, Y] = 0. \tag{7.7}$$

The local symmetry requires 4 gauge boson DoF–3 in the adjoint representation of $SU(2)$ and 1 related to the $U(1)$ symmetry:

$$W_a^\mu(3)_0, \qquad B^\mu(1)_0. \tag{7.8}$$

The corresponding field strengths are given by (see equations (2.28) in chapter 2 and (4.21) in chapter 4)

$$W_a^{\mu\nu} = \partial^\mu W_a^\nu - \partial^\nu W_a^\mu - g\varepsilon_{abc} W_b^\mu W_c^\nu, \qquad B^{\mu\nu} = \partial^\mu B^\nu - \partial^\nu B^\mu. \tag{7.9}$$

The covariant derivative is

$$D^\mu = \partial^\mu + igW_a^\mu T_a + ig' B^\mu Y. \tag{7.10}$$

Thus, there are two independent coupling constants in $\mathcal{L}_{\text{kin}}$: there is a single g for all the $SU(2)$ couplings and a different one, g', for the $U(1)$ coupling. (We set both to be positive.) The $SU(2)$ couplings must all be the same because they mix with one another under $SU(2)$ rotations. The $U(1)$ coupling can be different from that of $SU(2)$ because the generator Y never appears as a commutator of $SU(2)$ generators.

We define $\mathcal{L}_{\text{kin}}$ to include the kinetic terms of all the fields:

$$\mathcal{L}_{\text{kin}} = -\frac{1}{4} W_a^{\mu\nu} W_{a\mu\nu} - \frac{1}{4} B^{\mu\nu} B_{\mu\nu} + i\overline{L_L^i} \not{D} L_L^i + i\overline{E_R^i} \not{D} E_R^i + (D^\mu \phi)^\dagger (D_\mu \phi). \tag{7.11}$$

For the $SU(2)_L$ doublets, $T_a = \sigma_a/2$ (σ_a are the Pauli matrices), while for the $SU(2)_L$ singlets, $T_a = 0$. Explicitly,

$$D^\mu \phi = \left(\partial^\mu + \frac{i}{2} g W_a^\mu \sigma_a + \frac{i}{2} g' B^\mu\right) \phi,$$

$$D^\mu L_L^i = \left(\partial^\mu + \frac{i}{2} g W_a^\mu \sigma_a - \frac{i}{2} g' B^\mu\right) L_L^i,$$

$$D^\mu E_R^i = \left(\partial^\mu - i g' B^\mu\right) E_R^i. \tag{7.12}$$

7.2.2 $\mathcal{L}_\psi$

There are no mass terms for the fermions in the LSM. We cannot write Dirac mass terms for the fermions because they are assigned to chiral representations of the gauge symmetry. We cannot write Majorana mass terms for the fermions because they all have $Y \neq 0$. Hence,

$$\mathcal{L}_\psi = 0. \tag{7.13}$$

We learn that the LSM is a chiral theory; that is, a theory without bare mass terms for the fermions.

7.2.3 $\mathcal{L}_{\text{Yuk}}$

The Yukawa part of the Lagrangian is given by

$$-\mathcal{L}_{\text{Yuk}} = Y_{ij}^e \overline{L_L^i} E_R^j \phi + \text{h.c.}, \tag{7.14}$$

where $i, j = 1, 2, 3$. The Yukawa matrix Y^e is a general, complex 3×3 matrix of dimensionless couplings.

7.2.4 $\mathcal{L}_\phi$ and SSB

The Higgs potential, which leads to the SSB, is given by

$$-\mathcal{L}_\phi = \mu^2 \left(\phi^\dagger \phi\right) + \lambda \left(\phi^\dagger \phi\right)^2. \tag{7.15}$$

This discussion follows the same lines as the $U(1)$ and $SO(3)$ models presented in chapter 6. The quartic coupling λ is dimensionless and real, and we consider the case where it is positive, so the potential is bounded from below. The quadratic coupling μ^2 has a mass-dimension of 2 and is real. If the gauge symmetry is to be spontaneously broken, as in equation (7.4), we must take $\mu^2 < 0$. Defining

$$v^2 = -\frac{\mu^2}{\lambda}, \tag{7.16}$$

we can rewrite equation (7.15) as follows (up to a constant term):

$$-\mathcal{L}_\phi = \lambda\left(\phi^\dagger\phi - \frac{v^2}{2}\right)^2. \tag{7.17}$$

The scalar potential in equation (7.17) implies that the scalar field acquires a vacuum expectation value (VEV), $|\langle\phi\rangle| = v/\sqrt{2}$. We choose the VEV to lie in the real direction of the $T_3 = -1/2$ component:

$$\langle\phi\rangle = \begin{pmatrix} 0 \\ v/\sqrt{2} \end{pmatrix}. \tag{7.18}$$

This VEV corresponds to the following breaking pattern:

$$SU(2) \times U(1) \to U(1). \tag{7.19}$$

This statement corresponds to the fact that there is one (and only one) linear combination of generators that annihilates the vacuum state. With our specific choice, equation (7.18), it is $T_3 + Y$ (you are asked to prove it in question 7.2 at the end of this chapter). The unbroken subgroup is identified with $U(1)_{\mathrm{EM}}$, and hence its generator, Q, is identified as

$$Q = T_3 + Y. \tag{7.20}$$

(We could equally well choose the VEV to lie in the direction of the $T_3 = +1/2$ component. In this case, we would identify $Q = T_3 - Y$, and the physics would remain the same.)

Let us denote the four real components of the scalar doublet as three phases, $\theta_a(x)$ $(a = 1, 2, 3)$, and one magnitude, $h(x)$. We choose the three phases to be the three would-be Goldstone bosons, in a way that is similar to the case discussed in section 6.6 in chapter 6. In our case, the broken generators are T_1, T_2, and $T_3 - Y$, and thus we write

$$\phi(x) = \frac{1}{\sqrt{2}} \exp\left[\frac{i\left(\sigma_a\theta_a(x) - I\theta_3(x)\right)}{2v}\right]\begin{pmatrix} 0 \\ v + h(x) \end{pmatrix}. \tag{7.21}$$

The spontaneously broken $SU(2)_L \times U(1)_Y$ symmetry allows one to rotate away the explicit dependence on the three $\theta_a(x)$. They represent the three would-be Goldstone bosons that are eaten by the three gauge bosons that acquire masses as a result of the SSB. See the discussion in section 6.6. In this gauge, $\phi(x)$ has 1 DoF:

$$\phi(x) = \frac{1}{\sqrt{2}}\begin{pmatrix} 0 \\ v + h(x) \end{pmatrix}. \tag{7.22}$$

7.2.5 Summary

The renormalizable part of the LSM Lagrangian is given by

$$\mathcal{L}_{\text{LSM}} = -\frac{1}{4}W_a^{\mu\nu}W_{a\mu\nu} - \frac{1}{4}B^{\mu\nu}B_{\mu\nu} + (D^\mu\phi)^\dagger(D_\mu\phi) + i\overline{L_L^i}\not{D}L_L^i + i\overline{E_R^i}\not{D}E_R^i$$
$$- \left(Y_{ij}^e\overline{L_L^i}E_R^j\,\phi + \text{h.c.}\right) - \lambda\left(\phi^\dagger\phi - v^2/2\right)^2, \qquad (7.23)$$

where $i, j = 1, 2, 3$.

7.3 The Spectrum

7.3.1 Scalars: Back to $\mathcal{L}_\phi$

The scalar sector contains one real scalar field, which we denote by h. It is electromagnetically neutral. This is the Higgs boson of the LSM. Its mass can be obtained by plugging equation (7.22) into equation (7.17), and it is given by

$$m_h = \sqrt{2\lambda}v. \qquad (7.24)$$

Experiments give

$$m_h = 125.25 \pm 0.17 \text{ GeV}. \qquad (7.25)$$

7.3.2 Vector Bosons: Back to $\mathcal{L}_{\text{kin}}(\phi)$

Since the symmetry that is related to three of the four generators is spontaneously broken, three of the four vector bosons acquire masses, while one remains massless. Taking into account the SSB, the term that leads to gauge boson masses can be read from the kinetic term of ϕ (see equation (7.11)):

$$\mathcal{L}_{M_V} = (D_\mu\langle\phi\rangle)^\dagger(D^\mu\langle\phi\rangle). \qquad (7.26)$$

Using equation (7.12) for $D^\mu\phi$, we obtain

$$D^\mu\langle\phi\rangle = \frac{i}{\sqrt{8}}\left(gW_a^\mu\sigma_a + g'B^\mu\right)\begin{pmatrix}0\\ v\end{pmatrix} = \frac{i}{\sqrt{8}}\begin{pmatrix} gW_3^\mu + g'B^\mu & g(W_1^\mu - iW_2^\mu)\\ g(W_1^\mu + iW_2^\mu) & -gW_3^\mu + g'B^\mu\end{pmatrix}\begin{pmatrix}0\\ v\end{pmatrix}$$
$$= \frac{iv}{\sqrt{8}}\begin{pmatrix} g(W_1^\mu - iW_2^\mu)\\ -gW_3^\mu + g'B^\mu\end{pmatrix}. \qquad (7.27)$$

The mass terms for the vector bosons are thus given by

$$\mathcal{L}_{M_V} = \frac{v^2}{8}\begin{pmatrix} g(W_1 + iW_2)_\mu & -gW_{3\mu} + g'B_\mu\end{pmatrix}\begin{pmatrix} g(W_1 - iW_2)^\mu\\ -gW_3^\mu + g'B^\mu\end{pmatrix}. \qquad (7.28)$$

We define an angle θ_W via

$$\tan\theta_W \equiv \frac{g'}{g}. \tag{7.29}$$

We define four gauge boson states as follows:

$$W^{\pm}_{\mu} = \frac{1}{\sqrt{2}}(W_1 \mp iW_2)_{\mu}, \qquad Z^0_{\mu} = \cos\theta_W W_{3\mu} - \sin\theta_W B_{\mu},$$
$$A^0_{\mu} = \sin\theta_W W_{3\mu} + \cos\theta_W B_{\mu}, \tag{7.30}$$

such that $W^{\pm}_{\mu}$ form a complex field, while A_{μ} and Z_{μ} are real fields. In terms of the vector-boson fields of equation (7.30), the mass term, equation (7.28), reads as

$$\mathcal{L}_{M_V} = \frac{1}{4}g^2 v^2 W^{+\mu} W^{-}_{\mu} + \frac{1}{8}(g^2 + g'^2)v^2 Z^{\mu} Z_{\mu}. \tag{7.31}$$

We learn that the four states of equation (7.30) are the mass eigenstates, with masses:

$$m_W^2 = \frac{1}{4}g^2 v^2, \qquad m_Z^2 = \frac{1}{4}(g^2 + g'^2)v^2, \qquad m_A^2 = 0. \tag{7.32}$$

(Recall that for a complex bosonic field ϕ with mass m, the mass term is $m^2|\phi|^2$, while for a real field, it is $m^2\phi^2/2$.) Furthermore, these four states have well-defined charges under the unbroken $U(1)_{\rm EM}$ symmetry:

$$q(W^{+}_{\mu}) = +1, \qquad q(W^{-}_{\mu}) = -1, \qquad q(Z_{\mu}) = 0, \qquad q(A_{\mu}) = 0. \tag{7.33}$$

Thus, the super-indexes in equation (7.30) stand for the respective electromagnetic charges. In most cases, unless there is ambiguity, we do not write these indices.

Four points are worth emphasizing:

- As anticipated, three vector bosons acquire masses.
- $m_A^2 = 0$ provides a consistency check on our calculation.
- When a symmetry is partially broken, $G \to H$, mass eigenstates can be linear combinations of states that come from different representations of G, so long as they transform in the same way under the unbroken subgroup H. The mass eigenstates Z_{μ} and A_{μ} provide an explicit example.
- The angle θ_W represents the rotation angle of the two neutral vector bosons from the interaction basis, where fields have well-defined transformation properties under the full gauge symmetry, (W_3, B), into the mass basis for the vector bosons, (Z, A).

In chapter 6, it is emphasized that SSB leads to relations between observables that would have been independent in the absence of symmetry. One important such relation involves the vector-boson masses and their couplings:

$$\frac{m_W^2}{m_Z^2} = \frac{g^2}{g^2 + g'^2}. \tag{7.34}$$

This relation is testable. The left side can be derived from the measured spectrum and the right side from interaction rates. It is conventional to express this relation in terms of θ_W, as defined in equation (7.29):

$$\rho \equiv \frac{m_W^2}{m_Z^2 \cos^2\theta_W} = 1. \tag{7.35}$$

The $\rho = 1$ relation is a consequence of the SSB by $SU(2)$-doublets. (See question 7.6 for other possibilities.) It thus tests this specific ingredient of the LSM.

The experimental values of the weak gauge boson masses are given by

$$m_W = 80.377 \pm 0.012 \text{ GeV}; \qquad m_Z = 91.1876 \pm 0.0021 \text{ GeV}. \tag{7.36}$$

We can then use the $\rho = 1$ relation to determine $\sin^2\theta_W$:

$$\frac{m_W}{m_Z} = 0.8815 \pm 0.0001 \quad \Longrightarrow \quad \sin^2\theta_W = 1 - (m_W/m_Z)^2 = 0.2230 \pm 0.0003. \tag{7.37}$$

In sections 7.4.3 and 7.4.5, we describe the determination of $\sin^2\theta_W$ by various interaction rates. We will see that $\rho = 1$ is indeed realized in nature (within experimental errors, and up to calculable quantum corrections, which we discuss in detail in chapter 12).

7.3.3 *Fermions: Back to $\mathcal{L}_{\text{Yuk}}$*

Since the fermions in the LSM are chiral and charged under the $SU(2)_L \times U(1)_Y$ symmetry, the model has $\mathcal{L}_\psi = 0$, and fermion masses can arise from the Yukawa interaction only as a result of the SSB. The terms that lead to fermion masses can be read from $\mathcal{L}_{\text{Yuk}}$ (see equation (7.14)) as

$$-\mathcal{L}_{M_\ell} = Y^e_{ij}\overline{L^i_L}E^j_R\ \langle\phi\rangle + \text{h.c.} \tag{7.38}$$

We write explicitly the two components of the $SU(2)_L$ doublets L^i_L:

$$L^i_L = \begin{pmatrix} N^i_L \\ E^i_L \end{pmatrix}. \tag{7.39}$$

The various states have well-defined charges under the unbroken $U(1)_{\text{EM}}$ symmetry, as follows:

- E^i_L have $T_3 = -1/2$ and $Y = -1/2$, while E^i_R have $T_3 = 0$ and $Y = -1$. Hence, their electromagnetic charges are

$$q(E^i_L) = q(E^i_R) = -1. \tag{7.40}$$

- N_L^i have $T_3 = +1/2$ and $Y = -1/2$. Hence, they are neutral under $U(1)_{\rm EM}$:

$$q(N_L^i) = 0. \tag{7.41}$$

If $SU(2)_L \times U(1)_Y$ were an exact symmetry of nature, there would be no way of distinguishing particles that are members in the same $SU(2)_L$ multiplet. The SSB makes N_L^i distinguishable from E_L^i.

Replacing ϕ by its VEV, equation (7.18) in equation (7.38), and using the definitions of equation (7.39), we obtain the following mass terms:

$$-\mathcal{L}_{M_\ell} = Y_{ij}^e \begin{pmatrix} \overline{N_L^i} & \overline{E_L^i} \end{pmatrix} \begin{pmatrix} 0 \\ v/\sqrt{2} \end{pmatrix} E_R^j + \text{h.c.} = \frac{Y_{ij}^e v}{\sqrt{2}} \overline{E_L^i} E_R^j + \text{h.c.} \tag{7.42}$$

Without loss of generality (see the discussion in section 7.5.2, later in this chapter), we can choose a basis where Y^e is diagonal and real:

$$\hat{Y}^e = \begin{pmatrix} y_e & 0 & 0 \\ 0 & y_\mu & 0 \\ 0 & 0 & y_\tau \end{pmatrix}, \tag{7.43}$$

with the convention that $y_e < y_\mu < y_\tau$. In the basis defined by equation (7.43), we denote the left-handed fermion fields as follows:

$$L_{Le} = \begin{pmatrix} \nu_{eL} \\ e_L \end{pmatrix}, \qquad L_{L\mu} = \begin{pmatrix} \nu_{\mu L} \\ \mu_L \end{pmatrix}, \qquad L_{L\tau} = \begin{pmatrix} \nu_{\tau L} \\ \tau_L \end{pmatrix}, \tag{7.44}$$

and the right-handed fermion fields as follows:

$$e_R, \qquad \mu_R, \qquad \tau_R. \tag{7.45}$$

The charged lepton mass terms are given in this basis by

$$-\mathcal{L}_{m_\ell} = \frac{y_e v}{\sqrt{2}} \overline{e_L}\, e_R + \frac{y_\mu v}{\sqrt{2}} \overline{\mu_L}\, \mu_R + \frac{y_\tau v}{\sqrt{2}} \overline{\tau_L}\, \tau_R + \text{h.c.} \tag{7.46}$$

We conclude that the charged leptons, e, μ, and τ, are Dirac fermions of masses:

$$m_e = \frac{y_e v}{\sqrt{2}}, \qquad m_\mu = \frac{y_\mu v}{\sqrt{2}}, \qquad m_\tau = \frac{y_\tau v}{\sqrt{2}}. \tag{7.47}$$

We often refer to them as the three lepton flavors. We also see that the mass basis for the fermions is the one where Y^e is diagonal and real (equation (7.43)). The charged lepton masses have been measured as follows:

$$m_e = 0.51099895000(15)\ \text{MeV}, \qquad m_\mu = 105.6583755(23)\ \text{MeV},$$
$$m_\tau = 1776.86(12)\ \text{MeV}. \tag{7.48}$$

The crucial point is that while the leptons are in a chiral representation of the full gauge group $SU(2)_L \times U(1)_Y$, the charged leptons—e, μ, τ—are in a vectorial representation of $U(1)_{\rm EM}$, the subgroup that is not spontaneously broken. This situation is the key to the possibility of acquiring masses as a result of the SSB, as realized in equation (7.46).

Unlike the charged leptons, the neutrinos ν_e, ν_μ, and ν_τ are massless Weyl fermions:

$$m_{\nu_e} = m_{\nu_\mu} = m_{\nu_\tau} = 0. \tag{7.49}$$

(We drop the explicit index L from the neutrino fields.) Their masslessness in the LSM can be understood as follows. The LSM has no right-handed fields in the $(2)_{-1/2}$ representation, so there are no Dirac mass terms for the neutrinos. The LSM has neither $(1)_0$ fermion fields nor $(3)_1$ scalar fields, so the neutrinos can acquire neither a Dirac nor a Majorana mass as a result of the SSB. A priori, since the neutrinos have no charge under $U(1)_{\rm EM}$, the possibility of acquiring Majorana masses is not closed, but the neutrinos do not acquire Majorana masses from renormalizable terms. Thus, the lepton number is an accidental symmetry of the theory (see section 7.5.1). Since the masslessness of the neutrinos is protected by a symmetry, the neutrinos are exactly massless, and not only at the tree level.

Experimentally, the neutrinos were found to have very small masses—much smaller than all the other masses in the LSM spectrum. Thus, the LSM prediction of massless neutrinos is violated, and yet it constitutes a very good approximation to nature. We discuss neutrino masses and possible ways to generate them in chapter 14.

7.3.4 Summary

We presented the details of the spectrum of the LSM in this section. These are summarized in table 7.1. All masses are proportional to the VEV of the scalar field, v. For the three massive gauge bosons and the three charged leptons, this must be the case: In the absence of SSB, the former would be protected against acquiring a mass by the gauge symmetry

Table 7.1. The LSM particles

Particle	Spin	Charge	Mass (theoretical)
$W^\pm$	1	± 1	$gv/2$
Z^0	1	0	$\sqrt{g^2+g'^2}\,v/2$
A^0	1	0	0
h	0	0	$\sqrt{2\lambda}\,v$
e	1/2	-1	$y_e v/\sqrt{2}$
μ	1/2	-1	$y_\mu v/\sqrt{2}$
τ	1/2	-1	$y_\tau v/\sqrt{2}$
ν_e	1/2	0	0
ν_μ	1/2	0	0
ν_τ	1/2	0	0

and the latter by their chiral nature. For the Higgs boson, the situation is different, as a mass term does not violate any symmetry. Here, it is just a manifestation of the fact that the LSM has a single dimensionful parameter, which can be taken to be v, and therefore all masses must be proportional to this parameter.

7.4 The Interactions

In this section, we obtain the interactions among the mass eigenstates of the LSM. The scalar potential of equation (7.15) leads to Higgs boson self-interactions. The Yukawa terms of equation (7.14) lead to Higgs-mediated Yukawa interactions among the charged leptons. The kinetic terms of equation (7.11) lead to three types of interactions that are mediated by vector bosons: The photon-mediated electromagnetic interactions (Quantum ElectroDynamics (QED)), the Z-mediated weak interactions (neutral current weak interactions), and the $W^{\pm}$-mediated weak interactions (charged current weak interactions). To obtain these three types of interactions, we need to rewrite the covariant derivative given in equation (7.10) in terms of the vector boson mass eigenstates, A, Z, and $W^{\pm}$, as defined in equation (7.30):

$$D_\mu = \partial_\mu + ig(W_\mu^+ T^+ + W_\mu^- T^-) \qquad (7.50)$$
$$+ i(g \sin\theta_W\, T_3 + g' \cos\theta_W\, Y)A_\mu + i(g\cos\theta_W\, T_3 - g' \sin\theta_W\, Y)Z_\mu,$$

where

$$T^{\pm} = (T_1 \pm iT_2)/\sqrt{2}. \qquad (7.51)$$

7.4.1 The Higgs Boson

The interactions involving the Higgs boson, h, are given by

$$\mathcal{L}^h_{\text{int}} = \mathcal{L}^h_h + \mathcal{L}^h_V + \mathcal{L}^h_f$$
$$-\mathcal{L}^h_h = \frac{m_h^2}{2v}h^3 + \frac{m_h^2}{8v^2}h^4,$$
$$\mathcal{L}^h_V = \left(m_W^2 W_\mu^- W^{\mu +} + \frac{1}{2}m_Z^2 Z_\mu Z^\mu\right)\left(\frac{2h}{v} + \frac{h^2}{v^2}\right),$$
$$-\mathcal{L}^h_f = \frac{h}{v}\left(m_e\, \overline{e_L}\, e_R + m_\mu\, \overline{\mu_L}\, \mu_R + m_\tau\, \overline{\tau_L}\, \tau_R + \text{h.c.}\right). \qquad (7.52)$$

We write $\mathcal{L}^h_{\text{int}}$ in a way that demonstrates that all of the Higgs couplings can be expressed in terms of the masses of the particles to which it couples.

The Higgs boson self-couplings are proportional to the square of its mass, $m_h^2 = 2\lambda v^2$. The dimensionless h^4 coupling,

$$\frac{m_h^2}{8v^2} = \frac{\lambda}{4}, \qquad (7.53)$$

is unchanged from the quartic coupling of ϕ in equation (7.15). The h^3 coupling,

$$\frac{m_h^2}{2v} = \lambda v, \tag{7.54}$$

arises as a consequence of the SSB.

The Higgs couplings to the weak interaction vector bosons are proportional to their masses squared. The dimensionless $hhVV$ couplings,

$$\frac{m_W^2}{v^2} = \frac{g^2}{4}, \qquad \frac{m_Z^2}{2v^2} = \frac{g^2 + g'^2}{8}, \tag{7.55}$$

are unchanged from those of ϕ in equation (7.11). The hVV couplings,

$$\frac{2m_W^2}{v} = \frac{g^2 v}{2}, \qquad \frac{m_Z^2}{v} = \frac{(g^2 + g'^2)v}{4}, \tag{7.56}$$

arise as a consequence of the SSB.

There is neither hAA nor $hhAA$ coupling. One can understand the absence of these couplings in two ways. First, the Higgs boson is electromagnetically neutral, so it should not couple to the electromagnetic force carrier. Second, the photon is massless, so it should not couple to the Higgs boson.

The Yukawa couplings of the Higgs bosons to the charged leptons are proportional to their masses. The $h\ell^+\ell^-$ couplings,

$$\frac{m_\ell}{v} = \frac{y_\ell}{\sqrt{2}}, \tag{7.57}$$

are unchanged from equation (7.14).

There is no coupling of the Higgs to the neutrinos. This is related to the masslessness of the neutrinos.

7.4.2 QED: Electromagnetic Interactions

Here, we study the photon interactions to fermions and show that we recover QED. Using equation (7.50) and the definition of θ_W in equation (7.29), we find that the photon coupling is proportional to

$$g \sin\theta_W \, T_3 + g' \cos\theta_W \, Y = \frac{gg'}{\sqrt{g^2 + g'^2}}(T_3 + Y) = \frac{gg'}{\sqrt{g^2 + g'^2}} Q. \tag{7.58}$$

This is what we should have obtained. The coupling is proportional to $T_3 + Y$, which we defined as Q, the generator of $U(1)_{EM}$. We conventionally define

$$e = \frac{gg'}{\sqrt{g^2 + g'^2}}, \tag{7.59}$$

such that

$$g = \frac{e}{\sin\theta_W}, \qquad g' = \frac{e}{\cos\theta_W}. \tag{7.60}$$

Thus, the electromagnetic interactions are described by

$$\mathcal{L}_{A\bar{f}f} = e\,\overline{\ell_i}\not{A}\ell_i. \tag{7.61}$$

The $\ell_{1,2,3} = e, \mu, \tau$ fields are the Dirac fermions with $Q = -1$ that are formed from the $T_3 = -1/2$ component of a left-handed lepton doublet and from a right-handed lepton singlet (e.g., $\tau = (\tau_L, \tau_R)^T$). Note that the interaction in equation (7.61) is the same as the one from the QED Lagrangian, equation (3.20) in chapter 3. QED is now understood as the part of the LSM Lagrangian that involves the photon field, A, and the massive charged Dirac fermions, ℓ_i.

QED interactions are discussed in chapter 3. Here, we only reemphasize some important features that arise from equation (7.61):

1. The photon couplings are vector-like: Since the left-handed and right-handed fields carry the same charge, the photon couples to them in the same way.
2. The electromagnetic interactions are *P, C, and T conserving.*
3. *Diagonality:* The photon couples to e^+e^-, $\mu^+\mu^-$, and $\tau^+\tau^-$, but not to $e^\pm\mu^\mp$, $e^\pm\tau^\mp$, or $\mu^\pm\tau^\mp$ pairs. Thus, electromagnetic interactions do not change leptonic flavor. This is a result of the unbroken local $U(1)_{\rm EM}$ symmetry.
4. *Universality*: The couplings of the photon to the different generations are universal. This is a result of the $U(1)_{\rm EM}$ gauge invariance.

7.4.3 Neutral Current Weak Interactions

Here, we study the Z-boson interactions with fermions. Using equation (7.50) and the definition of θ_W in equation (7.29), we find that the Z-boson coupling is proportional to

$$g\cos\theta_W\, T_3 - g'\sin\theta_W\, Y = \frac{g}{\cos\theta_W}(T_3 - \sin^2\theta_W Q). \tag{7.62}$$

This leads to the following interactions between the Z-boson and the fermions:

$$\mathcal{L}_{Z\bar{f}f} = \frac{g}{\cos\theta_W}\sum_{\ell=e,\mu,\tau}\left[-\left(\frac{1}{2} - \sin^2\theta_W\right)\overline{\ell_L}\not{Z}\ell_L + \sin^2\theta_W\,\overline{\ell_R}\not{Z}\ell_R + \frac{1}{2}\,\overline{\nu_{\ell L}}\not{Z}\nu_{\ell L}\right]. \tag{7.63}$$

Interactions mediated by Z-exchange are called *neutral current weak interactions.* Equation (7.63) reveals some important features of the neutral current weak interactions:

1. Unlike the photon, the Z-boson couples to neutrinos.
2. *Parity violation:* The Z-boson couplings are chiral. Since left-handed and right-handed fields carry different T_3 values, the Z-boson couples to them with different strengths. Thus, the Z-interactions violate parity.

3. *Diagonality:* The Z-boson couplings are diagonal. For example, it couples to e^+e^-, $\mu^+\mu^-$, $\overline{\nu_{eL}}\nu_{eL}$, and $\overline{\nu_{\mu L}}\nu_{\mu L}$, but not to $e^\pm\mu^\mp$ and $\overline{\nu_{eL}}\nu_{\mu L}$ pairs. The diagonality holds to all orders in perturbation theory due to an accidental $[U(1)]^3$ symmetry of the LSM; for more details, see section 7.5.1.
4. *Universality*: The couplings of the Z-boson to the three fields within each of the three sectors ($\nu_{\ell L}$, ℓ_L, and ℓ_R) are universal. This is a result of a special feature of the LSM: all fermions of given chirality and given charge come from the same $SU(2) \times U(1)$ representation (more details are given in section 9.4 in chapter 9).

The above points have been experimentally tested. Below are a few examples for such tests:

1. The branching ratio of Z decays into invisible final states (which, in the LSM, are interpreted as the decay into final neutrinos) is measured to be

$$\mathrm{BR}(Z \to \nu\bar{\nu}) = (20.00 \pm 0.06)\%. \tag{7.64}$$

From equation (7.63), we obtain

$$\frac{\mathrm{BR}(Z \to \ell^+\ell^-)}{\mathrm{BR}(Z \to \nu_\ell\bar{\nu}_\ell)} = \frac{(1/2 - \sin^2\theta_W)^2 + \sin^4\theta_W}{1/4} = 1 - 4\sin^2\theta_W + 8\sin^4\theta_W. \tag{7.65}$$

2. As an example of parity violation, we consider the τ polarization P_τ (also denoted as A_τ) in the $Z \to \tau^+\tau^-$ decay. It is given by

$$P_\tau \equiv \frac{\sigma_L - \sigma_R}{\sigma_L + \sigma_R}, \tag{7.66}$$

where $\sigma_R(\sigma_L)$ is the cross section of producing a right-handed (left-handed) τ^- in Z decay. Experimentally,

$$P_\tau = 0.143 \pm 0.004. \tag{7.67}$$

In a parity-invariant theory, $P_\tau = 0$, and thus the measurement of $P_\tau \neq 0$ demonstrates parity violation. From equation (7.63), we obtain

$$P_\tau = \frac{(1/2 - \sin^2\theta_W)^2 - (\sin^2\theta_W)^2}{(1/2 - \sin^2\theta_W)^2 + (\sin^2\theta_W)^2}. \tag{7.68}$$

3. Diagonality has been tested by the following experimental searches:

$$\begin{aligned} \mathrm{BR}(Z \to e^+\mu^-) &< 7.5 \times 10^{-7}, \\ \mathrm{BR}(Z \to e^+\tau^-) &< 5.0 \times 10^{-6}, \\ \mathrm{BR}(Z \to \mu^+\tau^-) &< 6.5 \times 10^{-6}. \end{aligned} \tag{7.69}$$

4. Universality has been tested by Z-boson decays into charged lepton pairs:

$$\begin{aligned} \text{BR}(Z \to e^+e^-) &= (3.363 \pm 0.004)\%\,, \\ \text{BR}(Z \to \mu^+\mu^-) &= (3.366 \pm 0.007)\%\,, \\ \text{BR}(Z \to \tau^+\tau^-) &= (3.370 \pm 0.008)\%\,. \end{aligned} \tag{7.70}$$

These results confirm universality:

$$\begin{aligned} \Gamma(\mu^+\mu^-)/\Gamma(e^+e^-) &= 1.0009 \pm 0.0028, \\ \Gamma(\tau^+\tau^-)/\Gamma(e^+e^-) &= 1.0020 \pm 0.0032. \end{aligned} \tag{7.71}$$

Z decays provide additional ways to determine $\sin^2\theta_W$, as demonstrated by the following two examples. Using equation (7.65) and the central values from equations (7.64) and (7.70), we obtain

$$\sin^2\theta_W = 0.226. \tag{7.72}$$

Using equation (7.68) and the central value from equation (7.67), we obtain

$$\sin^2\theta_W = 0.232. \tag{7.73}$$

Both determinations are in agreement with equation (7.37). An additional determination, from W-mediated decay, is given in equation (7.88). We discuss the various determinations of $\sin^2\theta_W$ in more detail in chapter 12.

We end this subsection by giving some definitions. There are several ways to write the coupling of the Z-boson to the fermions. Equation (7.63) gives the couplings of the Z-boson to left-handed and right-handed charged leptons:

$$\begin{aligned} \mathcal{L}_{Z\ell\bar{\ell}} &= \frac{g}{\cos\theta_W}\left[g_L^e \overline{\ell_{iL}} \not{Z} \ell_{iL} + g_R^e \overline{\ell_{iR}} \not{Z} \ell_{iR}\right], \\ g_L^e &= -\frac{1}{2} + \sin^2\theta_W, \qquad g_R^e = \sin^2\theta_W. \end{aligned} \tag{7.74}$$

One could write these Z-boson couplings in terms of the Dirac fermions:

$$\begin{aligned} \mathcal{L}_{Z\ell\bar{\ell}} &= \frac{g}{2\cos\theta_W}\left[\bar{\ell}(g_V^e + g_A^e\gamma_5)\not{Z}\ell\right], \\ g_V^e &= g_R^e + g_L^e = -\frac{1}{2} + 2\sin^2\theta_W, \qquad g_A^e = g_R^e - g_L^e = \frac{1}{2}. \end{aligned} \tag{7.75}$$

In chapter 8, we consider additional fermions of various Q and T_3 assignments. The generalization of the definitions of $g_{L,R}$ and $g_{V,A}$ to such fermions is straightforward:

$$g_R = \frac{g_V + g_A}{2} = T_3^R - Q\sin^2\theta_W, \qquad g_L = \frac{g_V - g_A}{2} = T_3^L - Q\sin^2\theta_W, \tag{7.76}$$

where the flavor index is implicit, and

$$g_V = g_R + g_L = T_3^R + T_3^L - 2Q\sin^2\theta_W, \qquad g_A = g_R - g_L = T_3^R - T_3^L. \tag{7.77}$$

7.4.4 Charged Current Weak Interactions

Here, we study $W^\pm$-boson interactions with fermions. Using equation (7.50), and the explicit form of the $T^\pm$ matrices for doublets,

$$T^+ = \frac{1}{\sqrt{2}}\begin{pmatrix} 0 & 1 \\ 0 & 0 \end{pmatrix}, \qquad T^- = \frac{1}{\sqrt{2}}\begin{pmatrix} 0 & 0 \\ 1 & 0 \end{pmatrix}, \tag{7.78}$$

we find that the W-boson couplings to fermion pairs are given by

$$\mathcal{L}_{W\bar{f}f} = -\frac{g}{\sqrt{2}} \sum_{\ell=e,\mu,\tau} \left(\overline{\nu_{\ell L}}\, W\!\!\!/^{+} \ell_L^- + \overline{\ell_L^-}\, W\!\!\!/^{-} \nu_{\ell L} \right). \tag{7.79}$$

The interactions mediated by the $W^\pm$ vector-bosons are called *charged current weak interactions*. They are unique among the interactions of the LSM, as the fermion pairs to which the W-boson couples consist of two different fermions, a neutrino and a charged lepton. This must be the case, as the W-bosons are charged, so they must change the identity of the particle with which they interact.

Equation (7.79) reveals important features of the charged current weak interactions.

1. *Parity violation*: The W-boson couplings are chiral. More specifically, only left-handed fields take part in the charged current weak interactions. Thus, the W-interactions violate parity.
2. *Universality*: The couplings of the W-boson to $\tau\bar{\nu}_\tau$, $\mu\bar{\nu}_\mu$, and $e\bar{\nu}_e$ are equal. This is a result of the local nature of the imposed $SU(2)$: a global symmetry would have allowed an independent coupling to each lepton pair.

These predictions have been experimentally tested as follows:

1. Consider polarized $\mu^- \to e^- \nu_\mu \overline{\nu_e}$ decay. We define θ to be the angle between the muon polarization and the electron. Parity transformation does not change the spin direction, but it does reverse the momentum direction of the electron, leading to $\theta \to \pi - \theta$. We learn that a parity-invariant spectrum must be even in $\cos\theta$. Experiments are consistent with the LSM prediction for the differential decay rate in the muon rest frame:

$$\frac{d^2\Gamma}{dx\, d(\cos\theta)} \propto x^2 \left[3 - 2x + P_\mu(2x-1)\cos\theta\right], \tag{7.80}$$

where $x \equiv E_e/E_e^{\max}$ and P_μ is the polarization of the muon. The $\cos\theta$ dependence proves that parity is violated.

2. Universality has been tested by W-boson decays into lepton pairs:

$$\begin{aligned}
\mathrm{BR}(W^+ \to e^+\nu_e) &= (10.71 \pm 0.16) \times 10^{-2}, \\
\mathrm{BR}(W^+ \to \mu^+\nu_\mu) &= (10.63 \pm 0.15) \times 10^{-2}, \\
\mathrm{BR}(W^+ \to \tau^+\nu_\tau) &= (11.38 \pm 0.21) \times 10^{-2}.
\end{aligned} \tag{7.81}$$

These results confirm universality:

$$\begin{aligned}
\Gamma(\mu^+\nu)/\Gamma(e^+\nu) &= 0.993 \pm 0.015, \\
\Gamma(\tau^+\nu)/\Gamma(e^+\nu) &= 1.062 \pm 0.024.
\end{aligned} \tag{7.82}$$

7.4.5 *The Fermi Constant*

When discussing low-energy effects of the weak interaction, it is useful to define the Fermi constant:

$$G_F \equiv \frac{\pi\alpha}{\sqrt{2}\sin^2\theta_W m_W^2} = \frac{g^2}{4\sqrt{2}m_W^2} = \frac{1}{\sqrt{2}v^2}, \tag{7.83}$$

where $g^2 = 4\pi\alpha/\sin^2\theta_W$ based on equations (3.10) in chapter 3 and (7.60), and $m_W^2 = g^2v^2/4$ based on equation (7.32).

To demonstrate the usefulness of this definition, consider the $\mu^- \to e^-\nu_\mu\bar{\nu}_e$ decay. At low energy ($q^2 \ll m_W^2$), the W-mediated interaction is well approximated via a four-fermion coupling:

$$\mathcal{A}_{\mu^-\to e^-\nu_\mu\bar{\nu}_e} \propto \frac{g^2}{m_W^2 - q^2} \approx \frac{g^2}{m_W^2} = \frac{4\pi\alpha}{\sin^2\theta_W m_W^2} \equiv 4\sqrt{2}G_F. \tag{7.84}$$

The measured muon lifetime,

$$\tau_\mu = (2.196981 \pm 0.000002) \times 10^{-6}\ \mathrm{s}, \tag{7.85}$$

determines G_F via

$$\Gamma_\mu = \frac{1}{\tau_\mu} = \frac{G_F^2 m_\mu^5}{192\pi^3} f(m_e^2/m_\mu^2)(1+\delta_{\mathrm{RC}}), \qquad f(x) = 1 - 8x + 8x^3 - x^4 - 12x^2\log x, \tag{7.86}$$

where $f(x)$ is the phase-space function for a three-body decay with two massless final particles (it is normalized to 1 in the case when all final particles are massless) and δ_{RC} encodes radiative corrections and was calculated to $\mathcal{O}(\alpha^2)$. One gets

$$G_F = 1.1663787(6) \times 10^{-5}\ \mathrm{GeV}^{-2}. \tag{7.87}$$

We note the following points:

- The Fermi constant G_F can be used as yet another independent way to determine $\sin^2\theta_W$. Using equation (7.83) and the central values of α in equation (3.11), m_W in equation (7.36), and G_F in equation (7.87), we obtain

$$\sin^2\theta_W = 0.215, \tag{7.88}$$

in good agreement with equations (7.37) and (7.72). Differences among the three values are accounted for by higher-order radiative corrections, which are discussed in chapter 12.

- The Fermi constant G_F also determines the VEV of the Higgs field. Using the expression for G_F in equation (7.83), we obtain

$$v = (\sqrt{2}G_F)^{-1/2} \approx 246 \text{ GeV}. \tag{7.89}$$

- Historically, G_F was used as the effective four-fermion coupling well before the LSM was conceived. The LSM reveals that G_F is a not a fundamental constant, but rather a quantity that is derived from the parameters of the LSM.

7.4.6 Gauge Boson Self-interactions

The gauge boson self-interactions that are presently most relevant to experiments are the W^+W^-V ($V = Z, A$) couplings which, in the LSM, take the following form:

$$\begin{aligned}\mathcal{L}_{WWV} = {} & ie\cot\theta_W\left[(W^+_{\mu\nu}W^{-\mu} - W^-_{\mu\nu}W^{+\mu})Z^\nu + W^+_\mu W^-_\nu Z^{\mu\nu}\right] \\ & + ie\left[(W^+_{\mu\nu}W^{-\mu} - W^-_{\mu\nu}W^{+\mu})A^\nu + W^+_\mu W^-_\nu F^{\mu\nu}\right].\end{aligned} \tag{7.90}$$

Here, $W^\pm_{\mu\nu} = \partial_\mu W^\pm_\nu - \partial_\nu W^\pm_\mu$, $Z_{\mu\nu} = \partial_\mu Z_\nu - \partial_\nu Z_\mu$, and $F_{\mu\nu} = \partial_\mu A_\nu - \partial_\nu A_\mu$. (Do not confuse the definitions that we use here for the mass eigenstates to the ones that we use in equation (7.9) for the interaction eigenstates.)

The WWV interactions depend on only two parameters, e and θ_W. Moreover, these parameters can be measured from other sectors of the theory. Thus, they can be used to test the theory. To date, the LSM has passed all such tests.

Finally, we present the quartic vector-boson couplings within the LSM:

$$\begin{aligned}\mathcal{L}_{4V} = {} & g^2\cos^2\theta_W\left(W^+_\mu W^-_\nu Z^\mu Z^\nu - W^+_\mu W^{-\mu}Z_\nu Z^\nu\right) \\ & + g^2\left(W^+_\mu W^-_\nu A^\mu A^\nu - W^+_\mu W^{-\mu}A_\nu A^\nu\right) \\ & + \frac{g^2}{2}\left(W^+_\mu W^-_\nu\right)\left(W^{+\mu}W^{-\nu} - W^{+\nu}W^{-\mu}\right) \\ & + e^2\cot\theta_W\left[\left(W^+_\mu W^-_\nu\right)\left(Z^\mu A^\nu - Z^\nu A^\mu\right) - 2W^+_\mu W^{-\mu}Z_\nu A^\nu\right].\end{aligned} \tag{7.91}$$

The experimental precision is not yet good enough to probe these couplings significantly.

Table 7.2. The LSM lepton interactions

Interaction	Force Carrier	Coupling	Fermions
Electromagnetic	γ	eQ	e
Neutral current weak	Z^0	$(g/\cos\theta_W)(T_3 - \sin^2\theta_W Q)$	e, ν
Charged current weak	$W^{\pm}$	$(g/\sqrt{2})$	$\bar{\nu}e$
Yukawa	h	y_ℓ	e

7.4.7 Summary

Leptons have four types of interactions, which are summarized in table 7.2.

The name "weak interactions" is somewhat misleading. In fact, the weak coupling g is larger than the electromagnetic coupling e. The more important feature is that the weak interactions are mediated by massive vector bosons; and consequently, they are short range, while the electromagnetic interactions are mediated by the massless photon, and hence they are long range. It is the short range of the weak interactions that suppresses the related cross sections.

7.5 Global Symmetries and Parameters

In section 7.5.1, we identify the accidental global symmetries of the kinetic terms and of the full Lagrangian, and give examples of the resulting predictions. Preparing for the identification of the physical parameters, we describe the interaction and mass bases and the relation between them in section 7.5.2, and explain the symmetry-related method for counting the number of physical parameters in section 7.5.3. The LSM parameters are counted and identified in section 7.5.4.

7.5.1 Accidental Symmetries

If we set the Yukawa couplings to zero ($\mathcal{L}_{\rm Yuk}=0$), the fermion sector of the LSM gains a large accidental global symmetry:

$$G_{\rm LSM}^{\rm global}(Y^e=0) = U(3)_L \times U(3)_E = SU(3)_L \times SU(3)_E \times U(1)_L \times U(1)_E. \quad (7.92)$$

Under this symmetry, the L_L^i fields transform as $(3,1)_{q_L,0}$, the E_R^i fields transform as $(1,3)_{0,q_E}$, and all other fields are singlets, $(1,1)_{0,0}$. Concerning the $U(1)$ factors, the choice of q_L and q_E is arbitrary (except that both must not equal zero). It is customary to normalize these charges to $+1$.

The Yukawa couplings break this symmetry into the following subgroup:

$$G_{\rm LSM}^{\rm global} = U(1)_e \times U(1)_\mu \times U(1)_\tau. \quad (7.93)$$

The $U(1)$ factors are called the *electron number, muon number*, and *tau number*, respectively. The charges of ν_e and e are $(1,0,0)$, the charges of ν_μ and μ are $(0,1,0)$, and the charges of ν_τ and τ are $(0, 0, 1)$. Thus, the electron number, muon number, and tau number are conserved charges in the LSM. This situation allows, for example, the muon decay mode $\mu^- \to e^- \overline{\nu_e} \nu_\mu$, but it forbids $\mu^- \to e^- \gamma$ and $\mu^- \to e^- e^+ e^-$. Also, scattering processes such as $e^+ e^- \to \mu^+ \mu^-$ are allowed, but $e^+ \mu^- \to \mu^+ e^-$ is forbidden.

It is useful to define the total lepton number, which is the sum of the three lepton flavor numbers and corresponds to a $U(1)_L$ symmetry. (This $U(1)_L$ symmetry is different from the $U(1)_L$ of equation (7.92). From here on, $U(1)_L$ refers to the total lepton number.) Clearly, the total lepton number is also an accidental symmetry of the LSM. Since all neutrinos are charged under $U(1)_L$, the conservation of the total lepton number explains why Majorana masses for them are not allowed within the LSM.

These accidental symmetries, however, are all broken by nonrenormalizable terms, such as

$$\frac{1}{\Lambda} L_{Li}^T L_{Lj} \phi^T \phi. \tag{7.94}$$

(As explained in section 1.2.2 in chapter 1, for the sake of simplicity, we write L_L^T instead of $\overline{L_L^c}$.) If the scale Λ is high enough, these breaking effects are very small. It means that the forbidden processes mentioned in this discussion are expected to occur, but at very low rates. It also implies that we should expect very small Majorana masses. We discuss these points in detail in chapter 14.

7.5.2 The Interaction Basis and the Mass Basis

- An interaction basis is defined to be one where all fields have well-defined transformation properties under the imposed symmetries of the Lagrangian. In particular, in this basis, the gauge interactions are universal.
- A mass basis is defined to be one where all fields have well-defined masses. In this basis, all fields have well-defined transformation properties under the symmetries that are not spontaneously broken. The fields in this basis correspond to the particles that are eigenstates of free propagation in spacetime.

For the LSM, the interaction eigenstates that have well-defined transformation properties under the $SU(2)_L \times U(1)_Y$ symmetry are the following:

$$W_a(3)_0, \quad B(1)_0, \quad L_L^i(2)_{-1/2}, \quad E_R^i(1)_{-1}, \quad \phi(2)_{+1/2}, \tag{7.95}$$

with $i = 1, 2, 3$. The mass eigenstates that have well-defined electromagnetic charge and mass are the following:

$$W^\pm,\ Z^0,\ A^0,\ e^-,\ \mu^-,\ \tau^-,\ \nu_e,\ \nu_\mu,\ \nu_\tau,\ h^0. \tag{7.96}$$

The number of DoF is the same in both bases. To verify this statement, one has to take into account the following features:

- W_a and B have only transverse components, while $W^\pm$ and Z^0 also have a longitudinal one.

- L_L and E_R are Weyl fermions, while e, μ, τ are Dirac fermions.
- ϕ is a complex scalar, while h is a real one.

In section 7.3, we showed how to transform from the interaction basis to the mass basis for the LSM bosons. Here, we discuss this basis transformation for the LSM fermions or, more generally, for cases where there are several copies of fields with the same representation.

If there are several fields with the same quantum numbers, f_i, then the interaction basis is not unique. The kinetic and gauge terms are invariant under a global unitary transformation among these fields. On the other hand, the Yukawa terms and the fermion mass terms are generally not invariant under a unitary transformation among fermion fields with the same quantum numbers, $f_i \to U_{ji} f_i$. Thus, by performing such transformations, we are changing the interaction basis.

In the LSM, there are three copies of $(2)_{-1/2}$ left-handed fermions, $L_L = (L_L^1, L_L^2, L_L^3)^T$, and three copies of $(1)_{-1}$ right-handed fermions, $E_R = (E_R^1, E_R^2, E_R^3)^T$. Let us rewrite $\mathcal{L}_{\rm Yuk}$ of equation (7.14) in the following way:

$$\overline{L_L} Y^e E_R\, \phi = \overline{L_L}(V_{eL}^\dagger V_{eL}) Y^e (V_{eR}^\dagger V_{eR}) E_R\, \phi = (\overline{L_L} V_{eL}^\dagger)(V_{eL} Y^e V_{eR}^\dagger)(V_{eR} E_R)\, \phi, \quad (7.97)$$

where V_{eL} and V_{eR} are unitary matrices; and in the first equality, we simply inserted unit matrices. We can choose V_{eL} and V_{eR} to be the ones that are used in the bi-unitary transformation that would make Y^e real and diagonal:

$$Y^e \to V_{eL} Y^e V_{eR}^\dagger = Y^e_{\rm diag} = {\rm diag}(y_e, y_\mu, y_\tau). \quad (7.98)$$

While in a generic basis, Y^e is a 3×3 complex matrix and thus generally has nine complex parameters, equation (7.97) demonstrates that, without loss of generality, one can find a basis where the Yukawa matrix depends on only three real parameters. Often, one chooses a basis where the number of Lagrangian parameters is minimal, as is the case with the diagonal basis of equation (7.98). One could work in any other basis. However, when calculating physical observables, only the eigenvalues of $Y_e^\dagger Y_e$ would play a role. Using the diagonal basis just provides a shortcut to this result.

As explained in section 7.3.3, the basis where the Yukawa matrix is diagonal constitutes the mass basis. We thus identify $(L_{Le}, L_{L\mu}, L_{L\tau})^T = V_{eL} L_L$ and $(e_R, \mu_R, \tau_R)^T = V_{eR} E_R$ as the charged lepton mass eigenstate fields.

The case for the neutrinos is different since, at the renormalizable level, they are massless and, in particular, they are degenerate. Thus, there is freedom in choosing the mass basis for the neutrinos. We choose the basis where the $W^\pm$ couplings to the charged lepton mass eigenstates are diagonal. One could choose a different mass basis, related to the one that we chose by a unitary transformation of the three neutrino fields:

$$\begin{pmatrix} \nu_e \\ \nu_\mu \\ \nu_\tau \end{pmatrix} \to \begin{pmatrix} \nu_1 \\ \nu_2 \\ \nu_3 \end{pmatrix} = U^\dagger \begin{pmatrix} \nu_e \\ \nu_\mu \\ \nu_\tau \end{pmatrix}. \quad (7.99)$$

Let us see how the decay rate of the W-boson into an electron and a neutrino is calculated in this basis. Since the experiment does not distinguish between ν_1, ν_2, and ν_3, one has to sum over all three species:

$$\Gamma(W^+ \to e^+\nu) = \sum_{i=1,2,3} \Gamma(W^+ \to e^+\nu_i) = \Gamma(W^+ \to e^+\nu_e)(|U_{e1}|^2 + |U_{e2}|^2 + |U_{e3}|^2)$$
$$= \Gamma(W^+ \to e^+\nu_e). \tag{7.100}$$

Thus, if the neutrinos are degenerate, the elements of the matrix U have no physical significance and cannot appear in any physical observable. Our choice of basis $(\nu_e, \nu_\mu, \nu_\tau)$ provides a shortcut to this result.

In chapter 14, it is shown that nonrenormalizable terms provide the neutrinos with nondegenerate masses, and then the neutrino mass basis becomes unique.

7.5.3 Parameter Counting

Before we discuss the LSM parameters in detail, we explain the basics of identifying the number of physical parameters. The Lagrangian written in a general interaction basis might include a number of parameters that is larger than the number of physical parameters. This means that when we express physical observables in terms of Lagrangian parameters, only a subset of these parameters (or combinations of them) appears. It also means that there exists a specific basis where the number of Lagrangian parameters is minimal and equals the number of physical parameters. For the purpose of testing a model, it is important to count and identify its physical parameters. In this subsection, we explain how to determine the number of physical parameters.

We start with a simple example: the hydrogen atom. It is invariant under spatial rotations, which are described by the $SO(3)$ group. Furthermore, there is an energy eigenvalue degeneracy of the Hamiltonian: states with different angular momenta have the same energy. This degeneracy is a consequence of the symmetry of the system.

Switching on a magnetic field changes the situation. Without loss of generality, we can define the direction of the magnetic field as the positive z-direction. Consider this choice more carefully. A generic uniform magnetic field is described by three real numbers: the three components of the magnetic field (B_x, B_y, B_z). The magnetic field breaks the $SO(3)$ symmetry of the hydrogen atom system to an $SO(2)$ symmetry of rotations in the plane perpendicular to the magnetic field. Now the one generator of the $SO(2)$ symmetry is the only valid symmetry generator; the remaining two $SO(3)$ generators in the orthogonal plane are broken. These broken symmetry generators allow us to rotate the system such that the magnetic field points in the z-direction:

$$O_{xz}O_{yz}(B_x, B_y, B_z) = (0, 0, B_z'), \tag{7.101}$$

where O_{xz} and O_{yz} are rotations in the xz and yz planes, respectively. The two broken generators were used to rotate away two unphysical parameters, leaving us with one physical parameter, the magnitude of the magnetic field. We learn that when turning on the magnetic field, all measurable quantities in the system depend on only one new parameter rather than the naive three.

These results apply more generally. In particular, they are useful in studying the flavor physics of quantum field theories. Consider a gauge theory with matter content. The kinetic and gauge terms ($\mathcal{L}_{\text{kin}}$) have a certain global symmetry, G_f. In adding terms ($\mathcal{L}_\psi + \mathcal{L}_\phi + \mathcal{L}_{\text{Yuk}}$) that respect the imposed gauge symmetries, the global symmetry may be broken into a smaller symmetry group. In breaking the global symmetry, there is an added freedom to use the broken G_f generators to transform from one interaction basis to another and, in particular, rotate away unphysical parameters, as when a magnetic field is added to the hydrogen atom system.

We are interested in obtaining the number of parameters affecting physical measurements, N_{phys}. In a general basis, the added terms depend on N_{general} parameters. The global symmetry of $\mathcal{L}$, H_f, has fewer generators than the global symmetry of $\mathcal{L}_{\text{kin}}$, G_f. We call the difference in the number of generators N_{broken}. Then N_{phys} is given by

$$N_{\text{phys}} = N_{\text{general}} - N_{\text{broken}}. \tag{7.102}$$

Furthermore, the rule in equation (7.102) applies separately to real parameters and phases. A general $n \times n$ complex matrix can be parameterized by n^2 real parameters and n^2 phases. Imposing restrictions like hermiticity or unitarity reduces the number of parameters required to describe the matrix. A Hermitian matrix can be described by $n(n+1)/2$ real parameters and $n(n-1)/2$ phases. A unitary $n \times n$ matrix, which corresponds to $U(N)$ symmetry, has $n(n-1)/2$ real parameters and $n(n+1)/2$ phases. Thus, a $U(1)$ transformation (that is not a symmetry) can be used to remove a single phase, while an $SU(N)$ transformation (that is not a symmetry) can be used to remove $n(n-1)/2$ real parameters and $n(n+1)/2 - 1$ phases.

7.5.4 The LSM Parameters

We are now ready to discuss the parameters of the LSM. As a first step, we count the number of physical parameters. The situation is rather simple with regard to the gauge ($\mathcal{L}_{\text{kin}}$) and the scalar ($\mathcal{L}_\phi$) sectors: Each of them is described by two real parameters. As for the Yukawa interactions ($\mathcal{L}_{\text{Yuk}}$), the rule given in equation (7.102) should be applied. The kinetic terms for the fermions have a global $[U(3)]^2$ symmetry, defined in equation (7.92). A $U(3)$ algebra has 9 generators, so the total number of generators of G is 18. The Yukawa matrix Y^e, defined in equation (7.14), is a 3×3 complex matrix, which contains a total of 18 parameters. These parameters break the $[U(3)]^2$ to the $[U(1)]^3$ accidental symmetry of equation (7.93). Using equation (7.102), the number of physical parameters in the Yukawa sector is given by

$$N_{\text{phys}} = 18 - (18 - 3) = 3. \tag{7.103}$$

A more detailed calculation implies that the three parameters are real.

We conclude that the LSM has seven parameters. This implies that in principle, we need to perform seven appropriate measurements, and then we can make predictions for any other processes involving the leptons and the Higgs boson that are mediated by the electromagnetic, weak, or Yukawa interactions. It is convenient to think of these experiments as measurements of the seven parameters.

There are various ways in which we can choose the seven independent parameters. For example, we can choose the parameters that appear in the Lagrangian (equation (7.23)) in the basis with the diagonal Yukawa matrix (equation (7.43)):

$$g, \quad g', \quad v, \quad \lambda, \quad y_e, \quad y_\mu, \quad y_\tau. \tag{7.104}$$

Another example would be

$$m_W, \quad m_Z, \quad m_h, \quad m_e, \quad m_\mu, \quad m_\tau, \quad \alpha. \tag{7.105}$$

This example shows that by measuring the spectrum of the LSM and the fine structure constant, all other interaction rates are predicted.

A good choice of parameters would be one where the experimental errors in their determination are the smallest. As of now, this set is the following:

$$\alpha, \quad G_F, \quad m_e, \quad m_\mu, \quad m_\tau, \quad m_Z, \quad m_h. \tag{7.106}$$

By now, all seven parameters have been measured, with m_h (or, equivalently, λ in the list given in equation (7.104)) the latest addition.

Finally, the fact that the seven LSM parameters can be chosen to be real, as demonstrated explicitly in equations (7.104) or (7.106), implies that *CP* is conserved by the LSM.

7.6 Low-Energy Tests

Nowadays, experiments produce the W and Z bosons and measure their properties directly. It is interesting to understand, however, how the LSM was tested at the time before experiments achieved sufficiently high energies. It is not only the historical aspect that is interesting, it is also an important demonstration of how we can use low-energy data to understand shorter distances.

We consider neutrino–electron-scattering processes, which are useful for our purpose, as they are mediated by the weak interaction only (i.e., by W and/or Z exchanges). The strong and electromagnetic interactions do not contribute at the tree level, as the neutrino carries neither color nor electromagnetic charge. We distinguish four classes of processes:

- Processes that are forbidden due to the accidental symmetries—for example, $(\nu_\mu e^- \to \nu_e e^-)$.
- Processes that are mediated by W exchange only—for example, $(\nu_\mu e^- \to \nu_e \mu^-)$.
- Processes that are mediated by Z exchange only—for example, $(\nu_\mu e^- \to \nu_\mu e^-)$.
- Processes that are mediated by both W and Z exchange—for example $(\nu_e e^- \to \nu_e e^-)$.

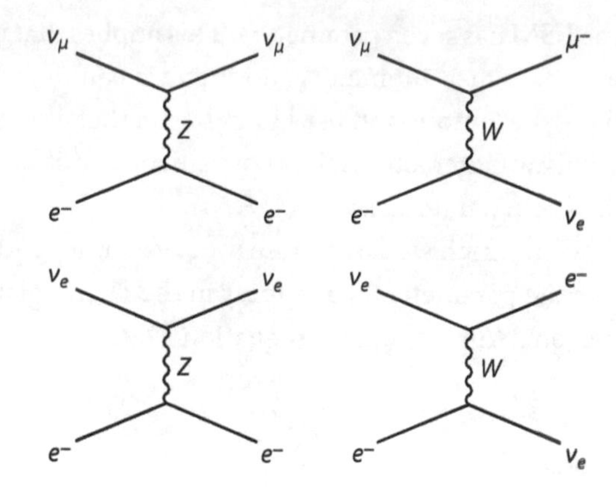

FIGURE 7.1. Tree-level diagrams for neutrino–electron scattering.

The relevant tree-level diagrams are given in figure 7.1. We work in the electron rest frame and define

$$s=(p_\nu+p_e)^2, \qquad y=\frac{p_e\cdot(p_\nu-p_\ell)}{p_e\cdot p_\nu}, \tag{7.107}$$

such that p_ℓ is the momentum of the outgoing charged lepton and p_ν is the momentum of the incoming neutrino or antineutrino. We consider only the case where

$$m_\mu^2 \ll s \ll m_W^2, \tag{7.108}$$

and keep terms that are first-order in s/m_W^2 and zeroth-order in m_μ^2/s. The results for the differential and total cross sections are given by

$$\frac{d\sigma}{d\Omega}=\frac{G_F^2 s}{\pi^2}\left[a^2+b^2(1-y)^2\right], \qquad \sigma=\frac{4G_F^2 s}{\pi}\left(a^2+\frac{b^2}{3}\right). \tag{7.109}$$

The dimensionless constants a and b depend on the process as follows:

$$\begin{aligned}
\nu_\mu e^- &\to \nu_e e^- : a=0, \quad b=0, \\
\nu_\mu e^- &\to \nu_e \mu^- : a=\frac{1}{2}, \quad b=0, \\
\nu_\mu e^- &\to \nu_\mu e^- : a=\frac{1}{2}-\sin^2\theta_W, \quad b=-\sin^2\theta_W, \\
\nu_e e^- &\to \nu_e e^- : a=-\frac{1}{2}-\sin^2\theta_W, \quad b=-\sin^2\theta_W,
\end{aligned} \tag{7.110}$$

and the same applies to $\mu\to\tau$. For the corresponding $\bar\nu-e$ scattering processes, $a\leftrightarrow b$:

$$\begin{aligned}
\bar\nu_\mu e^- &\to \bar\nu_e e^- : b=0, \quad a=0, \\
\bar\nu_e e^- &\to \bar\nu_\mu \mu^- : b=\frac{1}{2}, \quad a=0,
\end{aligned} \tag{7.111}$$

$$\bar{\nu}_\mu e^- \to \bar{\nu}_\mu e^- : b = \frac{1}{2} - \sin^2\theta_W, \quad a = -\sin^2\theta_W,$$

$$\bar{\nu}_e e^- \to \bar{\nu}_e e^- : b = -\frac{1}{2} - \sin^2\theta_W, \quad a = -\sin^2\theta_W.$$

These results can be used to test the LSM in many ways. Next, we give two examples.

7.6.1 *Charged Current Neutrino–Electron Scattering*

A comparison between the two W-mediated scattering processes, $\nu_\mu e^- \to \nu_e \mu^-$ and $\bar{\nu}_e e^- \to \bar{\nu}_\mu \mu^-$, provides a test of the vectorial nature of charged current weak interactions. (Here, "vectorial" means that the processes are mediated by a spin-one boson.) Using equations (7.110) and (7.111), we obtain

$$\frac{d\sigma(\bar{\nu}_e e^- \to \bar{\nu}_\mu \mu^-)/d\Omega}{d\sigma(\nu_\mu e^- \to \nu_e \mu^-)/d\Omega} = (1-y)^2, \tag{7.112}$$

$$\frac{\sigma(\bar{\nu}_e e^- \to \bar{\nu}_\mu \mu^-)}{\sigma(\nu_\mu e^- \to \nu_e \mu^-)} = \frac{1}{3}. \tag{7.113}$$

If the Dirac structure of the charged current weak interactions were not vectorial (e.g., if the mediator were a scalar), these ratios could be different. Here, we give the intuition behind this result.

In the center of mass frame $1-y=(1+\cos\theta)/2$, where θ as the angle between the outgoing muon and the incoming neutrino or antineutrino. Backward scattering of the beam particle corresponds to $1-y=0$ or, equivalently, $\cos\theta=-1$. Since only left-handed fields couple to the W boson and we consider the ultra-relativistic limit for the $\bar{\nu}_e e^-$ scattering, $\overline{\nu_L}$ and ℓ_L have positive and negative helicities, respectively. Thus, in the center of mass frame, their spins are in the same direction. Therefore, for the initial state, $(J_z)_i=+1$. When the scattering is backward, the respective momenta of the antineutrinos and the charged leptons are reversed, and consequently their helicities are reversed: $(J_z)_f=-1$. We conclude that backward $\bar{\nu}\ell$ scattering is forbidden by angular momentum conservation. This explains the $(1-y)^2=(1-\cos\theta)^2/4$ factor in the numerator of equation (7.112). In fact, the process $\bar{\nu}_e e \to \bar{\nu}_\mu \mu$ proceeds entirely in a $J=1$ state with a net helicity of $+1$. That is, only one of the three states is allowed. In contrast, in $\nu_\mu e \to \nu_e \mu$, backward scattering has $(J_Z)_i=(J_Z)_f=0$ and all helicity states are allowed. This explains the factor of $1/3$ in equation (7.113).

7.6.2 *Neutral Current Neutrino–Electron Scattering*

A comparison between the two Z-mediated scattering processes, $\nu_\mu e^- \to \nu_\mu e^-$ and $\bar{\nu}_\mu e^- \to \bar{\nu}_\mu e^-$, provides yet another way to determine $\sin^2\theta_W$. We first note a historical aspect of these processes. When the Standard Model was proposed, there were experimental pieces of evidence for the W-mediated charged current weak interactions but not for the Z-mediated neutral current weak interactions.

Using equations (7.110) and (7.111), we obtain

$$\frac{\sigma(\nu_\mu e \to \nu_\mu e)}{\sigma(\bar{\nu}_\mu e \to \bar{\nu}_\mu e)} = \frac{3(-1/2+\sin^2\theta_W)^2+\sin^4\theta_W}{(-1/2+\sin^2\theta_W)^2+3\sin^4\theta_W}. \tag{7.114}$$

The experimental result leads to

$$\sin^2\theta_W = 0.230 \pm 0.008, \tag{7.115}$$

in good agreement with other determinations—namely, equations (7.37), (7.72), (7.88), and (7.73).

For Further Reading

- For further details on vector-boson self-interactions, see Eboli and Gonzalez-Garcia [23].
- For reviews and more details on low-energy tests, see Erler and Su [24].
- More details on charged lepton—neutrino scattering can be found in Marciano and Parsa [25].

Problems

Question 7.1: Algebra

1. Write explicitly the h.c.-term in equation (7.14).
2. Starting from equation (7.28) and using the definitions of equations (7.29) and (7.30), derive equation (7.31).
3. Using the definitions of θ_W in equation (7.29) and of Q in equation (7.20), derive equation (7.62).
4. Derive equation (7.65) and discuss what approximations are made in the derivation.
5. Using the gauge boson kinetic terms from equation (7.23) and the definitions in equations (7.29), (7.30), and (7.60), derive equation (7.90).

Question 7.2: The broken generators

Below equation (7.18), we made the statement that our choice of VEV keeps T_3+Y unbroken, while the other three generators are broken. Here, you are asked to prove this statement.

1. With the choice of VEV in equation (7.18), show that T_3+Y annihilates the vacuum. That is, write T_3+Y explicitly as a 2×2 matrix, apply it to vector $\langle\phi\rangle$, and

show that the result is zero. This result proves that this operator is not broken (see question 6.3 in chapter 6).

2. The other three generators correspond to T_1, T_2, and $T_3 - Y$. Show that any nonvanishing linear combination of them (with real coefficients) does not annihilate the vacuum. This is proof that there is no other unbroken generator.

Question 7.3: The would-be Nambu-Goldstone bosons

The spontaneous breaking of a global symmetry entails Nambu–Goldstone bosons. If we now modify the model such that the spontaneously broken symmetry is a local one, these scalar DoF become the longitudinal components of the gauge bosons that correspond to the broken generators. To see this explicitly, we consider the Higgs sector of a modified LSM, such that the spontaneously broken symmetry is global.

We consider the Higgs potential of equation (7.15). Instead of equation (7.21) we write ϕ as

$$\phi = \begin{pmatrix} g^+ \\ (v + h + ig^0)/\sqrt{2} \end{pmatrix}, \tag{7.116}$$

where $g^+ = (\phi_1 - i\phi_2)/\sqrt{2}$.

1. Write $\phi^\dagger$ in terms of g^-, g^0 and h. Recall that $g^- = (g^+)^\dagger$.
2. Plug ϕ and $\phi^\dagger$ of equation (7.116) into equation (7.15) and find the spectrum. One way to get the spectrum is to define the mass matrix as

$$M^2_{ij} = \left. \frac{\partial^2 V}{\partial \phi_i \partial \phi_j^\dagger} \right|_{\phi_i = \phi_j^\dagger = 0}. \tag{7.117}$$

Explicitly, for example,

$$M^2(g^\pm) = \left. \frac{\partial^2 V}{\partial g^+ \partial g^-} \right|_{g^+ = g^- = 0}. \tag{7.118}$$

3. Check that the spectrum has 3 massless DoF, the $g^\pm$ and g^0, and 1 massive DoF, h. Check that the mass of h is the same in the global and local cases.

Question 7.4: Lepton decays

1. Find in the Particle Data Group (PDG) the main decay mode of the muon, and obtain the corresponding decay width.
2. Draw the tree-level Feynman diagram for the leading muon decay.
3. Find in the PDG the bound on

$$\mathrm{BR}(\mu \to e\gamma). \tag{7.119}$$

What is the LSM prediction for this mode? Explain.

4. Based on lepton universality, predict the following ratio:

$$\frac{\Gamma(\tau \to e\nu\bar{\nu})}{\Gamma(\tau \to \mu\nu\bar{\nu})}. \tag{7.120}$$

Compare your results with the PDG and explain any deviations.

5. Show that the LSM predicts (up to small phase-space factors)

$$\frac{\Gamma(\mu \to e\nu\bar{\nu})}{\Gamma(\tau \to e\nu\bar{\nu})} = \frac{m_\mu^5}{m_\tau^5}. \tag{7.121}$$

Compare the prediction to the data.

Question 7.5: A modified LSM

Consider the following model:

(*i*) The symmetry is a local

$$SU(2)_L \times U(1)_Y, \tag{7.122}$$

spontaneously broken to $U(1)_{\mathrm{EM}}$.

(*ii*) There is a single fermion generation:

$$L_L(3)_{-1}, \qquad E_R(2)_{-3/2}, \qquad . \tag{7.123}$$

(*iii*) There is a single scalar multiplet:

$$\phi(2T+1)_{+1/2}. \tag{7.124}$$

1. Can we have a bare mass term for the fermions?
2. What values of T allow a Yukawa term of the form $\overline{L_L}E_R\,\phi$? Write all the possibilities.
3. From now on, we assume $T = 1/2$. By considering the components of the fields and the Yukawa term, argue that the fermionic content of the model consists of a massless $Q = 0$ state (the neutrino ν), a massive $Q = -1$ state (the electron e), and a massive $Q = -2$ state (that was not observed in nature) that we denote as λ.
4. Show that $m_\lambda = \sqrt{2}m_e$.
5. Write the fermionic charged current weak interaction terms in the Lagrangian; that is, the coupling of the W boson to the fermions. Keep track of factors of 2.
6. Is λ stable? Explain.
7. Calculate the ratio

$$\frac{\Gamma(W^- \to \overline{e_R}^{+}\lambda_R^{--})}{\Gamma(W^- \to \overline{e_L}^{+}\lambda_L^{--})}. \tag{7.125}$$

8. Is this a viable model that can replace the LSM? Explain.

Question 7.6: ρ for general scalar representations

In the LSM, the Higgs field transforms under $SU(2)_L \times U(1)_Y$ as $(2)_{+1/2}$. However, any scalar that is charged under the gauge group and acquires a VEV will break the LSM gauge symmetry. We assume that the scalar potential is given by equation (7.17).

1. Consider a scalar ϕ that transforms as $(2T+1)_Y$. Since $SU(2)$ is a non-Abelian group, $2T+1$ has to be a positive integer, and thus T is a nonnegative half-integer. Since $U(1)$ is Abelian, a priori Y can assume any real value. Yet we require that $\langle\phi\rangle$ breaks $SU(2)_L \times U(1)_Y \to U(1)_{\rm EM}$, with $Q = T_3 + Y$. This definition restricts the possible values for Y. Find these values.
2. Show that
$$\rho \equiv \frac{m_W^2}{m_Z^2 \cos^2\theta_W} = \frac{T(T+1) - Y^2}{2Y^2}. \tag{7.126}$$
Hint: Use the following $2T+1$–dimensional representation of $SU(2)$:
$$T_3 = {\rm diag}\{T, T-1, T-2, \ldots, -T\}, \tag{7.127}$$
$$T_1 = \begin{pmatrix} 0 & a_1 & 0 & \ldots & 0 \\ a_1 & 0 & a_2 & 0 & \vdots \\ 0 & a_2 & \ddots & \ddots & \vdots \\ \vdots & 0 & \ddots & 0 & a_{2T} \\ 0 & \ldots & \ldots & a_{2T} & 0 \end{pmatrix} \qquad T_2 = -i \begin{pmatrix} 0 & a_1 & 0 & \ldots & 0 \\ -a_1 & 0 & a_2 & 0 & \vdots \\ 0 & -a_2 & \ddots & \ddots & \vdots \\ \vdots & 0 & \ddots & 0 & a_{2T} \\ 0 & \ldots & \ldots & -a_{2T} & 0 \end{pmatrix},$$
where
$$a_k = \frac{\sqrt{T(T+1) - (T-k)(T-k+1)}}{2}. \tag{7.128}$$
3. For $T > 0$ and $Y = 0$, one can see from equation (7.126) that $\rho \to \infty$, independent of T. Explain this result using symmetry arguments.
4. Suppose that there exist several scalar representations ($i = 1, \ldots, N$) whose neutral members acquire VEVs v_i. Find ρ in terms of v_i, T_i, and Y_i.
5. Assume that, in addition to the standard Higgs doublet $\{T = 1/2, Y = 1/2\}$ with VEV v_W, there exists one other multiplet $\{T_2, Y_2\}$, whose electromagnetic-neutral member acquires a much smaller VEV, $v_2 \ll v_W$. Find $\delta\rho \equiv \rho - 1$ to first-order in $(v_2/v_W)^2$.
6. Assume that experimentally, $-0.01 \le \delta\rho \le +0.005$. Find the constraint on $(v_i/v_W)^2$ for the following multiplets: $(3)_{+1}$, $(5)_{-1}$ and $(4)_{+3/2}$.
7. From equation (7.126), it is clear that $\rho = 1$ for all $3Y^2 = T(T+1)$ multiplets. The fact that experimentally, $\rho \simeq 1$ motivates the choice of one of these multiplets to spontaneously break the symmetry, but from the consideration of ρ alone, there is no difference which multiplet we take. Yet, in the LSM, we make the specific choice of $T = 1/2$ and $Y = 1/2$. What is the advantage of the LSM Higgs field over the other possible choices?

Question 7.7: Forward-backward asymmetry

We consider $e^+e^- \to \mu^+\mu^-$ scattering. This process is mediated at tree level by both photon and Z-boson exchanges. The photon exchange is parity conserving, while the Z exchange is parity violating. The interference between them leads to a forward-backward asymmetry, which is a manifestation of parity violation. The forward-backward asymmetry is defined as follows:

$$A_{\rm FB} = \frac{\sigma_{\rm F} - \sigma_{\rm B}}{\sigma_{\rm F} + \sigma_{\rm B}}, \quad \sigma_{\rm F} = 2\pi \int_0^1 d(\cos\theta)\, \frac{d\sigma}{d\cos\theta}, \quad \sigma_{\rm B} = 2\pi \int_{-1}^0 d(\cos\theta)\, \frac{d\sigma}{d\cos\theta}, \tag{7.129}$$

where θ is the angle between the incoming e^- and the outgoing μ^- and σ represent a cross section.

1. Draw the photon-exchange and Z-exchange diagrams and evaluate the corresponding amplitudes.
2. We define $s = (p_{e^+} + p_{e^-})^2$, where p_{e^+} (p_{e^-}) is the four-momentum of the incoming e^+ (e^-). Consider the range $m_\mu^2 \ll s \ll m_Z^2$ and neglect contributions that are proportional to $1 - 4\sin^2\theta_W$. Show that, to leading order in s/m_Z^2,

$$\frac{d\sigma}{d\Omega} = \frac{\alpha^2}{4s}\left[1 + \cos^2\theta + \frac{1}{4c_W^2 s_W^2}\frac{s}{s - m_Z^2}\cos\theta\right], \tag{7.130}$$

   where $c_W^2 \equiv \cos^2\theta_W$ and $s_W^2 \equiv \sin^2\theta_W$.
3. Show that, within our approximations,

$$A_{\rm FB} = \frac{3}{32\cos^2\theta_W \sin^2\theta_W} \times \frac{s}{s - m_Z^2}. \tag{7.131}$$

4. One of the data points is given by $A_{\rm FB}[s = (35\ \text{GeV})^2] = -(9.9 \pm 1.6) \times 10^{-2}$. What is the LSM prediction for this point, and how does it stand in comparison to experiment?

Question 7.8: Flavor symmetries

In this question, you are asked to use broken generators to transform to a basis with fewer parameters.

Consider a theory where we impose a global $U(1)$ symmetry and have two Dirac fermions that have the same charge under that $U(1)$. The most general Lagrangian is

$$\mathcal{L} = \overline{\psi_i}(i\partial\!\!\!/\,\delta_{ij} - m_{ij})\psi_j, \tag{7.132}$$

where $i, j = 1, 2$. The matrix m_{ij} must be Hermitian, and thus it depends on four real parameters.

1. The flavor symmetry of this Lagrangian is $U(1) \times U(1)$. However, this symmetry is not manifest. To make it so, proceed as follows. Write

$$m_{\text{diag}} \equiv \text{diag}(m_1, m_2) = UmU^\dagger. \tag{7.133}$$

Define $\psi' = U\psi$ and show that $\mathcal{L}$ in equation (7.132) is equal to

$$\mathcal{L} = \overline{\psi_i'}[i\partial\!\!\!/\delta_{ij} - (m_{\text{diag}})_{ij}]\psi_j'. \tag{7.134}$$

2. In equation (7.134), the two $U(1)$ symmetries are manifest. They are the independent rotations of each flavor:

$$\psi_i' \to e^{i\alpha_i}\psi_i' \qquad \text{for} \quad i = 1, 2, \tag{7.135}$$

which we denote as $U(1)_1$ and $U(1)_2$. Show that the $U(1)_1 \times U(1)_2$ symmetry can also be written as a $U(1)_0 \times U(1)_3$ symmetry, with

$$\psi_i' \to \left(e^{i\alpha_0}\right)_{ij}\psi_j' \qquad \psi_i' \to \left(e^{i\alpha_3\sigma_3}\right)_{ij}\psi_j', \tag{7.136}$$

and find α_0 and α_3 in terms of α_1 and α_2.
3. Identify the imposed $U(1)$ in terms of $U(1)_0$ and $U(1)_3$.
4. In the case that m is proportional to the unit matrix, the flavor symmetry is $SU(2) \times U(1)$. In the most general case, the symmetry is $U(1) \times U(1)$. In the basis where $m = m_{\text{diag}}$, the two broken generators correspond to σ_1 and σ_2. Show by explicit calculation that they are broken (to do this, it is enough to show it for only one of them).
5. The two broken symmetry generators can be used to generate U of equation (7.133):

$$U = e^{i\alpha_2\sigma_2}e^{i\alpha_1\sigma_1}, \tag{7.137}$$

with α_1 and α_2 given in terms of the parameters of m_{ij}. Find α_1 and α_2.

Hint: It may be easier to perform the two rotations one after the other. First, use $U_1 = e^{i\alpha_1\sigma_1}$ to make m real and symmetric, and then use $U_2 = e^{i\alpha_2\sigma_2}$ to diagonalize it.

Question 7.9: 2HDM

The LSM includes a single Higgs doublet. Here, we study the scalar sector that has two Higgs doublets. We return to other aspects of the Two Higgs Doublet Model (2HDM) in question 9.9.

We define the 2HDM as follows:

(*i*) The symmetry is

$$SU(2)_L \times U(1)_Y \times Z_2. \tag{7.138}$$

(*ii*) There are two scalar multiplets:

$$\phi_1(2)_{+1/2}, \qquad \phi_2(2)_{+1/2}. \tag{7.139}$$

Under the Z_2 symmetry, ϕ_1 is even and ϕ_2 is odd. For simplicity, we impose *CP* conservation.

The Higgs potential is given by

$$V = m_1^2\phi_1^\dagger\phi_1 + m_2^2\phi_2^\dagger\phi_2 + \frac{\lambda_1}{2}(\phi_1^\dagger\phi_1)^2 + \frac{\lambda_2}{2}(\phi_2^\dagger\phi_2)^2 \\ + \lambda_3(\phi_1^\dagger\phi_1)(\phi_2^\dagger\phi_2) + \lambda_4(\phi_1^\dagger\phi_2)(\phi_2^\dagger\phi_1) + \frac{\lambda_5}{2}\left[(\phi_1^\dagger\phi_2)^2 + (\phi_2^\dagger\phi_1)^2\right]. \tag{7.140}$$

We are interested in a model in which both doublets acquire a VEV.

1. We need to ensure that the potential is bounded from below (the equivalent of the $\lambda > 0$ condition in the LSM), and that at the origin, we have a maximum (the equivalent of the $\mu^2 < 0$ condition in the LSM). Argue that if all $\lambda_i > 0$ $(i = 1, \ldots, 5)$ and all $m_i^2 < 0$ $(i = 1, 2)$, the symmetry is spontaneously broken.
2. Assume that

$$\langle\phi_1\rangle = \begin{pmatrix} 0 \\ v_1/\sqrt{2} \end{pmatrix}, \qquad \langle\phi_2\rangle = \begin{pmatrix} 0 \\ v_2/\sqrt{2} \end{pmatrix}. \tag{7.141}$$

Show that the minimum equations, $\partial V/\partial v_1 = 0$ and $\partial V/\partial v_2 = 0$, give

$$m_1^2 v_1 + \frac{\lambda_1}{2}v_1^3 + \frac{\lambda_{345}}{2}v_1 v_2^2 = 0, \\ m_2^2 v_2 + \frac{\lambda_2}{2}v_2^3 + \frac{\lambda_{345}}{2}v_2 v_1^2 = 0, \tag{7.142}$$

where $\lambda_{345} = \lambda_3 + \lambda_4 + \lambda_5$.
3. Find v_1^2, v_2^2 and $\tan\beta \equiv v_2/v_1$ as a function of the model parameters. What extra assumptions on the model parameters are needed to ensure that both VEVs are real?
4. In the limit where $\lambda_{345} = 0$, equations (7.142) decouple. In that case, each of the doublets should have a potential like the one in the LSM. Show that the solution reduces to the LSM one in that case.

We now obtain the scalar spectrum. Since each doublet has 4 real DoF, the 2HDM has a total of 8. We denote them as follows (see question 7.3):

$$\phi_1 = \begin{pmatrix} \eta_1^+ \\ (v_1 + h_1 + i\chi_1^0)/\sqrt{2} \end{pmatrix}, \qquad \phi_2 = \begin{pmatrix} \eta_2^+ \\ (v_2 + h_2 + i\chi_2^0)/\sqrt{2} \end{pmatrix}, \tag{7.143}$$

where $\eta^{+}_{1,2}$ are complex and $h_{1,2}$ and $\chi_{1,2}$ are real fields. To obtain the spectrum, we use

$$M^2_{ij} = \left. \frac{\partial^2 V}{\partial\phi_i \partial\phi_j^\dagger} \right|_{\phi_i=\phi_j^\dagger=0}. \tag{7.144}$$

The mass matrix M^2 can be divided into three 2×2 matrices: one for the charged fields, $\eta_i^\pm$ (denoted as M^2_C), one for the h_i fields (denoted as M^2_S), and one for the χ_i fields (denoted as M^2_A).

5. Write M^2_C, M^2_A, and M^2_S.
6. Show that each of M^2_C and M^2_A has a zero eigenvalue. Show that the corresponding eigenstates are given by

$$g^\pm = c_\beta \eta_1^\pm + s_\beta \eta_2^\pm, \qquad g^0 = c_\beta \chi_1 + s_\beta \chi_2. \tag{7.145}$$

   Explain why we should have expected 3 massless and 5 massive DoF.
7. The orthogonal combinations are a physical charged scalar and a physical *CP*-odd neutral scalar:

$$H^\pm = -s_\beta \eta_1^\pm + c_\beta \eta_2^\pm, \qquad A = -s_\beta \chi_1 + c_\beta \chi_2. \tag{7.146}$$

   Find their masses, m_{H^+} and m_A.
8. The eigenstates of M^2_S are two neutral, *CP*-even scalars. Write them as $H = c_\alpha \eta_1 + s_\alpha \eta_2$ and $h = -s_\alpha \eta_1 + c_\alpha \eta_2$, with $m_h < m_H$. Find the two masses and express α in terms of the model parameters.

We return to this model in question 9.9 in chapter 9.

Question 7.10: The Leptonic Left-Right Symmetric Model

Here, we study the Leptonic Left-Right Symmetric Model (LLRSM), which extends both the gauge group and the lepton content of the LSM. It is motivated partly by allowing parity to be a spontaneously broken symmetry of the Lagrangian. More important, in contrast to the LSM, it predicts small neutrino masses.

The model is defined as follows:

(*i*) The symmetry is a local

$$G_{\rm LRS} = SU(2)_L \times SU(2)_R \times U(1)_X. \tag{7.147}$$

(*ii*) There are N_g fermion generations, each consisting of two different representations:

$$L_L(2,1)_{-1}, \qquad L_R(1,2)_{-1}. \tag{7.148}$$

(*iii*) There are three scalar multiplets:

$$\Phi(2,2)_0, \qquad \Delta_L(3,1)_{+2}, \qquad \Delta_R(1,3)_{+2}. \tag{7.149}$$

We start with $N_g = 1$. We label the various components of the doublets as

$$L_L = \begin{pmatrix} \nu_L \\ e_L \end{pmatrix}, \qquad L_R = \begin{pmatrix} \nu_R \\ e_R \end{pmatrix}. \tag{7.150}$$

The gauge fields are denoted by $W_L(3,1)_0$, $W_R(1,3)_0$, and $C(1,1)_0$, and the respective coupling constants by g_L, g_R, and g_X.

1. In what spaces are the doublets in equation (7.150) written?
2. Write the commutation relations between the various generators (the analog of equation (7.6)).
3. Write the covariant derivative of a general field and of all the fermions and scalars of the model (the analog of equation (7.10)).
4. Let us assume that only Δ_R acquires a VEV, and assign the VEV to its $T_3^R = -1$ component. What is $G_{\rm LRS}$ broken into (the analog of equation (7.19))?
5. Find the linear combinations of the LLRSM generators that give the LSM generator Y and the EM generator Q (the analog of equation (7.20)).
6. What are the $SU(2)_L \times U(1)_Y$ quantum numbers of ν_R?
7. Write Φ in components that are distinguishable when $\langle \Delta_R \rangle \neq 0$. Denote the two components as ϕ_1 and ϕ_2 and write their $SU(2)_L \times U(1)_Y$ quantum numbers.
8. The B^μ field of the LSM must be a linear combination of W_{R3}^μ and C^μ. The orthogonal linear combination is called Z'^μ. Write the expression of B^μ and Z'^μ in terms of W_{R3}^μ and C^μ (the analog of equation (7.30)). Denote the mixing angle by θ_R, and express it in terms of the LLRSM couplings.
9. Find the masses of the massive gauge bosons, $W_R^\pm$ and Z', after this stage of breaking (the analog of equations (7.31) and (7.32)).
10. Find the electric charges of the massive gauge bosons, $W_R^\pm$ and Z' (the analog of equation (7.33)).
11. Express g_R and g_X in terms of g' and θ_R (the analog of equation (7.60)). Here, g' is the coupling constant of $U(1)_Y$.
12. Find the coupling of the Z'-boson to fermions in terms of T_{3R} and Y (the analog of equation (7.63)).
13. Write down explicitly the charged current interactions of the leptons with the massive $W_R^\pm$-bosons (the analog of equation (7.79)).

We now assume that all three scalars acquire VEVs, with the following hierarchy:

$$\langle \Delta_R \rangle \gg \langle \Phi \rangle \gg \langle \Delta_L \rangle. \tag{7.151}$$

Then we can think about breaking the symmetry as a two-stage process:

$$SU(2)_L \times SU(2)_R \times U(1)_X \to SU(2)_L \times U(1)_Y \to U(1)_{\rm EM}. \tag{7.152}$$

The previous items referred to the first stage of the breaking. We now consider the second stage of the breaking due to $\langle \Phi \rangle \neq 0$.

14. Explain why we must require that only the neutral components of Φ acquire VEVs.
15. We denote $\langle \phi_1 \rangle = k_1$ and $\langle \phi_2 \rangle = k_2$. Calculate the mass of the $W_L^\pm$ bosons in terms of k_1 and k_2 (provisionally putting $\langle \Delta_L \rangle = 0$), and relate it to the LSM parameter v. We identify $W_L^\pm$ with $W^\pm$ of the LSM.

Our last step is to understand why we choose the hierarchies of VEVs as in equation (7.151).

16. Experiments established that $\rho = 1$ (ρ is defined in equation (7.35)) to a very good approximation. Show that this implies that $|\langle \Delta_L \rangle| \ll |\langle \Phi \rangle|$.

What you have shown is that the LLRSM constitutes a viable extension of the LSM. We return to this model in question 14.5 in chapter 14.

8

The Standard Model

The Standard Model accounts for the strong, weak, electromagnetic, and Yukawa interactions of the fermions. It combines the Leptonic Standard Model (LSM), presented in chapter 7, and Quantum ChromoDynamics (QCD), presented in chapter 5. In this chapter, we define the Standard Model and obtain its spectrum, interactions, accidental symmetries, and parameters. The rest of this book is dedicated to the study of the Standard Model beyond the tree and beyond the renormalizable levels.

8.1 Defining the Standard Model

The Standard Model is defined as follows:

1. The symmetry is a local
$$SU(3)_C \times SU(2)_L \times U(1)_Y. \tag{8.1}$$
2. There are three fermion generations, each consisting of five different representations:
$$Q_{Li}(3,2)_{+1/6}, \quad U_{Ri}(3,1)_{+2/3}, \quad D_{Ri}(3,1)_{-1/3}, \quad L_{Li}(1,2)_{-1/2},$$
$$E_{Ri}(1,1)_{-1}, \quad i=1,2,3. \tag{8.2}$$
3. There is a single scalar multiplet:
$$\phi(1,2)_{+1/2}. \tag{8.3}$$

Similar to the LSM, the presence of a scalar allows spontaneous symmetry breaking (SSB). We define the Standard Model as a model that exhibits the following breaking pattern:
$$SU(3)_C \times SU(2)_L \times U(1)_Y \to SU(3)_C \times U(1)_{\rm EM}. \tag{8.4}$$

We use the notation $(A, B)_Y$, where A is the representation under $SU(3)_C$, B is the representation under $SU(2)_L$, and Y is the hypercharge. The fermions that transform as triplets of $SU(3)_C$ are called *quarks*, while those that transform as singlets of $SU(3)_C$ are called *leptons*.

8.2 The Lagrangian

As explained in previous chapters, the most general renormalizable Lagrangian with scalar and fermion fields can be decomposed into

$$\mathcal{L} = \mathcal{L}_{\text{kin}} + \mathcal{L}_{\psi} + \mathcal{L}_{\phi} + \mathcal{L}_{\text{Yuk}}. \tag{8.5}$$

It is now our task to find the specific form of the Lagrangian made of the fermion fields Q_{Li}, U_{Ri}, D_{Ri}, L_{Li}, and E_{Ri} (equation (8.2)), and the scalar field ϕ (equation (8.3)), subject to the gauge symmetry (equation (8.1)) and leading to the SSB of equation (8.4).

8.2.1 $\mathcal{L}_{\text{kin}}$ and the Gauge Symmetry

The gauge group is given in equation (8.1). It has 12 generators: 8 L_as that form the $SU(3)$ algebra, 3 T_bs that form the $SU(2)$ algebra, and a single Y that corresponds to the $U(1)$ group:

$$[L_a, L_b] = if_{abc}L_c, \qquad [T_a, T_b] = i\varepsilon_{abc}T_c, \qquad [L_a, T_b] = [L_a, Y] = [T_b, Y] = 0. \tag{8.6}$$

The local symmetry requires 12 gauge boson degrees of freedom (DoF): 8 in the adjoint representation of $SU(3)_C$, 3 in the adjoint representation of $SU(2)_L$, and 1 related to the $U(1)_Y$ symmetry:

$$G^{\mu}_a(8,1)_0, \qquad W^{\mu}_a(1,3)_0, \qquad B^{\mu}(1,1)_0. \tag{8.7}$$

The corresponding field strengths are given by

$$\begin{aligned}
G^{\mu\nu}_a &= \partial^{\mu}G^{\nu}_a - \partial^{\nu}G^{\mu}_a - g_s f_{abc}G^{\mu}_b G^{\nu}_c,\\
W^{\mu\nu}_a &= \partial^{\mu}W^{\nu}_a - \partial^{\nu}W^{\mu}_a - g\varepsilon_{abc}W^{\mu}_b W^{\nu}_c,\\
B^{\mu\nu} &= \partial^{\mu}B^{\nu} - \partial^{\nu}B^{\mu},
\end{aligned} \tag{8.8}$$

where f_{abc} are the structure constants of $SU(3)$ and ε_{abc} are the structure constants of $SU(2)$. The covariant derivative is given by

$$D^{\mu} = \partial^{\mu} + ig_s G^{\mu}_a L_a + igW^{\mu}_b T_b + ig'YB^{\mu}. \tag{8.9}$$

There are three independent coupling constants in $\mathcal{L}_{\text{kin}}$: g_s, related to the $SU(3)_C$ subgroup; g, related to the $SU(2)_L$ subgroup; and g', related to the $U(1)_Y$ subgroup. For the $SU(3)_C$ triplets, $L_a = \frac{1}{2}\lambda_a$ (λ_a are the Gell-Mann matrices), while for the $SU(3)_C$ singlets, $L_a = 0$. For the $SU(2)_L$ doublets, $T_b = \frac{1}{2}\sigma_b$ (σ_b are the Pauli matrices), while for the $SU(2)_L$ singlets, $T_b = 0$. Explicitly, the covariant derivatives acting on the scalar and various fermion fields are given by

$$\begin{aligned}
D^\mu \phi &= \left(\partial^\mu + \frac{i}{2} g W_b^\mu \sigma_b + \frac{i}{2} g' B^\mu\right) \phi, \\
D^\mu Q_L &= \left(\partial^\mu + \frac{i}{2} g_s G_a^\mu \lambda_a + \frac{i}{2} g W_b^\mu \sigma_b + \frac{i}{6} g' B^\mu\right) Q_L, \\
D^\mu U_R &= \left(\partial^\mu + \frac{i}{2} g_s G_a^\mu \lambda_a + \frac{2i}{3} g' B^\mu\right) U_R, \\
D^\mu D_R &= \left(\partial^\mu + \frac{i}{2} g_s G_a^\mu \lambda_a - \frac{i}{3} g' B^\mu\right) D_R, \\
D^\mu L_L &= \left(\partial^\mu + \frac{i}{2} g W_b^\mu \sigma_b - \frac{i}{2} g' B^\mu\right) L_L, \\
D^\mu E_R &= \left(\partial^\mu - i g' B^\mu\right) E_R.
\end{aligned} \tag{8.10}$$

$\mathcal{L}_{\text{kin}}$ includes the kinetic terms of all the fields:

$$\begin{aligned}
\mathcal{L}_{\text{kin}} = &-\frac{1}{4} G_a^{\mu\nu} G_{a\mu\nu} - \frac{1}{4} W_b^{\mu\nu} W_{b\mu\nu} - \frac{1}{4} B^{\mu\nu} B_{\mu\nu} \\
&+ i\overline{Q_{Li}} \not{D} Q_{Li} + i\overline{U_{Ri}} \not{D} U_{Ri} + i\overline{D_{Ri}} \not{D} D_{Ri} + i\overline{L_{Li}} \not{D} L_{Li} + i\overline{E_{Ri}} \not{D} E_{Ri} \\
&+ (D^\mu \phi)^\dagger (D_\mu \phi).
\end{aligned} \tag{8.11}$$

8.2.2 $\mathcal{L}_\psi$

There are no mass terms for the fermions of the Standard Model:

$$\mathcal{L}_\psi = 0. \tag{8.12}$$

In chapter 7, we explained that this is the case for leptons. A larger symmetry means stronger constraints; hence, it is impossible that lepton masses would become allowed when the gauge symmetry is extended to include $SU(3)_C$. As for the quarks, we cannot write Dirac mass terms because they are assigned to chiral representations of the $SU(2)_L \times U(1)_Y$ gauge symmetry. We cannot write Majorana mass terms for the quarks because they all have $Y \neq 0$ and they are in a complex representation of $SU(3)_C$.

8.2.3 $\mathcal{L}_\phi$ and SSB

The scalar field is a singlet of the $SU(3)_C$ group. Thus, the form of $\mathcal{L}_\phi$ is the same as in the LSM (see equation (7.17)):

$$-\mathcal{L}_\phi = \lambda \left(\phi^\dagger \phi - \frac{v^2}{2}\right)^2. \tag{8.13}$$

Since ϕ is an $SU(3)_C$ singlet, the $SU(3)_C$ subgroup remains unbroken and the pattern of spontaneous symmetry breaking (SSB) is as required by equation (8.4).

It is useful to define

$$\widetilde{\phi}_a = \varepsilon_{ab}\phi_b^*, \tag{8.14}$$

where a and b are the $SU(2)$-indexes. The $\widetilde{\phi}$ field transforms as $(1,2)_{-1/2}$. When considering the vacuum expectation value (VEV), we can write equation (7.18) as

$$\langle\phi\rangle = \begin{pmatrix} 0 \\ v/\sqrt{2} \end{pmatrix}, \qquad \langle\widetilde{\phi}\rangle = \begin{pmatrix} v/\sqrt{2} \\ 0 \end{pmatrix}. \tag{8.15}$$

8.2.4 $\mathcal{L}_{\text{Yuk}}$

The Yukawa part of the Lagrangian is given by

$$-\mathcal{L}_{\text{Yuk}} = Y_{ij}^u \overline{Q_{Li}} U_{Rj}\, \widetilde{\phi} + Y_{ij}^d \overline{Q_{Li}} D_{Rj}\, \phi + Y_{ij}^e \overline{L_{Li}} E_{Rj}\, \phi + \text{h.c.}, \tag{8.16}$$

where $i, j = 1, 2, 3$ represent flavor indexes. Note that the Yukawa couplings of U_R involve $\widetilde{\phi}$. The Yukawa matrices Y^u, Y^d, and Y^e are general complex 3×3 matrices of dimensionless couplings.

8.2.5 Summary

The renormalizable Standard Model Lagrangian is given by

$$\begin{aligned}
\mathcal{L}_{\text{SM}} = & -\frac{1}{4}G_a^{\mu\nu}G_{a\mu\nu} - \frac{1}{4}W_b^{\mu\nu}W_{b\mu\nu} - \frac{1}{4}B^{\mu\nu}B_{\mu\nu} - (D^\mu\phi)^\dagger(D_\mu\phi) \\
& + i\overline{Q_{Li}}\not{D}Q_{Li} + i\overline{U_{Ri}}\not{D}U_{Ri} + i\overline{D_{Ri}}\not{D}D_{Ri} + i\overline{L_{Li}}\not{D}L_{Li} + i\overline{E_{Ri}}\not{D}E_{Ri} \\
& - \left(Y_{ij}^u \overline{Q_{Li}} U_{Rj}\, \widetilde{\phi} + Y_{ij}^d \overline{Q_{Li}} D_{Rj}\, \phi + Y_{ij}^e \overline{L_{Li}} E_{Rj}\, \phi + \text{h.c.}\right) \\
& - \lambda\left(\phi^\dagger\phi - v^2/2\right)^2,
\end{aligned} \tag{8.17}$$

where $i, j = 1, 2, 3$.

8.3 The Spectrum

8.3.1 Bosons

Given the spontaneous breaking of the $SU(2)_L \times U(1)_Y$ symmetry to the $U(1)_{\text{EM}}$ subgroup, the spectrum of the electroweak gauge bosons remains the same as in the LSM: three massive vector bosons, $W^\pm$ and Z^0, and a massless photon, A^0. Furthermore, since the breaking is induced by an $SU(2)_L$-doublet, the $\rho \equiv m_W^2/(m_Z^2 \cos^2\theta_W) = 1$ relation holds.

The new ingredient is QCD, and the existence of a gluon in the octet representation of $SU(3)_C$. Since the $SU(3)_C$ gauge symmetry remains unbroken, the gluon is massless. The gluon particle is often denoted by g.

As for scalars, the three would-be Goldstone bosons become the longitudinal components of the three massive vector bosons. The fourth scalar DoF is the Higgs boson h, a real massive scalar field.

8.3.2 Fermions

Since the Standard Model allows no bare mass terms for the fermions, their masses can arise only from the Yukawa part of the Lagrangian, which is given in equation (8.16). The resulting lepton spectrum is the same as in the LSM, as discussed in chapter 7.

The terms that lead to quark masses are given by

$$-\mathcal{L}_{M_q} = Y^u_{ij}\overline{Q_{Li}}U_{Rj}\,\langle\widetilde{\phi}\rangle + Y^d_{ij}\overline{Q_{Li}}D_{Rj}\,\langle\phi\rangle + \text{h.c.} \tag{8.18}$$

In what follows, we use

$$Q^i_L = \begin{pmatrix} U^i_L \\ D^i_L \end{pmatrix}, \qquad U^i_R, \qquad D^i_R \tag{8.19}$$

to denote the quark interaction eigenstates in a generic basis. The various states have well-defined charges under the unbroken $U(1)_{\text{EM}}$ symmetry:

- U^i_L have $T_3 = +1/2$ and $Y = +1/6$, while U^i_R have $T_3 = 0$ and $Y = +2/3$. Hence, their electromagnetic charges are

$$q(U^i_L) = q(U^i_R) = +2/3. \tag{8.20}$$

- D^i_L have $T_3 = -1/2$ and $Y = +1/6$, while D^i_R have $T_3 = 0$ and $Y = -1/3$. Hence, their electromagnetic charges are

$$q(D^i_L) = q(D^i_R) = -1/3. \tag{8.21}$$

Using equation (8.15), we obtain

$$-\mathcal{L}_{M_q} = \frac{Y^u_{ij}v}{\sqrt{2}}\overline{U_{Li}}U_{Rj} + \frac{Y^d_{ij}v}{\sqrt{2}}\overline{D_{Li}}D_{Rj} + \text{h.c.} \tag{8.22}$$

Without loss of generality, we can use a bi-unitary transformation:

$$Y^u \to \hat{Y}_u = V_{uL}Y^uV^\dagger_{uR} \tag{8.23}$$

to transform the basis into one where $\hat{Y}^u$ is diagonal and real:

$$\hat{Y}^u = \begin{pmatrix} y_u & 0 & 0 \\ 0 & y_c & 0 \\ 0 & 0 & y_t \end{pmatrix}. \tag{8.24}$$

Equation (8.22) implies that the basis defined in equation (8.24) is the mass basis for up-type quarks. In this basis, we denote the components of the quark $SU(2)$-doublets and the up-type quark $SU(2)$-singlets as follows:

$$Q_{Lu} = \begin{pmatrix} u_L \\ d_{uL} \end{pmatrix}, \qquad Q_{Lc} = \begin{pmatrix} c_L \\ d_{cL} \end{pmatrix}, \qquad Q_{Lt} = \begin{pmatrix} t_L \\ d_{tL} \end{pmatrix}; \qquad u_R, \qquad c_R, \qquad t_R, \tag{8.25}$$

with $y_u < y_c < y_t$. The up-type quark mass terms are given in this basis by

$$-\mathcal{L}_{M_u} = \frac{y_u v}{\sqrt{2}} \overline{u_L} u_R + \frac{y_c v}{\sqrt{2}} \overline{c_L} c_R + \frac{y_t v}{\sqrt{2}} \overline{t_L} t_R + \text{h.c.} \tag{8.26}$$

We conclude that the up-type quarks u, c, and t are Dirac fermions of masses:

$$m_u = \frac{y_u v}{\sqrt{2}}, \qquad m_c = \frac{y_c v}{\sqrt{2}}, \qquad m_t = \frac{y_t v}{\sqrt{2}}. \tag{8.27}$$

We can use yet another bi-unitary transformation:

$$Y^d \to \hat{Y}_d = V_{dL} Y^d V_{dR}^\dagger \tag{8.28}$$

to transform the basis into one where $\hat{Y}^d$ is diagonal and real:

$$\hat{Y}^d = \begin{pmatrix} y_d & 0 & 0 \\ 0 & y_s & 0 \\ 0 & 0 & y_b \end{pmatrix}. \tag{8.29}$$

Equation (8.22) implies that the basis defined in equation (8.29) is the mass basis for down-type quarks. In this basis, we denote the components of the quark $SU(2)$-doublets and the down-type quark $SU(2)$-singlets as follows:

$$Q_{Ld} = \begin{pmatrix} u_{dL} \\ d_L \end{pmatrix}, \qquad Q_{Ls} = \begin{pmatrix} u_{sL} \\ s_L \end{pmatrix}, \qquad Q_{Lb} = \begin{pmatrix} u_{bL} \\ b_L \end{pmatrix}; \qquad d_R, \qquad s_R, \qquad b_R, \tag{8.30}$$

with the convention that $y_d < y_s < y_b$. The down-type quark mass terms are given in this basis by

$$-\mathcal{L}_{M_d} = \frac{y_d v}{\sqrt{2}} \overline{d_L} d_R + \frac{y_s v}{\sqrt{2}} \overline{s_L} s_R + \frac{y_b v}{\sqrt{2}} \overline{b_L} b_R + \text{h.c.} \tag{8.31}$$

We conclude that the down-type quarks d, s, and b are Dirac fermions of masses:

$$m_d = \frac{y_d v}{\sqrt{2}}, \qquad m_s = \frac{y_s v}{\sqrt{2}}, \qquad m_b = \frac{y_b v}{\sqrt{2}}. \tag{8.32}$$

We conclude that all charged fermions acquire Dirac masses as a result of the SSB. The key to this feature is that, while the charged fermions are in chiral representations of the full gauge group $SU(3)_C \times SU(2)_L \times U(1)_Y$, they are in vector like representations of the

$SU(3)_C \times U(1)_{EM}$ group:

- The left-handed and right-handed charged lepton fields, e, μ, and τ, are in the $(1)_{-1}$ representation.
- The left-handed and right-handed up-type quark fields, u, c, and t, are in the $(3)_{+2/3}$ representation.
- The left-handed and right-handed down-type quark fields, d, s, and b, are in the $(3)_{-1/3}$ representation.

On the other hand, as discussed in section 7.3.3, the neutrinos remain massless:

$$m_{\nu_e} = m_{\nu_\mu} = m_{\nu_\tau} = 0. \tag{8.33}$$

The experimental values of the charged fermion masses are

$$\begin{aligned}
&m_e = 0.51099895000(15)\text{ MeV}, \qquad m_\mu = 105.6583755(23)\text{ MeV},\\
&m_\tau = 1776.86(12)\text{ MeV},\\
&m_u = 2.2^{+0.5}_{-0.3}\text{ MeV}, \qquad m_c = 1.27 \pm 0.02\text{ GeV}, \qquad m_t = 172.7 \pm 0.3\text{ GeV},\\
&m_d = 4.7^{+0.5}_{-0.2}\text{ MeV}, \qquad m_s = 93^{+9}_{-3}\text{ MeV}, \qquad m_b = 4.18^{+0.03}_{-0.02}\text{ GeV}.
\end{aligned} \tag{8.34}$$

For $q = u, d, s$, we quote the $\overline{\text{MS}}$ masses $m_q(\mu = 2\text{ GeV})$. For $Q = c, b$, we quote the running mass $m_Q(\mu = m_Q)$, while m_t is extracted from the kinematics of $t\bar{t}$ events. The issue of how quark masses are defined and extracted are complicated by the fact that QCD is confining. We do not elaborate on this point here.

8.3.3 The CKM Matrix

In the derivation of the quark spectrum in section 8.3.2, there is an important difference from the analysis of the lepton spectrum in section 7.3.3. For the leptons, there exists a basis that is simultaneously an interaction basis and a mass basis for both the charged leptons and the neutrinos (i.e., the $\hat{Y}_e$ basis). In contrast, for the quarks, there generally is no interaction basis that is also a mass basis for both up-type and down-type quarks. To see that, we denote $u^i = (u, c, t)$ and $d^i = (d, s, b)$ and write the relation of these mass eigenstates to the interaction eigenstates:

$$u^i_L = (V_{uL})_{ij} U^j_L, \qquad u^i_R = (V_{uR})_{ij} U^j_R, \qquad d^i_L = (V_{dL})_{ij} D^j_L, \qquad d^i_R = (V_{dR})_{ij} D^j_R. \tag{8.35}$$

If $V_{uL} \neq V_{dL}$, as is the general case, then the interaction basis defined by equation (8.24) is different from the interaction basis defined by equation (8.29). In the former, Y^d can be written as a unitary matrix times a diagonal one:

$$Y^u = \hat{Y}^u, \qquad Y^d = V\hat{Y}^d. \tag{8.36}$$

Table 8.1. The Standard Model particles

Particle	Spin	Color	Q	Mass $[v]$
$W^{\pm}$	1	(1)	± 1	$\frac{1}{2}g$
Z^0	1	(1)	0	$\frac{1}{2}\sqrt{g^2+g'^2}$
γ	1	(1)	0	0
g	1	(8)	0	0
h	0	(1)	0	$\sqrt{2\lambda}$
e, μ, τ	1/2	(1)	-1	$y_{e,\mu,\tau}/\sqrt{2}$
ν_e, ν_μ, ν_τ	1/2	(1)	0	0
u, c, t	1/2	(3)	$+2/3$	$y_{u,c,t}/\sqrt{2}$
d, s, b	1/2	(3)	$-1/3$	$y_{d,s,b}/\sqrt{2}$

In the latter, Y^u can be written as a unitary matrix times a diagonal one:

$$Y^d = \hat{Y}^d, \qquad Y^u = V^\dagger \hat{Y}^u. \tag{8.37}$$

In either case, the unitary matrix V is given by

$$V = V_{uL} V_{dL}^\dagger, \tag{8.38}$$

where V_{uL} and V_{dL} are defined by equations (8.23) and (8.28), respectively. Note that each of V_{uL}, V_{uR}, V_{dL}, and V_{dR} depends on the basis from which we start the diagonalization. The combination $V = V_{uL} V_{dL}^\dagger$, however, does not. This is a hint that V is physical. The matrix V is called the *Cabibbo-Kobayashi-Maskawa* (CKM) *matrix*. Its physical significance will become clear in section 8.4.4.

We discuss additional aspects of the CKM matrix in chapter 9.

8.3.4 Summary

The mass eigenstates of the Standard Model, their $SU(3)_C \times U(1)_{EM}$ quantum numbers, and their masses in units of the VEV v are presented in table 8.1. As is the case in the LSM, all masses are proportional to the VEV of the scalar field, v.

8.4 The Interactions

In this section, we discuss the electromagnetic, strong, weak, and Yukawa interactions of the fermion and scalar mass eigenstates of the Standard Model.

8.4.1 Electromagnetic (QED) and Strong (QCD) Interactions

The photon-mediated electromagnetic interactions are described by Quantum Electro-Dynamics (QED), which is part of the Standard Model. The discussion in section 3.2.3 in chapter 3 leads in a straightforward way to the following Lagrangian terms for the Standard Model Dirac fermions:

$$\mathcal{L}_{Af\bar{f}} = e\overline{e_i}\not{A}e_i - (2/3)e\overline{u_i}\not{A}u_i + (1/3)e\overline{d_i}\not{A}d_i. \tag{8.39}$$

The gluon-mediated strong interactions are described by QCD, which is part of the Standard Model. The discussion in section 5.2 of chapter 5 leads in a straightforward way to the following Lagrangian terms for the Standard Model quarks:

$$\mathcal{L}_{Gq\bar{q}} = -g_s\overline{q_i}\not{G}_a(\lambda_a/2)q_i. \tag{8.40}$$

8.4.2 The Higgs Boson Interactions

The only novelty in the Higgs boson interactions with respect to the LSM lies in its couplings to quarks. As for its couplings to bosons, $\mathcal{L}_h^h$ and $\mathcal{L}_V^h$ are unchanged from equation (7.52) in chapter 7. Note that the Higgs boson does not couple (at tree level) to the gluon. As for its couplings to fermions, they are given by the following extension of $\mathcal{L}_f^h$ of equation (7.52):

$$-\mathcal{L}_f^h = \frac{h}{v}\left(m_e\,\overline{e_L}\,e_R + m_\mu\,\overline{\mu_L}\,\mu_R + m_\tau\,\overline{\tau_L}\,\tau_R\right.$$
$$\left. + m_u\,\overline{u_L}\,u_R + m_c\,\overline{c_L}\,c_R + m_t\,\overline{t_L}\,t_R + m_d\,\overline{d_L}\,d_R + m_s\,\overline{s_L}\,s_R + m_b\,\overline{b_L}\,b_R + \text{h.c.}\right). \tag{8.41}$$

Since the couplings of the Higgs boson are proportional to the mass of the particle to which it couples, the Higgs decay is dominated by the heaviest particle that can be pair-produced in the decay. For $m_h \sim 125$ GeV, this is the bottom quark. More generally, the Standard Model predicts the following branching ratios for the leading decay modes:

$$\text{BR}_{\bar{b}b} : \text{BR}_{WW^*} : \text{BR}_{gg} : \text{BR}_{\tau^+\tau^-} : \text{BR}_{ZZ^*} : \text{BR}_{c\bar{c}} = 0.58 : 0.21 : 0.09 : 0.06 : 0.03 : 0.03, \tag{8.42}$$

where $\text{BR}_f \equiv \text{BR}(h \to f)$. The following comments are in order with regard to equation (8.42):

- The WW^* and ZZ^* modes stand for the three-body tree-level decays ($Wf\bar{f}'$ and $Zf\bar{f}$, respectively), where one of the vector bosons is on-shell and the other off-shell.
- The Higgs boson does not have a tree-level coupling to gluons since it carries no color (and the gluon is massless). The decay into final gluons proceeds via loop diagrams. The dominant contribution comes from the top-quark loop.
- From the six branching ratios, three ($b\bar{b}$, $\tau^+\tau^-$, $c\bar{c}$) stand for two-body tree-level decays. Thus, at the tree level, the respective branching ratios obey $\text{BR}_{\bar{b}b} : \text{BR}_{\tau^+\tau^-} : \text{BR}_{c\bar{c}} = 3m_b^2 : m_\tau^2 : 3m_c^2$. QCD radiative corrections are significant and suppress the

two modes with the quark final states (b, c) compared to one with the lepton final state (τ).

- The Higgs boson does not have a tree-level coupling to photons since it carries no electric charge (and the photon is massless). The decay into final photons proceeds via loop diagrams with a small ($\mathrm{BR}_{\gamma\gamma} \sim 0.002$) but observable rate. The dominant contributions come from the W-boson and the top-quark loops, which interfere destructively.

Experimentally, the decays into final $\tau^+\tau^-$, $b\bar{b}$, ZZ^*, WW^* and $\gamma\gamma$ have been established with rates that are consistent with the Standard Model predictions.

8.4.3 Neutral Current Weak Interactions

As explained in chapter 7, interactions mediated by Z-exchange are called neutral current weak interactions. The expression for the Z-couplings to fermions is the same as in the LSM (equation (7.62)), and the quarks follow the same patterns as the leptons. Using the T_3 and Y assignments of the various fermion fields, we find the following types of Z-couplings:

$$\mathcal{L}_{Z,\text{fermions}} = \frac{g}{c_W}\left[-\left(\frac{1}{2} - s_W^2\right)\overline{e_L^i}\not{Z}e_L^i + s_W^2\,\overline{e_R^i}\not{Z}e_R^i + \frac{1}{2}\,\overline{\nu_L^i}\not{Z}\nu_L^i + \left(\frac{1}{2} - \frac{2}{3}s_W^2\right)\overline{u_L^i}\not{Z}u_L^i\right.$$

$$\left. - \frac{2}{3}s_W^2\,\overline{u_R^i}\not{Z}u_R^i - \left(\frac{1}{2} - \frac{1}{3}s_W^2\right)\overline{d_L^i}\not{Z}d_L^i + \frac{1}{3}s_W^2\,\overline{d_R^i}\not{Z}d_R^i\right], \tag{8.43}$$

where $s_W \equiv \sin\theta_W$, $c_W \equiv \cos\theta_W$, $e^i = e, \mu, \tau$, $\nu^i = \nu_e, \nu_\mu, \nu_\tau$, $u^i = u, c, t$, and $d^i = d, s, b$. We learn that the Z-couplings are chiral, parity-violating, diagonal, and universal.

Equation (8.43) implies that the Z-boson couplings in each fermion sector are universal. While universality is a necessary feature of the couplings of gauge bosons corresponding to unbroken generators, it is not an obvious feature of the couplings of vector bosons corresponding to spontaneously broken generators. We discuss this point in section 9.4.2 in chapter 9.

Omitting common factors and phase-space factors, we obtain the following predictions for the Z-decays into a one-generation fermion pair of each type:

$$\begin{aligned}
\Gamma(Z \to \nu\bar{\nu}) &\propto 1, \\
\Gamma(Z \to \ell\bar{\ell}) &\propto 1 - 4s_W^2 + 8s_W^4, \\
\Gamma(Z \to u\bar{u}) &\propto 3\left(1 - \frac{8}{3}s_W^2 + \frac{32}{9}s_W^4\right), \\
\Gamma(Z \to d\bar{d}) &\propto 3\left(1 - \frac{4}{3}s_W^2 + \frac{8}{9}s_W^4\right).
\end{aligned} \tag{8.44}$$

With $s_W^2 = 0.225$, we obtain

$$\Gamma_\nu : \Gamma_\ell : \Gamma_u : \Gamma_d = 1 : 0.51 : 1.74 : 2.24. \tag{8.45}$$

Experiments measure the following average branching ratio into a single generation of each fermion species:

$$\begin{aligned} \text{BR}(Z \to \nu\bar{\nu}) &= (6.67 \pm 0.02)\%, \\ \text{BR}(Z \to \ell\bar{\ell}) &= (3.366 \pm 0.002)\%, \\ \text{BR}(Z \to u\bar{u}) &= (11.6 \pm 0.6)\%, \\ \text{BR}(Z \to d\bar{d}) &= (15.6 \pm 0.4)\%, \end{aligned} \tag{8.46}$$

which, using central values, gives

$$\Gamma_\nu : \Gamma_\ell : \Gamma_u : \Gamma_d = 1 : 0.505 : 1.74 : 2.34, \tag{8.47}$$

which has very nice agreement with the predictions.

8.4.4 Charged Current Weak Interactions

As explained in chapter 7, interactions mediated by W-exchange are called charged current weak interactions. We now study the couplings of the charged vector bosons, $W^\pm$, to fermion pairs. For the lepton mass eigenstates, things are simple because there is an interaction basis that is also a mass basis. Thus, the W interactions must be universal also in the mass basis:

$$\mathcal{L}_{W,\ell} = -\frac{g}{\sqrt{2}} \left(\overline{\nu_{eL}}\, \not{W}^+ e_L^- + \overline{\nu_{\mu L}}\, \not{W}^+ \mu_L^- + \overline{\nu_{\tau L}}\, \not{W}^+ \tau_L^- + \text{h.c.} \right). \tag{8.48}$$

As for quarks, things are more complicated since there is no interaction basis that is also a mass basis. In the interaction basis, the W interactions take the following form:

$$\mathcal{L}_{W,q} = -\frac{g}{\sqrt{2}} \overline{U_L^i}\, \not{W}^+ D_L^i + \text{h.c.} \tag{8.49}$$

Using equation (8.35) to write $U_L^i = (V_{uL}^\dagger)_{ij} u_L^j$ and $D_L^i = (V_{dL}^\dagger)_{ij} d_L^j$, we can rewrite $\mathcal{L}_{W,q}$ in terms of the mass eigenstates:

$$\mathcal{L}_{W,q} = -\frac{g}{\sqrt{2}} \overline{u_L^k} (V_{uL})_{ki}\, \not{W}^+ (V_{dL}^\dagger)_{il} d_L^l + \text{h.c.} = -\frac{g}{\sqrt{2}} \overline{u_L^k} V_{kl}\, \not{W}^+ d_L^l + \text{h.c.}, \tag{8.50}$$

where V is the CKM matrix defined in equation (8.38).

Equation (8.50) reveals some important features of the model:

- Only left-handed particles take part in charged current interactions. Consequently, parity is violated by these interactions.

- The W couplings to the quark mass eigenstates are not universal. The universality of gauge interactions is hidden in the unitarity of the CKM matrix, V.
- The W couplings to the quark mass eigenstates are not diagonal. This is a manifestation of the fact that no pair of up-type and down-type mass eigenstates fits into an $SU(2)_L$ doublet. For example, the d and u mass eigenstates are not members of a single $SU(2)_L$ doublet.

Omitting common factors and phase-space factors, we obtain the following predictions for the W-decays:

$$\Gamma(W^+ \to \ell^+ \nu_\ell) \propto 1,$$
$$\Gamma(W^+ \to u_i \overline{d_j}) \propto 3|V_{ij}|^2 \qquad (i=1,2;\ j=1,2,3). \tag{8.51}$$

The top quark is not included because it is heavier than the W-boson. Taking this fact into account, as well as the CKM unitarity relations,

$$|V_{ud}|^2 + |V_{us}|^2 + |V_{ub}|^2 = |V_{cd}|^2 + |V_{cs}|^2 + |V_{cb}|^2 = 1, \tag{8.52}$$

we obtain

$$\Gamma(W \to \text{quarks}) \approx 2\Gamma(W \to \text{leptons}). \tag{8.53}$$

Experiments give

$$\text{BR}(W \to \text{leptons}) = (32.58 \pm 0.27)\%, \qquad \text{BR}(W \to \text{quarks}) = (67.41 \pm 0.27)\%, \tag{8.54}$$

leading to

$$\Gamma(W \to \text{quarks}) / \Gamma(W \to \text{leptons}) = 2.07 \pm 0.02, \tag{8.55}$$

which, when including radiative corrections, is in good agreement with the Standard Model prediction. The hidden universality within the quark sector is tested by the prediction

$$\Gamma(W \to uX) = \Gamma(W \to cX) = \frac{1}{2}\Gamma(W \to \text{quarks}). \tag{8.56}$$

Experimentally,

$$\text{BR}(W \to cX) = (33.3 \pm 2.6)\%, \qquad \text{BR}(W \to \text{quarks}) = (67.41 \pm 0.27)\%, \tag{8.57}$$

which leads to

$$\Gamma(W \to cX) / \Gamma(W \to \text{quarks}) = 0.49 \pm 0.04. \tag{8.58}$$

8.4.5 *Gauge Boson Self-interactions*

The electroweak vector-boson self-interactions are the same as in the LSM, and the gluon self-interactions are the same as in QCD. There are no interactions that involve both the gluon and electroweak bosons. Thus, the vector boson self-interactions of the Standard

Model are given by the combination of the terms given in equations (5.13), (7.90), and (7.91).

8.4.6 Summary

The various types of interactions that fermions have in the Standard Model are summarized in table 8.2.

8.5 Global Symmetries and Parameters

8.5.1 Accidental Symmetries

If we set the Yukawa couplings to zero ($\mathcal{L}_{\rm Yuk}=0$), the Standard Model gains a large accidental global symmetry in the fermion sector:

$$G_{\rm SM}^{\rm global}(Y^{u,d,e}=0)=U(3)_Q\times U(3)_U\times U(3)_D\times U(3)_L\times U(3)_E, \tag{8.59}$$

where $U(3)_Q$ has (Q_{L1}, Q_{L2}, Q_{L3}) transforming as an $SU(3)_Q$ triplet, and all other fields as singlets; $U(3)_U$ has (U_{R1}, U_{R2}, U_{R3}) transforming as an $SU(3)_U$ triplet, and all other fields as singlets; $U(3)_D$ has (D_{R1}, D_{R2}, D_{R3}) transforming as an $SU(3)_D$ triplet, and all other fields as singlets; $U(3)_L$ has (L_{L1}, L_{L2}, L_{L3}) transforming as an $SU(3)_L$ triplet, and all other fields as singlets; and $U(3)_E$ has (E_{R1}, E_{R2}, E_{R3}) transforming as an $SU(3)_E$ triplet, and all other fields as singlets.

The Yukawa couplings break this symmetry into the following subgroup:

$$G_{\rm SM}^{\rm global}=U(1)_B\times U(1)_e\times U(1)_\mu\times U(1)_\tau. \tag{8.60}$$

Under $U(1)_B$, all quarks (antiquarks) carry the charge $+1/3$ $(-1/3)$, while all other fields are neutral. It explains why proton decay has not been observed. Possible proton decay modes, such as $p\to e^+\gamma$, are not forbidden by the $SU(3)_C\times U(1)_{\rm EM}$ symmetry. However, they violate $U(1)_B$, and therefore they do not occur within the Standard Model. The lesson here is general: the lightest particle that carries a conserved charge is stable.

Table 8.2. The Standard Model fermion interactions

Interaction	Force Carrier	Coupling	Fermions
Electromagnetic	γ	eQ	u, d, and e
Strong	g	g_s	u and d
Neutral current weak	Z^0	$(g/\cos\theta_W)(T_3-\sin^2\theta_W Q)$	u, d, e, and ν
Charged current weak for quarks	$W^\pm$	$(g/\sqrt{2})V_{ij}$	$\bar{u}_i d_j$
Charged current weak for leptons	$W^\pm$	$(g/\sqrt{2})$	$\bar{\nu} e$
Yukawa	h	y_f	u, d, and e

8.5.2 The Standard Model Parameters

On how many physical parameters does the Standard Model depend? The gauge, scalar, and lepton sectors are very similar to the case of the LSM, discussed in section 7.5.4 in chapter 7. Taking into account the one extra gauge coupling of the Standard Model, g_s, these sectors depend on eight physical parameters. What remains to be obtained is the number of physical parameters in the quark sector. To do so, we follow the procedure presented in section 7.5.3.

The two Yukawa matrices defined in equation (8.16), Y^u and Y^d, are 3×3 complex matrices, which contain a total of 36 parameters (18 real parameters and 18 phases) in a general basis. The kinetic terms for the quarks have a global symmetry:

$$G_q = U(3)_Q \times U(3)_U \times U(3)_D, \tag{8.61}$$

which has 27 generators. The Yukawa terms break this symmetry into a baryon number, $H_q = U(1)_B$, which has a single generator. Thus, $N_{\text{broken}} = 26$. This broken symmetry allows us to rotate away a large number of the parameters by moving to a more convenient basis. Using equation (7.102), the number of physical parameters in the quark sector is given by

$$N_{\text{phys}} = 36 - 26 = 10. \tag{8.62}$$

These parameters can be split into real parameters and phases. The three unitary matrices generating G_q have a total of 9 real parameters and 18 phases. The symmetry is broken down to a symmetry with only one phase generator. Thus,

$$N_{\text{phys}}^{(r)} = 18 - 9 = 9, \qquad N_{\text{phys}}^{(i)} = 18 - 17 = 1. \tag{8.63}$$

Let us now identify these parameters. Of the 9 real parameters, 6 are the quark masses and three are real parameters in the CKM matrix, which are called *mixing angles*. The one phase is a *CP*-violating phase in the CKM matrix, which is called the *Kobayashi–Maskawa* (KM) *phase*. We give examples of explicit parameterizations of the CKM matrix in section 9.2.1 in chapter 9.

We conclude that the Standard Model has 18 parameters: 3 gauge couplings, 2 parameters of the Higgs potential, 3 charged lepton masses, 6 quark masses, and 4 CKM parameters. Of these, 7 parameters are the LSM parameters, discussed in section 7.5.4, and 7 (g_s and the 6 quark masses) were introduced in previous sections of this chapter. The four CKM parameters are discussed in chapter 9.

8.5.3 "A Standard Model" versus "the Standard Model"

At this point, we would like to distinguish "*a* Standard Model" from "*the* Standard Model." The properties of the former depend only on the definition of the model (i.e., its imposed symmetry and particle content). The properties of the latter depend as well on the specific values of the parameters. Thus, our discussion so far has dealt with "a Standard Model."

Once we present the values of the parameters, we can move on to discuss "the Standard Model." This distinction is particularly useful when we discuss flavor physics in chapter 13.

8.5.4 Discrete Symmetries: P, C, and CP

Similarly to the LSM, the Standard Model violates C and P, as it is a chiral theory. As for CP, however, the Standard Model differs from the LSM: while CP is a good symmetry of the LSM, it is not a good symmetry of the Standard Model. The single phase of the CKM matrix is a source of CP violation within the Standard Model. CP violation in the Standard Model is closely related to the number of fermion generations: three is the minimal number that exhibits a CP-violating phase.

Note, however, that unlike the case of P and C, which are violated in the Standard Model (and in any chiral theory) already by the definition of the particle content, CP can be violated only by the interactions. For a generic case, $W^\pm$ interactions with the quarks violate CP. We specify the exact conditions on the model parameters for CP to be violated in chapter 9. Here, we just mention that it is experimentally established that CP is indeed violated in nature.

Appendix

8.A Anomalies and Nonperturbative Effects

The issues of anomalies and nonperturbative effects are usually discussed only in advanced quantum field theory (QFT) courses. Moreover, for almost all phenomenological implications of the Standard Model, these issues are irrelevant. Thus, we mention them in this appendix for the sake of completeness. Readers who do not have the required background can skip this appendix.

8.A.1 The Strong CP Parameter

The counting of parameters described in this chapter is done at the classical level. Usually, when quantizing a system, the number of parameters is not changed. Yet there are exceptions that are related to non-Abelian gauge groups. It turns out that in the Standard Model, there is one more renormalizable parameter that is unphysical at the classical level but physical at the quantum level. This parameter is called $\theta_{\rm QCD}$:

$$\mathcal{L}_{\theta_{\rm QCD}} = \frac{\theta_{\rm QCD}}{32\pi^2}\varepsilon_{\mu\nu\rho\sigma}G_a^{\mu\nu}G_a^{\rho\sigma}. \tag{8.64}$$

This term violates P and CP. In particular, it leads to an electric dipole moment (EDM) of the neutron d_n. The experimental upper bound on the EDM of the neutron,

$$d_n < 1.8 \times 10^{-26}\, e\,\text{cm}, \tag{8.65}$$

implies that

$$\theta_{\rm QCD} \lesssim 10^{-9}. \tag{8.66}$$

The question of why θ_{QCD} is so small is known as the "strong CP problem." We do not discuss it any further here. We just conclude that the number of independent parameters in the quantum Standard Model is 19: the 18 mentioned earlier and θ_{QCD}.

8.A.2 Anomalies

For a theory to be consistent, its gauge symmetry must be anomaly-free. This issue has not been raised thus far because an anomalous theory can be fixed by adding appropriate DoF to it. Yet it is interesting to note that QED, QCD, and the Standard Model are all free of gauge anomalies. In contrast, the LSM is an anomalous theory, and thus it could not be a full theory of nature. Indeed, the addition of the quark representations of the Standard Model removes the LSM anomalies.

A point to note is that vectorial theories are anomaly-free, and thus QED and QCD are as well. For chiral theories, some conditions on the fermion quantum numbers have to be satisfied. These conditions are satisfied within each generation of the Standard Model.

Concerning global symmetries, a theory is consistent even if its accidental symmetries are anomalous. An anomaly in a global symmetry implies that the symmetry is violated by nonperturbative effects. Each of the four accidental $U(1)$ symmetries of equation (8.60) is anomalous. However, one can always define three combinations, such as $U(1)_{B-L}$, $U(1)_{\mu-e}$ and $U(1)_{\tau-\mu}$, which are anomaly-free, but a fourth independent combination, such as $U(1)_{B+L}$, is unavoidably anomalous.

Due to the anomalies, baryon and lepton number–violating processes occur nonperturbatively. However, the nonperturbative effects obey a selection rule, $\Delta B = \Delta L = 3n$, with n integer, and thus do not lead to proton decay. Moreover, the nonperturbative B- and L-violating effects are negligibly small in all the processes that we study, except in the context of cosmology, as we discuss in section 15.3 in chapter 15.

The accidental symmetries of the renormalizable part of the Standard Model Lagrangian explain the vanishing of neutrino masses. A Majorana mass term violates not only the anomalous L symmetry, but also the nonanomalous $B-L$ symmetry. Thus, the symmetry prevents neutrino mass terms not only at the tree level, but also to all orders in perturbation theory, and even—due to the nonanomalous $B-L$ symmetry—at the nonperturbative level. We conclude that the renormalizable Standard Model gives the following *exact* prediction:

$$m_\nu = 0. \tag{8.67}$$

For Further Reading

- A collection of original papers that set the stage for the Standard Model can be found in Cahn and Goldhaber [26].
- More on the issue of anomalies (discussed in the appendix) can be found, for example, in chapter 30 of Schwartz [15].
- More on the strong CP problem (discussed in the appendix) can be found, for example, in Hook [27].

Problems

Question 8.1: Algebra

1. Using equation (7.18) in chapter 7, derive equation (8.15).
2. Using equation (8.18) and the definitions in this chapter, derive equations (8.22), (8.26), and (8.31).
3. Using equation (8.35), derive equations (8.36) and (8.37). Show that in both cases, V is indeed the one defined in equation (8.38).
4. Draw the Feynman diagrams for $W\overline{u_i}d_j$ interactions, and write the Feynman rules, making sure to include the relevant CKM factors.

Question 8.2: The N_g-generation Standard Model

Calculate the number of physical parameters related to $\mathcal{L}_{\text{Yuk}}$ in a Standard Model with N_g generations. Separate them into masses, mixing angles, and phases. In particular, show that CP is violated only with three or more generations.

Question 8.3: The four-generation Standard Model

Consider an extension of the Standard Model with an additional (fourth) generation of quarks and leptons:

$$L_4(1,2)_{-1/2}, \qquad E_4(1,1)_{-1}, \qquad Q_4(3,2)_{+1/6}, \qquad U_4(3,1)_{+2/3}, \qquad D_4(3,1)_{-1/3}. \tag{8.68}$$

We denote the fourth-generation mass eigenstates by t', b', τ', and $\nu_{\tau'}$ in an obvious notation.

1. Write the Yukawa interactions. Use matrix notation (e.g., use L_i, where $i = 1, 2, 3, 4$, to denote the lepton doublets).
2. What is the global symmetry of the model? What is the number of physical parameters related to the Yukawa terms? Separate them into masses, mixing angles, and phases.
3. Is CP violated in the quark sector? Is it violated in the lepton sector?
4. Assuming that $m_{\tau'} < m_t, m_{t'}, m_{b'}$ and $m_b \ll m_{\tau'} \ll m_W$, list the possible tree-level decay processes of the τ'-lepton. Estimate the branching ratio of each mode in terms of the elements of the 4×4 CKM matrix. Neglect the outgoing fermion masses.
5. Explain why the measured value of the invisible width of the Z-boson, $\Gamma_{\text{inv}} = 499.0 \pm 1.5$ MeV, rules out the model.

Question 8.4: Leptoquarks and lepton flavor violations

A leptoquark is a hypothetical bosonic field that transforms as a triplet of $SU(3)_C$, and consequently (with appropriate $SU(2)_L \times U(1)_Y$ quantum numbers) can couple to quark–lepton pairs. In this question, we add to the Standard Model a single scalar field, $F(3,2)_{Y_F}$.

1. We require that the following couplings are allowed:

$$\lambda^{FQE}\,\overline{Q_L}\,E_R\,F. \tag{8.69}$$

Determine Y_F.

2. Restore the color, $SU(2)_L$, and flavor indexes in equation (8.69).
3. We denote the components of the F-doublet as

$$F = \begin{pmatrix} F_u \\ F_d \end{pmatrix}. \tag{8.70}$$

What are the electric charges of F_u and F_d?

4. The model is ruled out if $\langle F\rangle \neq 0$. Explain why this is the case. In what follows, we then assume that $\langle F\rangle = 0$.
5. Based on equation (8.69), find the baryon and lepton numbers of F. (Recall that quarks carry $B = +1/3$ and leptons carry $L = +1$.)
6. Starting with equation (8.69), write explicitly the Yukawa terms of F_u and F_d in the mass basis. Denote the Yukawa matrices by $\lambda^{Fu\ell}_{ij}$ and $\lambda^{Fd\ell}_{ij}$, respectively.
7. Find the relation between $\lambda^{Fd\ell}_{ij}$ and $\lambda^{Fu\ell}_{ij}$ in terms of the CKM matrix.
8. The Higgs VEV introduces splitting between the masses of F_u and F_d. Write the mass-squared term for F and the couplings to the Higgs field. Calculate the masses of F_u and F_d and explicitly write the mass splitting. Note that there are three independent ways to contract the $SU(2)_L$ indexes in the terms that involve the ϕ and F fields. Be sure to include all of them.

From this point on, assume that the mass splitting between F_u and F_d is negligible.

9. In the Standard Model, the $b \to s\mu^+e^-$ decay is forbidden. Explain why.
10. F mediates the $b \to s\mu^+e^-$ decay. Draw the tree-level Feynman diagram for this decay and estimate the amplitude.
11. Estimate the F-mediated amplitude of $b \to se^+\mu^-$.

Next, we obtain a lower bound on m_F. To do so, we compare the rate of the F-mediated $b \to s\mu^+e^-$ decay rate to that of the W-mediated $b \to ce^-\bar{\nu}$ decay.

12. Draw the tree-level diagram for $b \to ce^-\bar{\nu}$.
13. Estimate the ratio

$$\frac{\Gamma(b \to s\mu^+e^-)}{\Gamma(b \to ce^-\bar{\nu})} \tag{8.71}$$

in terms of $\lambda^{Fd\ell}$, m_F, g, m_W, and the CKM matrix elements. Assume that $m_F \gg m_b$ and neglect phase-space effects. Be sure to write explicitly the flavor structure of the couplings.

14. We now assume that $\lambda^{FQE}_{ij} \sim g$ for all ij. Using the experimental data

$$\text{BR}(b \to ce^-\bar{\nu}) \sim 10^{-1}, \qquad \text{BR}(b \to s\mu^+e^-) \lesssim 10^{-5}, \tag{8.72}$$

and $|V_{cb}| \approx 0.04$, estimate a lower bound on m_F.

We conclude that the mass of such a hypothetical leptoquark cannot be at the weak scale (i.e., close to the W mass) if its Yukawa couplings are of the order of the weak coupling g.

Question 8.5: Leptoquarks and proton decay

We continue to study hypothetical leptoquarks, but here we encounter leptoquark fields with much stronger lower bounds on their mass scale than the one we considered in question 8.4. Specifically, we add to the Standard Model a single scalar field, $R(3,1)_{Y_R}$.

1. We require that the following couplings are allowed:

$$\lambda_{ij}^{RDE}(\overline{d_R})_i(e_R^c)_jR. \tag{8.73}$$

   (Recall from section 1.2.2 that ψ_R^c is a left-handed field with opposite charges compared to ψ_R. In particular, e_R^c is a left-handed field that transforms as $(1,1)_{+1}$ and carries a lepton number of -1.) Determine Y_R.
2. R also couples to pairs of quarks. For example, we can have

$$\lambda_{ij}^{RUU}(\overline{u_R^c})_i(u_R)_jR. \tag{8.74}$$

   Show that this term is indeed a singlet under the Standard Model gauge group. The field u_R^c is left-handed and transforms as $(\bar{3},1)_{-2/3}$. As for SU(3) algebra, recall that $3\times 3=6+\bar{3}$ and $3\times\bar{3}=1+8$.
3. Explain why λ_{ij}^{RUU} has to be antisymmetric with respect to i and j.
4. Adding R to the Standard Model breaks both the baryon and lepton numbers. That is, one cannot assign lepton and baryon numbers to R such that B and L are respected by the couplings of both equation (8.73) and equation (8.74). Show that this is the case.
5. One combination of B and L is still conserved. Find this combination and determine the charge of R under it.
6. R can mediate proton decay. Consider the $p^+\to e^+\pi^0$ decay. (π^0 is the neutral pion, which will be discussed in chapter 10.) In terms of quarks, this decay corresponds to the following transition:

$$uud\to e^+d\bar{d}. \tag{8.75}$$

   Draw an example of a tree-level Feynman diagram for this decay. Note that that the diagram involves both W- and R-propogators.
7. Give a rough estimate of the amplitude in terms of the couplings, m_W, and m_R. Write explicitly the flavor dependence of the couplings. Assume that $m_R\gg m_p\sim 1$ GeV, and thus neglect the momentum dependence of the W- and R-propagators.

8. Experiments put a lower bound on the proton lifetime, $\tau_p \gtrsim 10^{33}$ years. Taking all the couplings of R to be of order 1, estimate the lower bound on m_R. Express your result in units of GeV. Do not worry about factors of order 1. Recall that $\hbar = 1 = 6.58 \times 10^{-22}$ MeV s.

We learn that if R exists, either its mass is very large or its couplings to the Standard Model quarks and leptons are very small.

9

Flavor Physics

The appearance of the Cabibbo-Kobayashi-Maskawa (CKM) matrix in the interactions of the W-boson with quarks introduces two new qualitative ingredients in the Standard Model, as compared to the Leptonic Standard Model (LSM): Flavor-changing interactions and CP violation. Thus, the physics of the CKM matrix deserves a more detailed discussion, which is provided in this chapter.

9.1 Introduction

The term *flavors* is used to describe several mass eigenstates with the same quantum numbers. Within the Standard Model, each of the four types of fermions (namely, different $SU(3)_C \times U(1)_{EM}$ representations) comes in three flavors:

- Up-type quarks in the $(3)_{+2/3}$ representation: u, c, and t
- Down-type quarks in the $(3)_{-1/3}$ representation: d, s, and b
- Charged leptons in the $(1)_{-1}$ representation: e, μ, and τ
- Neutrinos in the $(1)_0$ representation: ν_1, ν_2, and ν_3

In this section, we discuss only quark flavor physics. (We discuss lepton flavor physics in chapter 14.)

The term *flavor physics* refers to interactions that distinguish among flavors. Within the Standard Model, these are W-mediated weak interactions and Yukawa interactions. The term *flavor parameters* refers to parameters that carry flavor indexes. Within the quark sector of the Standard Model, there are 10 flavor parameters: the 6 quark masses and the 4 CKM parameters. The term *flavor-universal* refers to interactions with couplings that are proportional to a unit matrix in flavor space. Within the Standard Model, the strong, electromagnetic, and Z-mediated weak interactions are flavor-universal. The term *flavor-diagonal* refers to interactions with couplings that are diagonal, but not necessarily universal, in flavor space (i.e., in the mass basis). Within the Standard Model, the Yukawa interactions are flavor-diagonal. Note that all flavor-universal interactions are also flavor-diagonal.

The term *flavor-changing* refers to processes in which the initial and final flavor-numbers (i.e., the number of particles of a certain flavor minus the number of antiparticles of the same flavor) are different. A central role in testing the flavor sector of the Standard Model is played by flavor-changing processes. In *flavor-changing charged current* (FCCC) processes, both up-type and down-type flavors are involved. Examples of FCCC decays include $s \to u\mu^-\bar{\nu}_\mu$ transition and $b \to c\bar{c}s$ transition. Within the Standard Model, these processes are mediated by the W-bosons and occur at the tree level. In *flavor-changing neutral current* (FCNC) processes, either up-type or down-type flavors, but not both, are involved. Examples of FCNC decays include $s \to d\mu^+\mu^-$ transition and $b \to s\bar{s}s$ transition.

In nature, FCNC processes are strongly suppressed compared to FCCC processes. This situation is well accounted for in the Standard Model, as discussed in detail in section 9.4 and in chapter 13.

9.2 The CKM Matrix

Among the Standard Model interactions, the W-mediated interactions are the only ones that are not diagonal in the mass basis. Consequently, all flavor-changing processes depend on the CKM parameters. The fact that there are only four independent CKM parameters, while the number of measured flavor-changing processes is much larger, allows stringent tests of the CKM mechanism for flavor-changing processes.

9.2.1 The Standard Parameterization

The CKM matrix is defined in equation (8.38) of chapter 8. Its explicit form is not unique. First, there is freedom in defining V, in that we can permute between the generations. This freedom is fixed by ordering the up quarks and the down quarks by their masses $(u_1, u_2, u_3) \to (u, c, t)$ and $(d_1, d_2, d_3) \to (d, s, b)$. We then write the W interaction of equation (8.50) as

$$\mathcal{L}_{W,q} = -\frac{g}{\sqrt{2}} \left(\overline{u_L}\, \overline{c_L}\, \overline{t_L}\right) V \not{W}^+ \begin{pmatrix} d_L \\ s_L \\ b_L \end{pmatrix} + \text{h.c.} \tag{9.1}$$

The elements of V, therefore, are written as follows:

$$V = \begin{pmatrix} V_{ud} & V_{us} & V_{ub} \\ V_{cd} & V_{cs} & V_{cb} \\ V_{td} & V_{ts} & V_{tb} \end{pmatrix}. \tag{9.2}$$

Second, we can redefine the phases of the quark fields in such a way that the masses remain real, but the phase structure of the CKM matrix changes. This freedom can be used to choose an explicit parameterization that depends on three real and one imaginary parameters. For example, the standard parameterization, used by the Particle Data Group (PDG), is given by

$$V = \begin{pmatrix} c_{12}c_{13} & s_{12}c_{13} & s_{13}e^{-i\delta} \\ -s_{12}c_{23} - c_{12}s_{23}s_{13}e^{i\delta} & c_{12}c_{23} - s_{12}s_{23}s_{13}e^{i\delta} & s_{23}c_{13} \\ s_{12}s_{23} - c_{12}c_{23}s_{13}e^{i\delta} & -c_{12}s_{23} - s_{12}c_{23}s_{13}e^{i\delta} & c_{23}c_{13} \end{pmatrix}, \tag{9.3}$$

where $c_{ij} \equiv \cos\theta_{ij}$ and $s_{ij} \equiv \sin\theta_{ij}$. The three θ_{ij} are the three mixing angles, while δ is the Kobayashi–Maskawa (KM) phase. With the fixed mass ordering explained here, we have $\theta_{ij} \in \{0, \pi/2\}$ and $\delta \in \{0, 2\pi\}$. The mixing angles θ_{ij} are often referred to as the "real parameters," and δ as the "imaginary one" or the "*CP*-violating one."

The fitted values of the four parameters are given by

$$\begin{aligned} \sin\theta_{12} &= 0.22500 \pm 0.00067, \\ \sin\theta_{23} &= 0.04182^{+0.00085}_{-0.00074}, \\ \sin\theta_{13} &= 0.00369 \pm 0.00011, \\ \delta &= 1.144 \pm 0.027. \end{aligned} \tag{9.4}$$

This translates into the following ranges for the magnitude of the CKM elements:

$$|V| = \begin{pmatrix} 0.97435 \pm 0.00016 & 0.22500 \pm 0.00067 & 0.00369 \pm 0.00011 \\ 0.22486 \pm 0.00067 & 0.97349 \pm 0.00016 & 0.04182^{+0.00085}_{-0.00074} \\ 0.00857^{+0.00020}_{-0.00018} & 0.04110^{+0.00083}_{-0.00072} & 0.999118^{+0.000031}_{-0.000036} \end{pmatrix}. \tag{9.5}$$

We discuss some of the ways in which these entries are determined in section 9.3.

9.2.2 The Wolfenstein Parameterization

Equation (9.5) implies that the CKM matrix is numerically close to a unit matrix, with small off-diagonal terms that obey the following hierarchy:

$$|V_{ub}|, |V_{td}| \ll |V_{cb}|, |V_{ts}| \ll |V_{us}|, |V_{cd}|. \tag{9.6}$$

This situation inspires an approximate parameterization, known as the *Wolfenstein parameterization*. The Wolfenstein parameters consist of the three real parameters, λ, A, and ρ, and the imaginary (*CP*-violating) parameter, $i\eta$. The expansion is in the small parameter:

$$\lambda = |V_{us}| \approx 0.23. \tag{9.7}$$

The order of magnitude of each element can be read from the power of λ. To $\mathcal{O}(\lambda^3)$, the CKM matrix is written in terms of the Wolfenstein parameters as follows:

$$V = \begin{pmatrix} 1 - \frac{1}{2}\lambda^2 & \lambda & A\lambda^3(\rho - i\eta) \\ -\lambda & 1 - \frac{1}{2}\lambda^2 & A\lambda^2 \\ A\lambda^3(1 - \rho - i\eta) & -A\lambda^2 & 1 \end{pmatrix}. \tag{9.8}$$

The relations between the standard parameters and the Wolfenstein parameters are given by

$$\lambda = s_{12}, \qquad A\lambda^2 = s_{23}, \qquad A\lambda^3(\rho - i\eta) = s_{13}e^{-i\delta}. \tag{9.9}$$

The fitted values of the four parameters can be read from equation (9.4):

$$\begin{aligned}\rho &= 0.159 \pm 0.010,\\ \eta &= 0.348 \pm 0.010,\\ A &= 0.826^{+0.018}_{-0.015},\\ \lambda &= 0.22500 \pm 0.00067.\end{aligned} \tag{9.10}$$

In section 8.5.3 in chapter 8, we distinguished *the* Standard Model from *a* Standard Model. The experimental fact that the CKM matrix is close to a unit matrix is one of the ingredients of *the Standard Model* that are far from a generic Standard Model. The hierarchy in the quark masses constitutes another such ingredient.

9.2.3 CP Violation

Various parameterizations differ in the way that the freedom of phase rotation is used to leave a single phase in V. One can define, however, a CP-violating quantity in V that is independent of the parameterization. This quantity, the Jarlskog invariant, $J_{\rm CKM}$, is defined through

$$\mathcal{I}m\left(V_{ij}V_{kl}V_{il}^*V_{kj}^*\right) = J_{\rm CKM}\sum_{m,n=1}^{3}\varepsilon_{ikm}\varepsilon_{jln}, \qquad (i,j,k,l=1,2,3). \tag{9.11}$$

(There is no sum over the i, j, k, l indexes.) In terms of the explicit parameterizations given in equations (9.3) and (9.8), the Jarlskog invariant is given by

$$J_{\rm CKM} = c_{12}c_{23}c_{13}^2 s_{12}s_{23}s_{13}\sin\delta \approx \lambda^6 A^2 \eta. \tag{9.12}$$

Note that $|J_{\rm CKM}|$ is bounded from above:

$$|J_{\rm CKM}| \le \frac{1}{6\sqrt{3}} \sim 0.1. \tag{9.13}$$

The current best fit for $J_{\rm CKM}$ is given by

$$J_{\rm CKM} = (3.00^{+0.15}_{-0.09}) \times 10^{-5}, \tag{9.14}$$

which is much smaller than the upper bound of equation (9.13). More significantly, the experimental value is much smaller than the value that it would have if all the relevant parameters were $O(1)$. This is one more demonstration that within the flavor sector, the Standard Model has nongeneric features.

While a generic Standard Model violates CP, specific realizations of it could still conserve CP. For *the Standard Model* to violate CP, the following necessary and sufficient condition must be fulfilled:

$$X_{CP} \equiv \Delta m_{tc}^2 \Delta m_{tu}^2 \Delta m_{cu}^2 \Delta m_{bs}^2 \Delta m_{bd}^2 \Delta m_{sd}^2 J_{\rm CKM} \neq 0, \tag{9.15}$$

where $\Delta m_{ij}^2 \equiv m_i^2 - m_j^2$. Equation (9.15) puts the following requirements on the Standard Model for CP to be violated:

- Within each quark sector, there must be no mass degeneracy.
- The Jarlskog invariant must not vanish.

These conditions can also be written as a single requirement on the quark mass matrices in any interaction basis:

$$X_{CP} = \mathcal{I}m \left\{ \det \left[M_d M_d^\dagger, M_u M_u^\dagger \right] \right\} \neq 0 \Leftrightarrow CP \text{ violation.} \tag{9.16}$$

This is a convention-independent condition.

9.2.4 Unitarity Triangles

A very useful concept with regard to CP violation is that of the unitarity triangle. The unitarity of the CKM matrix leads to various relations among its elements. Of particular interest are the six relations:

$$\sum_{i=u,c,t} V_{iq} V_{iq'}^* = 0 \qquad (qq' = ds, db, sb),$$

$$\sum_{i=d,s,b} V_{qi} V_{q'i}^* = 0 \qquad (qq' = uc, ut, ct). \tag{9.17}$$

Each of these relations requires the sum of three complex quantities to vanish. Therefore, they can be geometrically represented in the complex plane as triangles, which are called *unitarity triangles*. It is a feature of the CKM matrix that all six unitarity triangles are equal in area. Moreover, the area of each unitarity triangle equals $|J_{\text{CKM}}|/2$, while the sign of J_{CKM} gives the direction of the complex vectors around the triangles.

The triangle that corresponds to the relation

$$V_{ud} V_{ub}^* + V_{cd} V_{cb}^* + V_{td} V_{tb}^* = 0 \tag{9.18}$$

has three sides of roughly the same length, $\mathcal{O}(\lambda^3)$—see equation (9.8). Furthermore, both the length of its sides and the size of its angles are experimentally accessible. For these reasons, the term the *unitarity triangle* is reserved for equation (9.18).

We further define the *rescaled unitarity triangle* here. It is derived from equation (9.18) by choosing a phase convention such that $(V_{cd} V_{cb}^*)$ is real, and dividing the lengths of all the sides by $|V_{cd} V_{cb}^*|$. The rescaled unitarity triangle is similar to the unitarity triangle. Two vertices of the rescaled unitarity triangle are fixed at (0,0) and (1,0). The coordinates of the remaining vertex correspond to the Wolfenstein parameters (ρ, η). The rescaled unitarity triangle is shown in figure 9.1. The lengths of the two complex sides are

$$R_u \equiv \left| \frac{V_{ud} V_{ub}}{V_{cd} V_{cb}} \right| = \sqrt{\rho^2 + \eta^2}, \qquad R_t \equiv \left| \frac{V_{td} V_{tb}}{V_{cd} V_{cb}} \right| = \sqrt{(1-\rho)^2 + \eta^2}. \tag{9.19}$$

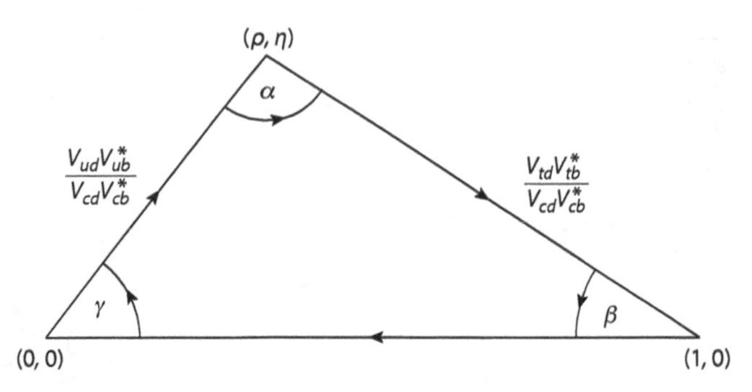

FIGURE 9.1. The rescaled unitarity triangle.

The three angles of the unitarity triangle are defined as follows:

$$\alpha \equiv \arg\left[-\frac{V_{td}V_{tb}^*}{V_{ud}V_{ub}^*}\right], \qquad \beta \equiv \arg\left[-\frac{V_{cd}V_{cb}^*}{V_{td}V_{tb}^*}\right], \qquad \gamma \equiv \arg\left[-\frac{V_{ud}V_{ub}^*}{V_{cd}V_{cb}^*}\right]. \quad (9.20)$$

They are physical quantities and can be independently measured, as we discuss next and in chapter 13. Another commonly used notation is $\phi_1 = \beta$, $\phi_2 = \alpha$, and $\phi_3 = \gamma$. Note that in the standard parameterization, $\gamma = \delta$.

9.3 Tree-Level Determination of the CKM Parameters

In this section, we describe the determination of CKM parameters from tree-level processes. There is an inherent difficulty about determining the CKM parameters: while the Standard Model Lagrangian has the quarks as its degrees of freedom (DoF), in nature, they appear only within bound states called hadrons. There are various tools to overcome this difficulty, some of which are discussed in chapter 10.

At the tree-level, the W-mediated interactions lead to only FCCC processes. These suffice, however, to overconstrain the CKM parameters. The most useful processes are semileptonic ones. We discuss some of these processes in more detail in chapter 10 (and in question 9.3 at the end of this chapter). Here, we give a short summary of the results:

- Processes related to $d \to u\ell^-\bar{\nu}$ transitions give $|V_{ud}| = 0.97373 \pm 0.00031$.
- Processes related to $s \to u\ell^-\bar{\nu}$ transitions give $|V_{us}| = 0.2243 \pm 0.0008$.
- Processes related to $c \to d\ell^+\nu$ or to $\nu_\mu + d \to c + \mu^-$ transitions give $|V_{cd}| = 0.221 \pm 0.004$.
- Processes related to $c \to s\ell^+\nu$ or to $c\bar{s} \to \ell^+\nu$ transitions give $|V_{cs}| = 0.975 \pm 0.006$.
- Processes related to $b \to c\ell^-\bar{\nu}$ transitions give $|V_{cb}| = 0.0408 \pm 0.0014$.
- Processes related to $b \to u\ell^-\bar{\nu}$ transitions give $|V_{ub}| = 0.00382 \pm 0.00020$.

There are two additional classes of tree-level processes that depend on the CKM parameters:

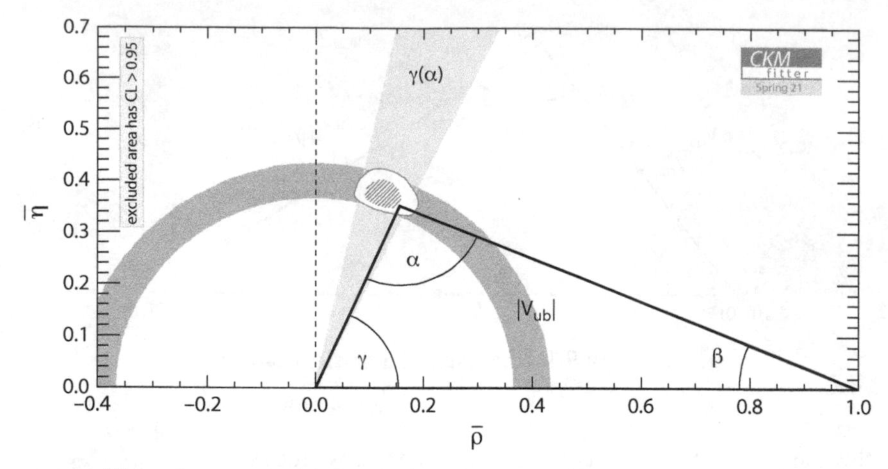

FIGURE 9.2. Allowed region in the (ρ, η) plane from Standard Model tree-level processes. Taken from CKMfitter Group [28].

- Processes related to single top production in hadron colliders give $|V_{tb}| = 1.014 \pm 0.029$.
- Processes related to $b \to sc\bar{u}$ and $b \to su\bar{c}$ transitions give $\gamma = (65.9^{+3.3}_{-3.5})^o$. (See section 13.C.1 in chapter 13 for details.)

These eight distinct classes of processes depend on only four CKM parameters. The system is thus overconstrained and tests the Standard Model.

The values of λ and A can be straightforwardly extracted from the measurements of $|V_{us}|$ and $|V_{cb}|$, respectively:

$$\lambda = 0.2245 \pm 0.0008, \qquad A = 0.837 \pm 0.016. \tag{9.21}$$

The values of ρ and η are extracted mainly from combining the measurements of $|V_{ub}|$ and γ, as shown in figure 9.2:

$$\rho = 0.13 \pm 0.03, \qquad \eta = 0.38 \pm 0.02. \tag{9.22}$$

The fact that the ranges of the four parameters in equations (9.21) and (9.22) are consistent with all the measurements means that the Standard Model passes the test.

Note that the error bars on the determination here, equations (9.21) and (9.22), are larger than those in equation (9.10). The reason is that here, we consider only tree-level processes.

9.4 No FCNC at Tree Level

Historically, the strong suppression of FCNC has played a very important role in constructing the Standard Model. At present, it continues to play a significant role in testing it. In this section, we explain why, in the Standard Model, there are no tree-level contributions to FCNC processes. Since there is no symmetry that forbids FCNC in the quark sector, there are loop contributions to these processes, which are discussed in chapter 13.

The W-boson cannot mediate FCNC processes at tree level, since it couples to up-down pairs or neutrino-charged lepton pairs. Only neutral bosons could mediate FCNC at tree level. The Standard Model has four neutral bosons: the gluon, the photon, the Z-boson, and the Higgs boson. As derived explicitly in chapter 8, within the Standard Model, all of them couple diagonally in the mass basis, and therefore they cannot mediate FCNC at tree level. Here, we explain the qualitative features of the Standard Model that lead to this situation.

9.4.1 *Photon- and Gluon-Mediated FCNC*

As for the massless gauge bosons, the gluon and the photon, their couplings are flavor-universal and, in particular, flavor-diagonal. This is guaranteed by gauge invariance. The universality of the kinetic terms in the canonical basis requires universality of the gauge couplings related to the unbroken symmetries. Hence, neither the gluon nor the photon can mediate flavor-changing processes at tree level. Since we require that extensions of the Standard Model respect the local $SU(3)_C \times U(1)_{EM}$ symmetry, this result holds for all such extensions.

9.4.2 *Z-Mediated FCNC*

The Z-boson, similar to the W-boson, corresponds to a broken gauge symmetry (as manifest in the fact that it is massive). Hence, there is no fundamental symmetry principle that forbids flavor-changing Z-couplings. Yet, as we explicitly find in section 8.4.3 of chapter 8, in the Standard Model, the Z-couplings are universal and diagonal.

The key point is defined as follows. The Z-couplings are proportional to $T_3 - Q \sin^2 \theta_W$. A sector of mass eigenstates is characterized by spin, $SU(3)_C$ representation, and $U(1)_{EM}$ charge. While Q must be the same for all the flavors in a given sector, there are two possibilities regarding T_3:

1. All mass eigenstates in this sector originate from interaction eigenstates in the same $SU(2)_L \times U(1)_Y$ representation, and thus they have the same T_3 and Y.
2. The mass eigenstates in this sector mix interaction eigenstates with the same $Q = T_3 + Y$ but different $SU(2)_L \times U(1)_Y$ representations, and in particular, different T_3 and Y values.

Let us examine the Z-couplings in the interaction and mass bases for several flavors of hypothetical fermions in the same $SU(3)_C \times U(1)_{EM}$ representation:

1. In the first class, the Z-couplings in the fermion interaction basis are universal; namely, they are proportional to the unit matrix (times $T_3 - Q \sin^2 \theta_W$ of the relevant interaction eigenstates). The rotation to the mass basis maintains the

universality as follows:

$$V_{fM} \times \mathbf{1} \times V_{fM}^{\dagger} = \mathbf{1}, \qquad (f = u, d; M = L, R). \qquad (9.23)$$

2. In the second class, the Z-couplings in the fermion interaction basis are diagonal but not universal. Each diagonal entry is proportional to the relevant $T_3 - Q\sin^2\theta_W$. In this case, the rotation to the mass basis generally does not maintain the diagonality:

$$V_{fM} \times \hat{G}_{\text{diagonal}} \times V_{fM}^{\dagger} = G_{\text{nondiagonal}}, \qquad (f = u, d; M = L, R). \qquad (9.24)$$

Question 9.6 provides a specific example of a model of this type.

The Standard Model fermions belong to the first class. All fermion mass eigenstates with a given chirality and in a given $SU(3)_C \times U(1)_{\text{EM}}$ representation come from the same $SU(3)_C \times SU(2)_L \times U(1)_Y$ representation. For example, all the left-handed up-quark mass eigenstates, which are in the $(3)_{+2/3}$ representation, come from interaction eigenstates in the $(3, 2)_{+1/6}$ representation. This is the reason that the Standard Model predicts universal Z-couplings to fermions. If, for example, nature had a fourth left-handed quark, in the $(3, 1)_{+2/3}$ representation, then the Z-couplings in the left-handed up-quark sector would be nonuniversal, and the Z-boson could mediate FCNC, such as $t \to cZ$ decay, at tree level.

9.4.3 Higgs-Mediated FCNC

The Yukawa couplings of the Higgs boson are not universal. In fact, in the interaction basis, they are given by completely general 3×3 matrices. Yet, as explained in section 8.4.2 in chapter 8, in the fermion mass basis, they are diagonal. The reason is that the fermion mass matrix is proportional to the corresponding Yukawa matrix and, consequently, the mass matrix and the Yukawa matrix are simultaneously diagonalized. The general condition for the absence of Higgs-mediated FCNC at tree level is that the only source of masses for any fermion type is a single Higgs field.

The relevant features of the Standard Model are the following:

1. All the Standard Model fermions are chiral, and therefore, there are no bare mass terms.
2. The scalar sector has a single Higgs doublet.

In contrast, either of the following possible extensions would lead to flavor-changing Higgs couplings:

1. There are quarks and/or leptons in vector-like representations, and thus there are bare mass terms.
2. There are more than one $SU(2)_L$-doublet scalar, which couple to a specific type of fermions.

Question 9.6 provides an example of the first case; question 9.9 provides an example of the second case.

We conclude that, within the Standard Model, all FCNC processes are loop-suppressed. However, in extensions of the Standard Model, FCNC can appear at the tree level, mediated by the Z-boson, the Higgs boson, or by new massive bosons.

For Further Reading

There are many useful reviews of the CKM matrix and flavor physics. See, for example, Branco, Lavoura, and Silva [29], Nir [30], Grossman and Tanedo [31], and the PDG review of the CKM matrix, Workman et al. [1].

Problems

Question 9.1: Algebra

1. Using equations (9.9) and (9.3) and working to $\mathcal{O}(\lambda^3)$, derive equation (9.8).
2. Using the definition of J_{CKM} in equation (9.11), show that

$$J_{\text{CKM}} = \mathcal{I}m\left(V_{ud}V_{cs}V_{us}^*V_{cd}^*\right). \tag{9.25}$$

3. Using equation (9.25) and the standard parameterization, equation (9.3), show that [see equation (9.12)]

$$J_{\text{CKM}} = c_{12}c_{23}c_{13}^2 s_{12}s_{23}s_{13}\sin\delta. \tag{9.26}$$

4. Explain why one cannot obtain J_{CKM} in terms of the Wolfenstein parameters using equation (9.25) and the approximate Wolfenstein parameterization of equation (9.8). Examine equation (9.8) and replace equation (9.25) with an equation that will allow you to show that [see equation (9.12)]

$$J_{\text{CKM}} \approx \lambda^6 A^2 \eta. \tag{9.27}$$

5. Using equation (9.26), show that [see equation (9.13)]

$$|J_{\text{CKM}}| \leq \frac{1}{6\sqrt{3}}. \tag{9.28}$$

Question 9.2: The bs unitarity triangle

In this chapter, we have discussed the unitarity triangle; that is, the *bd* triangle (equation (9.18)). We mention that each of its sides is of $O(\lambda^3)$, and each of its angles is of $O(1)$. Here, we discuss another unitarity triangle, the *bs* triangle.

1. Write the equation for the *bs* unitarity triangle (the equivalent of equation (9.18)).

2. Draw the rescaled bs unitarity triangle and label the angles as α_s, β_s and γ_s, in a similar way to our discussion of the rescaled bd triangle (the equivalent of figure 9.1).
3. Estimate, in terms of powers of λ, the size of each side and each angle of the bs unitarity triangle.
4. Equation (9.12) tells us that $J_{\text{CKM}} = \mathcal{O}(\lambda^6)$. Verify that the area of the bs unitarity triangle is also of the same order.
5. Using the standard parameterization, show explicitly that the bs and bd triangles have the same area.

Question 9.3: Semileptonic decays and the CKM matrix

A semileptonic decay is one where the final state has both hadrons and leptons. Here, we consider semileptonic b decays.

1. Draw the tree-level diagram for $b \to ue\bar{\nu}$ and estimate the size of the amplitude.
2. Estimate the ratio

$$\frac{\Gamma(b \to ue\bar{\nu})}{\Gamma(b \to ce\bar{\nu})} \tag{9.29}$$

as a function of CKM matrix elements. Neglect m_e and use

$$\frac{m_c}{m_b} \approx 0.3, \quad \frac{m_u}{m_b} \approx 0. \tag{9.30}$$

The phase-space function for a three-body decay with one massive and two massless final particles is given by

$$f(x_f) = 1 - 8x_f + 8x_f^3 - x_f^4 - 12x_f^2 \log x_f, \qquad x_f \equiv (m_f/m_b)^2. \tag{9.31}$$

3. The experimental value is

$$\frac{\Gamma(b \to u\ell\bar{\nu})}{\Gamma(b \to c\ell\bar{\nu})} \approx 2 \times 10^{-2}. \tag{9.32}$$

Estimate the relevant ratio of CKM matrix elements. Comment on the agreement between your estimate and the best-fit value for that ratio.

Question 9.4: τ decays to hadrons

The tau-lepton decays either leptonically or semileptonically.

1. List the possible tree-level decay processes of the tau-lepton in terms of leptons and quarks. (Note that, while $m_c < m_\tau$, what is relevant to this question is the mass of the lightest hadron containing a c-quark, the D-meson, and $m_D > m_\tau$.)
2. In terms of CKM elements, estimate the branching ratio of each mode. Remember to include color factors.

3. Estimate the Standard Model prediction for the branching ratio (BR) of the $\tau^- \to e^- \nu_\tau \bar{\nu}_e$ decay. Compare your prediction to the data and discuss any disagreement between the two.

Question 9.5: W decays

In this question, we study decays of the W-boson. Neglect all fermion masses except m_t. At tree level,

$$\Gamma_\mu \equiv \Gamma(W^+ \to \mu^+ \nu_\mu) = \frac{g^2 m_W}{48\pi} = \frac{G_F m_W^3}{6\sqrt{2}\pi} \approx 227 \text{ MeV}. \tag{9.33}$$

1. What is the Standard Model prediction for $\Gamma(W^+ \to e^+ \nu_e)$ and $\Gamma(W^+ \to \tau^+ \nu_\tau)$?
2. Write all the hadronic decays of the W-boson in terms of quarks (i.e., $W \to q\bar{q}'$), and estimate their widths in terms of Γ_μ, the CKM elements, and the number of colors ($N_c = 3$).
3. Estimate Γ_W, the total width of the W. Write your answer in GeV, to a precision of two decimal points.
4. Experimentally, $\Gamma_W = 2.085 \pm 0.042$ GeV. Explain the difference between your result and the experimental measurement.

Question 9.6: Exotic light quarks

In this question, we discuss a model of historical importance. Before the charm quark was discovered, only three quarks were known: the up, down, and strange quarks. One of the reasons to predict the existence of a fourth quark, the charm quark, was that a three-quark model predicts large rates for FCNC, in contradiction to experimental bounds.

Consider the following model

(*i*) The symmetry is a local

$$SU(3)_C \times SU(2)_L \times U(1)_Y. \tag{9.34}$$

(*ii*) The quark fields are

$$Q_L(3,2)_{+1/6}, \quad S_L(3,1)_{-1/3}, \quad U_R(3,1)_{+2/3}, \quad D_R(3,1)_{-1/3}, \quad S_R(3,1)_{-1/3}. \tag{9.35}$$

The lepton fields are

$$L_{Li}(1,2)_{-1/2}, \quad E_{Ri}(1,1)_{-1}, \quad i = 1,2. \tag{9.36}$$

(*iii*) There is a single scalar field:

$$\phi(1,2)_{+1/2}, \tag{9.37}$$

with $\langle \phi^0 \rangle = v/\sqrt{2}$.

We denote the components of the $SU(2)_L$-doublets as follows: $Q_L = (U_L, D_L)^T$, $L_{L1} = (\nu_{eL}, e_L)^T$, $L_{L2} = (\nu_{\mu L}, \mu_L)^T$, and $\phi = (\phi^+, \phi^0)^T$.

1. Consider the interaction basis of the quarks. Write the following:
   i. The covariant derivative for each of the quark fields
   ii. The charged current weak interactions
   iii. The Yukawa interactions of ϕ^0
   iv. The bare mass terms
   v. The quark mass matrices, taking into account the spontaneous symmetry breaking (SSB).
2. How many physical flavor parameters are there in the quark sector of this model? Separate the parameters into masses, mixing angles and phases.
3. Are there photon- or gluon-mediated FCNCs? Base your answer on symmetry-related arguments.
4. Write the Higgs couplings to the quarks in the mass basis. Are there Higgs-mediated FCNCs? Give a separate answer to each set of fermions with the same $SU(3)_C \times U(1)_{EM}$ quantum numbers.
5. Write the gauge interactions of the quark mass eigenstates with the W-boson.
6. In the Standard Model, the rate of $\bar{s} \to \bar{u}\mu^+\nu_\mu$ is used to determine $|V_{us}| \approx 0.2243$. Use this result to obtain the value for the mixing angle in $\mathcal{L}_W$.
7. Write down the gauge interactions of the quarks with the Z-boson in both the interaction and the mass bases. Are there tree-level Z-mediated FCNCs? Answer separately for each set of fermions with the same chirality and quantum numbers.
8. Consider the FCNC process $\bar{s} \to \bar{d}\mu^+\mu^-$. Explain why the Z-mediated contribution is quantitatively more significant than the h-mediated one.
9. We define

$$R_q \equiv \frac{\Gamma(\bar{s} \to \bar{d}\mu^+\mu^-)}{\Gamma(\bar{s} \to \bar{u}\mu^+\nu_\mu)}. \tag{9.38}$$

   Calculate R_q as a function of the model parameters, keeping only the leading effects.
10. Historically, the model was tested in kaon decays. While we did not yet discuss mesons, all you need to know for this question is that $R_q \sim R_H$, where

$$R_H \equiv \frac{\Gamma(K_L \to \mu^+\mu^-)}{\Gamma(K^+ \to \mu^+\nu)}. \tag{9.39}$$

   Find in the PDG the experimental value of R_H and explain why this result rules out the model. (While the data on R_H before the charm quark was discovered were not as good as they are today, they were good enough to exclude the model even then.)

The fact that, once a charm quark is added to the model, the tree-level value is $R_H = 0$ provided the motivation to suggest the existence of the charm quark.

Question 9.7: Even more exotic light quarks

Repeat question 9.6, items 1–5 and 7, for a model where we assign U_R and D_R to a doublet of $SU(2)_L$ that we denote as Q_R.

Question 9.8: The SM+χ

This is a generalization of the model presented in question 9.6. We add to the Standard Model a pair of quark fields, $\chi_L(3,1)_{-1/3}$ and $\chi_R(3,1)_{-1/3}$. The χ_R is distinguished from the three D_{Ri} by defining it as the only right-handed field that has a Dirac mass term with χ_L: $m_{\chi\chi}\overline{\chi_L}\chi_R$. We assume that $m_{\chi\chi} \gg v$.

1. Write for the quarks in the interaction basis:
   i. The covariant derivative for each of the quark fields
   ii. The Yukawa interactions
   iii. The bare mass terms
   iv. The quark mass matrices, taking into account the SSB
2. Write the gauge interactions of the quarks with the Z- and W-bosons in the interaction basis.
3. Write the gauge interactions of the quarks with the Z- and W-bosons in the mass basis. Write the W couplings using a suitably defined 3×4 mixing matrix $\hat{V}$, and write the Z-couplings using, for left-handed down quarks, a suitably defined 4×4 matrix $\hat{U}$.
4. Argue that, in general, there are off-diagonal Z couplings to the down-type quarks.
5. The bd unitarity relation is not represented by a triangle in this model. Write the generalization of equation (9.18) and plot it. What is the geometrical shape of it?

In question 13.3 of chapter 13, you are asked to derive bounds on the parameters of this model.

Question 9.9: FCNC in the 2HDM

Consider the Two Higgs Doublet Model (2HDM) extension of the Standard Model. In question 7.9 in chapter 7, we discussed the scalar sector. Here, we add the quark sector to the discussion. The model is defined as follows:

(*i*) The symmetry is

$$SU(3)_C \times SU(2)_L \times U(1)_Y. \tag{9.40}$$

(*ii*) There are three fermion generations, each consisting of three different representations:

$$Q_{Li}(3,2)_{+1/6}, \quad U_{Ri}(3,1)_{+2/3}, \quad D_{Ri}(3,1)_{-1/3}, \quad i=1,2,3. \tag{9.41}$$

(*iii*) There are two scalar multiplets:

$$\phi_1(2)_{+1/2}, \qquad \phi_2(2)_{+1/2}. \tag{9.42}$$

We make two simplifying assumptions with regard to the scalar potential:

I: *CP* conservation. Thus, all couplings are real.
II: The Higgs potential is invariant under the Z_2 symmetry defined in question 7.9. Thus, we can use the results of question 7.9 regarding the scalar spectrum.

These assumptions simplify the computations without affecting the main results.

1. Write (in matrix notation) the most general $\mathcal{L}_{\rm Yuk}$ for the quarks.
2. Find the masses of the quarks in terms of the Yukawa couplings and of the vacuum expectation values (VEVs) v_1 and v_2.
3. Find the couplings of the W- and Z-bosons to the quark mass eigenstates.
4. Find the couplings of h, H, and A to quarks in the mass basis for the quarks, in terms of the Yukawa couplings, and the angles α and β as defined in question 7.9. Show that the couplings are generally nondiagonal. This implies that the 2HDM has tree-level FCNCs.
5. Impose the Z_2 symmetry defined in question 7.9, with Q_L and D_R even and U_R odd under the symmetry. Show that there are no Higgs-mediated FCNCs.

Question 9.10: *b* and *c* lifetimes

In this question, we study decay rates and the lifetimes of the b and c quarks. We only consider tree-level decays. We denote by Q the b or c quark, and by q the u, d, and s quarks. We also approximate in the phase-space factors $m_q = m_e = m_\mu = 0$.

The phase-space factor for three-body decay with one massless particle is given by

$$\begin{aligned} f(x_1,x_2) =& \sqrt{\lambda_{12}}\left[1-7(x_1+x_2)-7(x_1^2+x_2^2)+x_1^3+x_2^3+x_1x_2(12-7(x_1+x_2))\right] \\ &+12\left[x_1^2\ln\frac{(1+x_1-x_2+\sqrt{\lambda_{12}})^2}{4x_1}+x_2^2\ln\frac{(1+x_2-x_1+\sqrt{\lambda_{12}})^2}{4x_2}\right] \\ &-12x_1^2x_2^2\ln\frac{(1-x_2-x_1+\sqrt{\lambda_{12}})^2}{4x_2x_1}, \end{aligned} \tag{9.43}$$

with

$$x_i \equiv \frac{m_i^2}{m_Q^2}, \qquad \lambda_{12} = 1-2(x_1+x_2)+(x_1-x_2)^2 . \tag{9.44}$$

For two massless particles, equation (9.43) reduces to the phase-space factor in equation (9.31).

1. Consider semileptonic decays, $Q \to q\ell\bar{\nu}$ (here $\ell = e$ or $\ell = \mu$). Draw the tree-level diagram and estimate the amplitude.

2. The decay width can be written as

$$\Gamma_Q = C m_Q^n |V_{ij}|^2 f(x_q, x_\ell), \tag{9.45}$$

such that C is a constant, V_{ij} is the relevant CKM matrix element, $x_f \equiv m_f^2/m_Q^2$, and f is a dimensionless phase-space factor (given for two massless and one massive final particles in equation (9.31)), with $f(0,0)=1$. Find n.
3. Draw the diagram of the $Q \to q_i \bar{q}_j q_k$ decay, and modify equation (9.45) to this case.
4. List the leading (in the CKM parameter λ) decay modes for c and b quarks.
5. Estimate the BRs of the semileptonic modes:

$$b \to ce\nu, \qquad c \to se\nu. \tag{9.46}$$

For the rough estimate in this question, use the masses from equation (8.34): $m_c = 1.27$ GeV, $m_b = 4.18$ GeV, and $m_\tau = 1.77$ GeV. For b decays, setting $x_e = x_\mu = x_u = x_d = x_s = 0$, the relevant phase-space factors are

$$f(x_c, x_\tau) \approx 0.06, \quad f(x_c, x_c) \approx 0.18, \quad f(x_c, 0) \approx 0.51, \quad f(0,0) = 1. \tag{9.47}$$

For c decays, setting $x_e = x_u = x_d = 0$, space factors are, the relevant phase-space factors are

$$f(x_s, 0) \approx 0.96, \quad f(x_s, x_\mu) \approx 0.91. \tag{9.48}$$

6. Using the value of the CKM elements from equation (9.5), estimate the ratio of the b and c lifetimes, τ_b/τ_c.
7. Compare your estimates for the lifetimes and for the semileptonic BRs to the PDG values. To do so, assume that the b and c lifetimes and decay modes are given by those of the corresponding mesons, B^+ and D^+. (We discuss the validity of this assumption in chapter 10.)

10

QCD at Low Energies

In chapter 5, we presented Quantum ChromoDynamics (QCD) in the high-energy regime, where the theory is perturbative. We mentioned that in the low-energy regime, QCD is strongly coupled. Consequently, on the theoretical side, there is no perturbative expansion. On the experimental side, we do not observe free quarks and gluons, but rather bound states which we call *hadrons*. (Bosonic hadrons are called *mesons* and fermionic hadrons are called *baryons*.) In this chapter, we discuss low-energy aspects of QCD and relate it to the theory that we define at the ultraviolet (UV) regime.

10.1 Introduction

To understand the implications of QCD at the infrared (IR), and to connect experiment to theory, we need to find ways to overcome the absence of a perturbative expansion. There are indeed ways that aim to solve QCD at the IR from first principles. We mention two examples here. First, *lattice QCD* is a nonperturbative approach based on discretizing spacetime. As of now, while providing relevant and useful results, lattice QCD is still not at the level of fully solving QCD. Second, *chiral perturbation theory* is an effective theory, with hadrons as its degrees of freedom (DoF), which is applicable at very low energy, $E \ll \Lambda_{\rm QCD}$. It provides a systematic expansion in power of $E/\Lambda_{\rm QCD}$. We do not discuss these approaches any further in this book.

One may use a less rigorous (but still useful) approach by employing a *model*. Indeed, in the next section, we introduce the quark model, but let us first comment on our use of the term *theory* versus *model*. A theory is based on first-principle axioms that are assumed to be exact. Any approximations made should be controlled (i.e., there is an expansion in a small parameter). (From this point of view, the Standard Model is in fact a theory.) For a model, on the other hand, the starting point is a set of assumptions that are known to be inexact. It is impossible to quantify the errors on the predictions of a model. A familiar example is that of the Bohr model for the hydrogen atom compared to the Schrödinger equation, where the latter is an output of a well-defined theory of quantum mechanics.

10.2 Hadronic Properties

Hadrons are formed due to the confining nature of QCD, so they cannot be treated perturbatively within it. In this section, we discuss some general properties of hadrons and define the quark model that enables us to understand more of the dynamics.

10.2.1 *General Properties*

There are some properties of hadrons that can be determined independent of our ability to describe their internal structure. What is measured directly by experiments is obviously generic. The measurable properties are the hadron mass, lifetime, spin, and electric charge. Other properties are not exact, but they are still generic. We describe three classes of hadronic quantum numbers next.

- **Exact quantum numbers**
  - The electromagnetic charge, Q
  - The spin, J
  - Baryon number, B
- **quantum numbers that are respected by QCD and QED but are broken by the weak interaction**
  - The charges under the global $[U(1)]^6$ flavor symmetry of the QCD and Quantum ElectroDynamics (QED) Lagrangians, discussed in section 5.8 in chapter 5, often referred to as *flavor quantum numbers*
  - Parity, P
  - For neutral mesons, charge conjugation, C
- **quantum numbers under approximate symmetries of QCD**
  - $SU(2)$-isospin and $SU(3)$-flavor, as discussed in section 10.4.1
  - Heavy quark symmetry, as discussed in section 10.4.2

10.2.2 *The Quark Model*

The basic idea behind the quark model is that hadrons are color-singlet combinations of quarks and antiquarks. In particular, we assume that the quantum numbers of the hadrons are dictated by the quantum numbers of the constituent quarks and antiquarks. This is clearly a model, as it assumes a minimum quark content of the hadron, while the structure of the hadron is unavoidably more complex. Yet the model works surprisingly well and provides many insights into the IR properties of QCD. It is difficult, however, to estimate the related uncertainties.

We first use the quark model to understand the spectroscopy of the hadrons. The fact that QCD is confining results in an infinite spectrum of bound states. Their names and quantum numbers are discussed in Appendix 10.A. In the quark model, the simplest hadrons belong to one of three types:

- *Mesons,* which have quark-antiquark constituents, $M = q\bar{q}$
- *Baryons,* which have three quark constituents, $B = qqq$
- *Antibaryons,* which have three antiquark constituents, $\overline{B} = \bar{q}\bar{q}\bar{q}$

The lightest meson are the pion:

$$\pi^+ = u\bar{d}, \qquad \pi^0 = \frac{1}{\sqrt{2}}(u\bar{u} - d\bar{d}), \qquad \pi^- = d\bar{u}. \tag{10.1}$$

The lightest baryons are the proton and the neutron:

$$p = uud, \qquad n = udd. \tag{10.2}$$

Mesons carry no baryon number, baryons carry a baryon number that we normalize to $B = +1$, and antibaryons carry a baryon number $B = -1$. Note that the letter B is used to denote a baryon and baryon number (as well as mesons that contain a b quark). Which meaning is being used should be clear from the context.

Some of the properties of the hadrons described in section 10.2.1 can be interpreted using the quark model. For example, the total spin, J, of a meson can be interpreted as a combination of the spins of the quark and the antiquark, S (that can be zero or 1), and the orbital angular momentum, L.

As an example of the use of the quark model, consider π^+, the charged pion, and ρ^+, the charged rho-meson. The pion is a pseudoscalar ($J^P = 0^-$) with a mass of about 140 MeV, while the ρ-meson is a vector ($J^P = 1^-$) with a mass of about 770 MeV. Within the quark model, we interpret both as $u\bar{d}$ states. They are the ground state ($n = 1$ and $L = 0$) and are distinguished by their spin. The pion is the $S = 0$ combination, while the rho-meson is the $S = 1$ combination. Using $J = S + L$ and $P = -(-1)^L$ (the extra minus sign in P comes from fermion pair exchange), we see how the spin and parity of a meson come from the properties of the constituent quarks.

One should be comfortable thinking about the plethora of mesons analogous to excited hydrogen atoms in quantum mechanics. The main difference is that for the hydrogen atom, we can analytically obtain the eigenvalues from the Schrödinger equation, while in QCD, since we do not know the potential, we cannot calculate the spectrum. Yet the basic formalism is similar, in that it is a two-body, bound state system.

The quark model contains states that involve more than two or three quarks and antiquarks, which are referred to as *exotics*. These include tetraquarks ($q\bar{q}q\bar{q}$) and pentaquarks ($qqqq\bar{q}$). While usually not described as part of the quark model, there are also states involving gluons, in particular the glueballs that are made of gluons only. Some of these states have recently been discovered. We do not elaborate on them any further in this book.

10.2.3 Hadron Masses

The mass of a hydrogen atom comes from the masses of its constituent proton and electron plus a potential energy ($V = -13.6$ eV). To make this statement, it is crucial that we are able to pull an electron and proton apart and measure their masses independent of their mutual electric field. For hadrons, the situation is more complicated. We can think about the mass of a hadron as a combination of the constituent quark masses and the binding energy, which is of the order of Λ_{QCD}. Confinement implies, however, that we cannot separate quarks and measure their masses in isolation. It is a priori not clear, then, how we can separate the contributions to the hadron mass from the constituent quark masses and the binding energy.

In principle, we can measure quark masses at high energy, where the couplings are small and can be treated perturbatively. The result, like any parameter in the Lagrangian, is scale dependent. However, while the effects of running are very small for lepton masses, they are dramatic for quark masses. In fact, we cannot calculate the running quark mass below the confinement scale. (For some purposes, it is useful to define a "valence quark mass," which is just the mass of the hadron divided by the number of constituent quarks. For example, the valence mass of the u and d can be extracted from the mass of the proton and the neutron, and it is roughly $m_p/3$.)

We then split the six quarks into three groups: light quarks, $q = u, d, s$; heavy quarks, $Q = c, b$; and a very heavy quark, t. The distinction between light and heavy quarks is based on $m_q \ll \Lambda_{\text{QCD}}$ (light quarks) or $m_Q \gg \Lambda_{\text{QCD}}$ (heavy and very heavy quarks). Very heavy quarks are distinguished by their large width ($\Gamma_t \gg \Lambda_{\text{QCD}}$).

- $q = u, d, s$: For hadrons made from only light quarks, the binding energy is the main contribution to the hadron mass.
- $Q = c, b$: Hadrons that have at least one heavy quark acquire their mass mainly from the heavy quark mass, and the binding energy is a small correction of relative order Λ_{QCD}/m_Q.
- t: The decay width of the top quark is very large ($\Gamma_t \gg \Lambda_{\text{QCD}}$). This situation is often described by saying that the top quark "does not form hadrons." Thus, to leading order, we treat the top quark as a free particle, and it is irrelevant to the discussion of the quark model of hadrons.

10.2.4 Hadron Lifetimes

With some stretch of language, we classify hadrons as either *stable particles* or *unstable resonances*. In this context, stability is not meant in the absolute sense. (The only truly stable hadron is the proton.) We can define what we mean by "stability" in two ways: experiment-related and theory-related.

Experimentally, we can determine the lifetime in two ways: by measuring a decay width (i.e., by the energy of the decay products), or directly, by measuring the distribution of the decay times. We characterize a particle as stable if its lifetime is large enough to be measured by the latter method (or even larger, so that it escapes the detector before decaying). Theoretically, the definition is that a stable particle does not decay through QCD or QED interactions, but only via weak interactions. These are the hadrons whose decays violate the accidental flavor symmetry of QCD and QED. Resonances are the particles that are not stable. In fact, some resonances are so broad, with a width of the order of their mass, that it becomes questionable whether it is a bound state at all. In the Particle Data Group (PDG), with a few exceptions, particles whose name contains their mass in parentheses, such as $\rho(770)$, are resonances, while those that do not are particles.

As an example of a resonance, consider the ρ^+ meson, which decays via QCD, almost always to two pions, $\rho^+ \to \pi^+\pi^0$. Its width is large, $\Gamma_\rho \approx 150$ MeV, just a factor of 5 less than its mass. As an example of a stable particle, consider the π^+ meson. Its main decay is via the weak interaction to leptons, $\pi^+ \to \mu^+\nu$. Its width is of order 10^{-8} eV, which is clearly much smaller than its mass. The ratio $\Gamma_\rho/\Gamma_\pi \sim 10^{16}$ demonstrates the difference between a resonance and a stable particle, and explains why the weakly decaying hadrons are called stable.

When discussing the difference between stable hadrons and resonances, tritium, which is a hydrogen isotope with two neutrons, provides a useful analogy. The excited $2P$ tritium state emits a photon and decays to the $1S$ ground state at a time scale of the order of 1 ns. The $1S$ state decays into helium-3, with a lifetime of about 18 years. The $2P$ state is the equivalent of a resonance, while the $1S$ is the equivalent of a stable particle.

10.3 Combining QCD with Weak Interactions

When we aim to extract from experiments the weak interaction parameters for quarks, we face the problem that our Lagrangian is written in terms of quarks and gluons, while experiments observe hadrons. When analyzing physical processes in quantum field theory (QFT), one uses asymptotic free states, but quarks and gluons cannot be free at infinity. The way that we deal with this problem is to "parameterize our ignorance." We isolate the parts of the amplitudes that are nonperturbative and then use whichever relevant method we have at our disposal (approximate symmetries, lattice QCD, models, etc.) to calculate them or relate them to another measurable process. The aim of this section is to describe this procedure.

10.3.1 Factorization

The starting point of factorization is the assumption that we can factorize a process to a QCD part and a non-QCD part. Then we treat the non-QCD parts (i.e., the leptonic and electromagnetic part) perturbatively.

Consider first a purely leptonic decay, such as $W \to \ell\nu$. The amplitude, or the matrix element, is written as

$$\mathcal{A}_{W\to\ell\nu} = g\langle\ell\nu|\mathcal{O}|W\rangle, \qquad \mathcal{O} \sim \overline{\ell_L}\gamma_\mu W^\mu \nu_L, \tag{10.3}$$

where $\mathcal{O}$ is the operator that mediates the process in the Standard Model at tree level. The calculation is simple, in that the operator creates and annihilates the particles of the external states and the amplitude is trivially $\mathcal{A} \propto g$.

Next, consider a hadron decay, such as $\pi^- \to \ell^- \bar{\nu}$. In the Standard Model, the matrix element is given by

$$\mathcal{A}_{\pi\to\ell\nu} = G_F V_{ud} \langle \ell\bar{\nu}|\mathcal{O}|\pi^-\rangle, \qquad \mathcal{O} \sim \overline{\ell_L}\gamma^\mu \nu_L \overline{u_L}\gamma_\mu d_L. \tag{10.4}$$

The leptonic part of this matrix element is simple. Just as in the purely leptonic case, we have asymptotic external states that are contracted with the creation and annihilation operators in $\mathcal{O}$. The complication is with the pion. In the quark model, the charged pion is made of $u\bar{d}$, and thus we can think of the $\bar{u}\gamma_\mu d$ combination in the operator as the part that annihilates the pion. However, this is not the whole story since the operator annihilates a free d quark and a free $\bar{u}$ antiquark, while inside the pion, the quark and the antiquark are bound and not free.

The factorization hypothesis states that we can treat the leptonic and hadronic parts of the matrix element separately and write the result as a product of the two terms. The leptons do not participate in the strong interaction, so they can be factored out. We then write

$$\langle \ell\bar{\nu}|\mathcal{O}|\pi^-\rangle = \langle \ell\bar{\nu}|\mathcal{O}_\ell|0\rangle \times \langle 0|\mathcal{O}_H|\pi^-\rangle \tag{10.5}$$

with

$$\mathcal{O}_\ell \sim \overline{\ell_L}\gamma^\mu \nu_L, \qquad \mathcal{O}_H \sim \overline{u_L}\gamma_\mu d_L. \tag{10.6}$$

The leptonic part is simple to calculate perturbatively, but the hadronic part (i.e., $\langle 0|\mathcal{O}_H|\pi^-\rangle$) is the one that we cannot calculate perturbatively. It is this *hadronic matrix element* that we parameterize and use nonperturbative methods to calculate.

The next example is semileptonic kaon decay, $K^+ \to \pi^0 \ell^+ \nu$. The matrix element in the Standard Model is

$$\mathcal{A}_{K\to\pi\ell\nu} = G_F V_{us}^* \langle \pi^0 \ell^+ \nu|\mathcal{O}|K^+\rangle, \qquad \mathcal{O} \sim \overline{\nu_L}\gamma^\mu \ell_L \overline{s_L}\gamma_\mu u_L. \tag{10.7}$$

Once again, we appeal to factorization and write

$$\langle \pi^0 \ell^+ \nu|\mathcal{O}|K^+\rangle = \langle \ell^+ \nu|\mathcal{O}_\ell|0\rangle \times \langle \pi^0|\mathcal{O}_H|K^+\rangle. \tag{10.8}$$

The result is similar to the previous example, but here, the hadronic part involves two mesons, as opposed to the one meson in equation (10.5).

Hadronic matrix elements with one hadron, such as in equation (10.5), are parameterized by *decay constants*, and they are discussed in section 10.3.2. Hadronic matrix elements with two hadrons, such as in equation (10.8), are parameterized by *form factors*, and those are discussed in section 10.3.3. We do not discuss matrix elements with more than two hadrons here.

Before we discuss these specific examples, we present some guidelines that apply to all hadronic matrix elements. In the most general case involving mesons, we can write

$$\langle H_f|\bar{q}\Gamma q'|H_i\rangle = f(p, \epsilon), \tag{10.9}$$

where H_f and H_i stand for hadronic states (including the vacuum), Γ represents some Dirac structure, p represents momenta, and ϵ stands for polarizations. We can always write

Γ as a combination of S, P, V, A, and T (scalar, pseudoscalar, vector, axial vector, and tensor, respectively) structures. The crucial point is that the two sides of equation (10.9) must transform in the same way under parity and Lorentz transformations. As we see next, there is usually a small number of combinations of momenta and polarizations that satisfy this requirement. This situation leads to considerable simplification in the analysis.

10.3.2 The Decay Constant

Considering pion decay (see equation (10.5)), we need to evaluate

$$\langle 0|\mathcal{O}_H|\pi^-(p_\pi)\rangle \sim \langle 0|V^\mu - A^\mu|\pi^-(p_\pi)\rangle, \qquad V^\mu = \bar{u}\gamma^\mu d, \qquad A^\mu = \bar{u}\gamma^\mu\gamma_5 d, \tag{10.10}$$

where we use

$$\mathcal{O}_H \sim \overline{u_L}\gamma^\mu d_L = \frac{1}{2}(V^\mu - A^\mu). \tag{10.11}$$

The pion is a pseudoscalar and the vacuum is parity-even. Thus, under Lorentz transformation, the matrix element of V^μ must transform as a pseudovector, and the matrix element of A^μ must transform as a vector. The only available physical observable is the pion momentum, p_μ, which is a vector. It is impossible to construct a product of any number of p_π^μ that transforms like a pseudovector. We conclude that

$$\langle 0|V^\mu|\pi\rangle = 0. \tag{10.12}$$

(Similar considerations imply that the matrix elements of $\Gamma = S, T$ vanish as well. The matrix element for $\Gamma = P$ is not zero. Since it is irrelevant to the Standard Model, we defer a discussion of that topic until question 10.5 at the end of the chapter.)

We do not know how to calculate the hadronic matrix element $\langle 0|A^\mu|\pi\rangle$. Instead, we parameterize it. From the considerations given here, we must have

$$\langle 0|A^\mu|\pi\rangle \propto p_\pi^\mu. \tag{10.13}$$

The proportionality constant could depend only on Lorenz scalar quantities. In our case, there is only one such quantity, $p_\pi^2 = m_\pi^2$, which is not a dynamical variable. We can then define the proportionality constant as a *decay constant*, f_π:

$$\langle 0|A^\mu|\pi\rangle \equiv -if_\pi p_\pi^\mu, \tag{10.14}$$

where $-i$ is a commonly used normalization factor. While the decay constant cannot be calculated, it can be extracted from measurements. For example, the decay rate for $\pi^- \to \mu^-\bar{\nu}_\mu$ is given by (see question 10.3)

$$\Gamma(\pi \to \mu\nu) = \frac{G_F^2|V_{ud}|^2}{8\pi} m_\mu^2 m_\pi \left(1 - \frac{m_\mu^2}{m_\pi^2}\right)^2 f_\pi^2. \tag{10.15}$$

From the measured decay rate, one extracts

$$f_\pi = 130.4 \pm 0.2 \text{ MeV}. \tag{10.16}$$

(In some of the research literature, a different normalization is used—namely, $F_\pi \equiv f_\pi/\sqrt{2} \approx 93$ MeV.) Similar decay constants are defined for other mesons.

To gain intuition with regard to the physics of the decay constant, we consider a perturbative system, such as the positronium, as a useful analog. Positronium is an e^+e^- bound state that decays into photons. One then can define a positronium decay constant. Unlike the pion, however, the positronium decay constant is calculable using QED. Semiclassically, the decay occurs when the electron and positron annihilate. For this to happen, they must be in the same place. Quantum mechanically, the decay amplitude is proportional to the wave function Ψ at $r=0$ (r is the distance between the electron and the positron). This intuition carries over to QCD bound states. The physics of a decay constant has to do with the wave function at the origin:

$$f_\pi \propto |\Psi(r=0)|. \tag{10.17}$$

While we cannot calculate the wave function perturbatively, we know general scaling properties of it in some cases. Moreover, lattice QCD provides precise nonperturbative calculations of decay constants.

10.3.3 Form Factors

Considering kaon decay (see equation (10.7)), we need to evaluate

$$\langle \pi^0(p_\pi)|V^\mu - A^\mu|K^+(p_K)\rangle, \qquad V^\mu = \bar{s}\gamma^\mu u, \qquad A^\mu = \bar{s}\gamma^\mu\gamma_5 u. \tag{10.18}$$

The available variables are p_K^μ and p_π^μ. Both K and π mesons are pseudoscalars. Thus, under Lorentz symmetry and parity, the matrix element of V^μ must transform as a vector, and the matrix element of A^μ must transform as an axial vector. It is impossible, however, to construct a combination of p_K^μ and p_π^μ that transforms like an axial vector. We conclude that

$$\langle \pi^0(p_\pi)|A^\mu|K^+(p_K)\rangle = 0. \tag{10.19}$$

For the vector current, we write

$$\langle \pi^0(p_\pi)|V^\mu|K^+(p_K)\rangle = a p_K^\mu + b p_\pi^\mu. \tag{10.20}$$

The coefficients a and b are called *form factors*, and they depend only on Lorentz scalars. In the case of $K \to \pi e \nu$, there are three relevant Lorentz scalars: $p_K^2 = m_K^2$, $p_\pi^2 = m_\pi^2$, and $p_K \cdot p_\pi$. The first two are mass parameters, so they are not dynamical. We are left with a single dynamical variable. It is customary to define

$$q \equiv p_K - p_\pi, \qquad q^2 = p_K^2 + p_\pi^2 - 2p_K \cdot p_\pi \tag{10.21}$$

and use the form factors $f_\pm(q^2)$, defined as follows:

$$\langle \pi^0(p_\pi)|V^\mu|K^+(p_K)\rangle = f_+(q^2)(p_K + p_\pi)^\mu + f_-(q^2)(p_K - p_\pi)^\mu. \tag{10.22}$$

A similar procedure applies to the other cases.

The form factors are where the nonperturbative physics plays a role. As noted here, they can be calculated by lattice QCD. In many cases, we can use approximate symmetries of QCD to learn about some properties of them and to relate them to each other.

The physical intuition with regard to the form factor is that it corresponds to the wave function overlap of the two hadrons. This intuition is borrowed from the sudden approximation in quantum mechanics. The probability of a fast transition between an initial and a final state depends on the overlap of the initial and final wave functions. In semileptonic hadron decay, there is a sudden transition due to the weak interaction. The form factors provide a formal way to encode the wave function overlap.

10.4 The Approximate Symmetries of QCD

Approximate symmetries of QCD are useful to relate hadronic matrix elements, or even to determine their value at a specific kinematic point. Results obtained from symmetry properties of QCD are model independent. Corrections to the symmetry limit (which we do not discuss here), however, are often evaluated in a model-dependent way (e.g., using the quark model).

10.4.1 Isospin Symmetry

Consider the Lagrangian for a two-flavor (up and down) QCD [see equation (5.11) in chapter 5]:

$$\mathcal{L}_{\rm QCD} = -\frac{1}{4} G_a^{\mu\nu} G_{a\mu\nu} + \sum_{q=u,d} (i\overline{q}\not{D}q - m_q\overline{q}q). \tag{10.23}$$

Since $m_u \neq m_d$, this Lagrangian has a $[U(1)]^2$ flavor symmetry. If, however, we had $m_u = m_d$, the flavor symmetry would be $U(2) = SU(2)_I \times U(1)_B$, where the $U(1)_B$ is the baryon number symmetry. The $SU(2)_I$ part is called *isospin symmetry*. Under isospin, the up and down quarks form a doublet:

$$Q = \begin{pmatrix} u \\ d \end{pmatrix}. \tag{10.24}$$

We emphasize that the isospin-$SU(2)_I$ is different from the gauge-$SU(2)_L$. The u and d quarks in the doublet are Dirac fermions, and not only left-handed components, as with $SU(2)_L$-doublet quarks. Furthermore, all quarks except up and down are singlets of $SU(2)_I$.

The up and down quarks are much lighter than $\Lambda_{\rm QCD}$ and thus $SU(2)_I$ is an approximate symmetry of the QCD interactions, broken by a small parameter:

$$\epsilon_I \equiv \frac{m_d - m_u}{\Lambda_{\rm QCD}} \sim 10^{-2}. \tag{10.25}$$

One might wonder why the QCD scale enters here. In general, we have to compare the mass difference of the quarks to the relevant energy in the event. Due to confinement, the energy is at least of the order of the QCD scale, and thus the symmetry-breaking parameter may be smaller, but it is not larger than ϵ_I.

In addition to ϵ_I, the isospin symmetry is broken by QED due to the different charges of the up and down quarks. The size of this breaking is of the order of $\alpha \sim 0.01$, similar to ϵ_I.

The approximate $SU(2)_I$ symmetry implies that the hadron mass eigenstates can be assigned into well-defined representations of isospin, where the hadrons in an isospin multiplet are approximately degenerate. This statement is based on a symmetry of the QCD Lagrangian and not on any quark model. We use the quark model to assign the hadron mass eigenstates into quark representations.

The expected approximate degeneracies are indeed manifest in the baryon spectrum. The two lightest baryons are the neutron and the proton, which are almost degenerate: $m_p = 938.272$ MeV and $m_n = 939.565$ MeV. It is natural to assume that they form an isospin doublet ($I = 1/2$). At a somewhat higher scale, $m_\Delta \approx 1.23$ GeV, there are the four nearly degenerate Δ states with charges $Q = -1, 0, +1, +2$, which fit into an isospin quadruplet ($I = 3/2$). Within the quark model, a baryon is made of three quarks. Thus, in terms of isospin, the baryons that are formed from u and d quarks are a product of three isospin doublets. This product of three doublets gives

$$\frac{1}{2} \times \frac{1}{2} \times \frac{1}{2} = \frac{1}{2} + \frac{1}{2} + \frac{3}{2}. \tag{10.26}$$

Only one combination of the two isospin doublets on the right-hand side can be combined with the spin-1/2 wave functions to provide the required overall symmetry properties of the three-quark state. Therefore, there are six light baryon states, which are identified as the nucleon doublet and the Δ quadruplet.

The quark model charge assignment fits as well: these baryons are made of the u and d quarks. The quark assignment for the nucleons is given in equation (10.2), while for the Δ baryons, it is given by

$$\Delta^- = ddd, \qquad \Delta^0 = ddu, \qquad \Delta^+ = duu, \qquad \Delta^{++} = uuu. \tag{10.27}$$

The quasi-degeneracy within each multiplet is a result of the approximate symmetry. The splitting of $O(\Lambda_{\rm QCD})$ between the masses of the nucleon doublet and the Δ quartet is a result of the very different binding energies from QCD interactions for different representations of the isospin group.

The approximate degeneracies are also manifest in the meson spectrum. Among the $J^P = 0^-$ mesons, we identify the three pions, which are quasi-degenerate ($m_{\pi^\pm} = 139.57$ MeV, $m_{\pi^0} = 134.98$ MeV), as an isospin-triplet ($I = 1$); and the heavier η meson ($m_\eta =$ 547.86 MeV) as an isospin singlet ($I = 0$). Indeed, the combination of two isospin doublets gives

$$\frac{1}{2} \times \frac{1}{2} = 0 + 1. \tag{10.28}$$

The quark model assignment of the pions is given in equation (10.1), while that of η is given by

$$\eta^0 = \frac{u\bar{u} + d\bar{d}}{\sqrt{2}}. \tag{10.29}$$

This assignment is valid in the limit where only the u and d quarks are light compared to $\Lambda_{\rm QCD}$. In the real world, where the s quark is also light, the η^0 meson includes an $s\bar{s}$ component.

In addition to spectroscopy, isospin plays an important role in weak decays. It relates form factors and even allows one to predict their values in specific kinematic points. In Appendix 10.B, we provide an example of a matrix element whose value is predicted in the isospin limit.

One can also treat the s quark as light compared to $\Lambda_{\rm QCD}$. In the limit of degenerate u, d, and s quarks, the QCD Lagrangian gains an approximate $SU(3)_F$ symmetry, which is usually called "flavor $SU(3)$," with (u, d, s) forming a triplet. The breaking of this symmetry is much larger than that of isospin, of the order of $(m_s - m_d)/\Lambda_{\rm QCD} \sim 0.3$, but the symmetry considerations are still very useful. We do not discuss it any further in this book.

10.4.2 Heavy Quark Symmetry

The term *heavy quark symmetry* applies to hadrons that contain one heavy quark. When the heavy quark mass goes to infinity, the theory gains extra symmetries. This symmetry does not manifest itself in $\mathcal{L}_{\rm QCD}$, as one cannot simply take parameters to infinity. Yet it is rigorously defined and gives definite and useful predictions.

Most of the essential physics associated with heavy quark symmetry can be demonstrated by the quantum mechanics of the hydrogen atom. Consider the difference between the hydrogen isotopes: hydrogen, deuterium, and tritium. Even though they have different masses and spins, their chemical properties are approximately the same. The reason for this similarity among isotopes is that, to a very good approximation, the electrons are largely insensitive to the mass and spin of the nucleus. To put it in more general terms, the light DoF are insensitive to the heavy DoF that source the potential, and it is the former that determine the chemistry. To the zeroth order in the inverse mass of the nucleus, the potential is insensitive to the mass and spin of the actual nuclear source of the electromagnetic field. At the order of m_e/m_A (m_A is the mass of the nucleus), there are two leading corrections:

- The difference in the reduced mass of the system:

$$\mu \equiv \frac{m_e m_A}{m_e + m_A} \approx m_e \left(1 - \frac{m_e}{m_A}\right). \tag{10.30}$$

- The hyperfine splitting:

$$\Delta E^{\rm hf} \sim m_e \alpha^4 \frac{m_e}{m_A}. \tag{10.31}$$

The extra m_e/m_A suppression of the hyperfine splitting is a manifestation of the fact that, in the limit of infinite nucleus mass, one cannot rotate the nucleus to reverse its dipole moment.

We now return to our discussion of hadrons. Consider, for example, mesons that, in the quark model, are made of a b quark and a light (d or u), quark. The analogy is that the heavy DoF, the b quark, is like the heavy nucleus, and the light DoF, the u or d quark, is like the electron. The binding, however, comes from QCD rather than from QED, and thus we need to overcome the confining nature of QCD. What is helpful here is the fact that $m_b \gg \Lambda_{QCD}$, and thus we can get rigorous results in the symmetry ($m_b \to \infty$) limit.

Specifically, consider the B $(J=0)$ and the B^* $(J=1)$ mesons, which in the quark model are the singlet and triplet states of the hyperfine structure. In the $m_b/\Lambda_{QCD} \to \infty$ limit, the B and B^* mesons are degenerate. The $m_{B^*} - m_B$ mass splitting is the analog of the hyperfine splitting in the hydrogen system. To the leading order, the splitting scales like the inverse mass of the heavy quark:

$$m_{B^*} - m_B \propto \frac{\text{const}}{m_{B^*} + m_B}. \tag{10.32}$$

Thus, the prediction is that for any heavy quark Q,

$$(m_{Q^*} - m_Q)(m_{Q^*} + m_Q) = m_{Q^*}^2 - m_Q^2 = \text{const}. \tag{10.33}$$

This prediction of the heavy quark symmetry is experimentally confirmed:

$$m_{B^*}^2 - m_B^2 \approx 0.47 \text{ GeV}^2, \qquad m_{D^*}^2 - m_D^2 \approx 0.55 \text{ GeV}^2. \tag{10.34}$$

The small difference between the two is attributed to higher-order effects.

The situation is similar in other mesons, like the D, D_s and B_s, and their excited states. You may wonder if indeed $m_c \gg \Lambda_{QCD}$. While m_c is not much larger than Λ_{QCD} and corrections can be large, we still formally work in this limit.

Another useful feature that arises in the heavy quark limit is that the wave function representing the internal structure of the B and the D mesons does not depend on the color source. Consequently, for $B \to D\ell\nu$ decay, in the kinematic point where the two mesons are relatively at rest, the wave function of the light DoF is not affected by the $B \to D$ decay and we have (up to some normalization)

$$\langle B(p)|\mathcal{O}|D(p)\rangle = 1. \tag{10.35}$$

We use this result in Appendix 10.C.

10.5 Hadrons in High-Energy QCD

In the previous section, we explained how we can interpret hadronic processes in terms of quark transitions using form factors and decay constants. This method is most useful when we have a small number of hadrons. Specifically, we discussed processes involving one or two hadrons. In this section, we explore the opposite limit: processes that involve many hadrons. In such processes, the relative momenta between the underlying quarks and gluons are very large (i.e., much larger than $\Lambda_{\rm QCD}$). In such processes, which we call *high-energy QCD processes,* interpretation in terms of quarks and gluons is possible if one integrates over large parts of the parameter space. It is this integration that makes it possible to connect the calculation in terms of quarks and gluons to the experiments that measure hadronic rates.

10.5.1 Quark-Hadron Duality

The notion of *quark-hadron duality* states that inclusive hadronic observables at high energies, when integrated over a large enough part of phase space, can be described by the calculation in terms of quarks and gluons. (The generic name for a quark or a gluon in such processes is a *parton.*)

The quark-hadron duality works well in many cases. Despite this success, the notion of the quark-hadron duality is often vague. Yet the intuition is clear: at short distance, QCD is perturbative. The strong interaction results in energy shifts of order $\Lambda_{\rm QCD}$ between the various partons. By integrating over a large-enough interval (i.e., much larger than $\Lambda_{\rm QCD}$), we average over these hadronic processes and recover the underlying results. In practice, there are instances where instead of performing a simple integration, we can smooth out the hadronic distribution at scale of the order of $\Lambda_{\rm QCD}$ or greater.

To demonstrate this idea, consider the following two cross sections, which can be measured:

$$\begin{aligned}\sigma_h(s) &\equiv \sigma(e^+e^- \to \text{hadrons})[s],\\ \sigma_\mu(s) &\equiv \sigma(e^+e^- \to \mu^+\mu^-)[s];\end{aligned} \tag{10.36}$$

and the following two cross sections, which can be calculated:

$$\begin{aligned}\sigma_a(s) &\equiv \int_0^\infty \sigma(e^+e^- \to \text{hadrons})[s']F(s,s')ds',\\ \sigma_q(s) &\equiv \sigma(e^+e^- \to q\bar{q})[s],\end{aligned} \tag{10.37}$$

where $s \equiv (p_+ + p_-)^2$, p_+ (p_-) are the four momenta of the positron (electron); and $F(s,s')$ is a function that smooths $\sigma_h(s)$ over a scale of the order of $\Lambda_{\rm QCD}$. ($\sigma_a(s)$ is calculable given the measured $\sigma_h(s)$.) We further define the three ratios of cross sections:

$$R_h(s) \equiv \frac{\sigma_h(s)}{\sigma_\mu(s)}, \qquad R_a(s) \equiv \frac{\sigma_a(s)}{\sigma_\mu(s)}, \qquad R_q(s) \equiv \frac{\sigma_q(s)}{\sigma_\mu(s)}. \tag{10.38}$$

The ratio $R_q(s)$ can be calculated perturbatively for $\sqrt{s} \gg \Lambda_{\rm QCD}$:

$$R_q(s) = 3\sum_q Q_q^2. \tag{10.39}$$

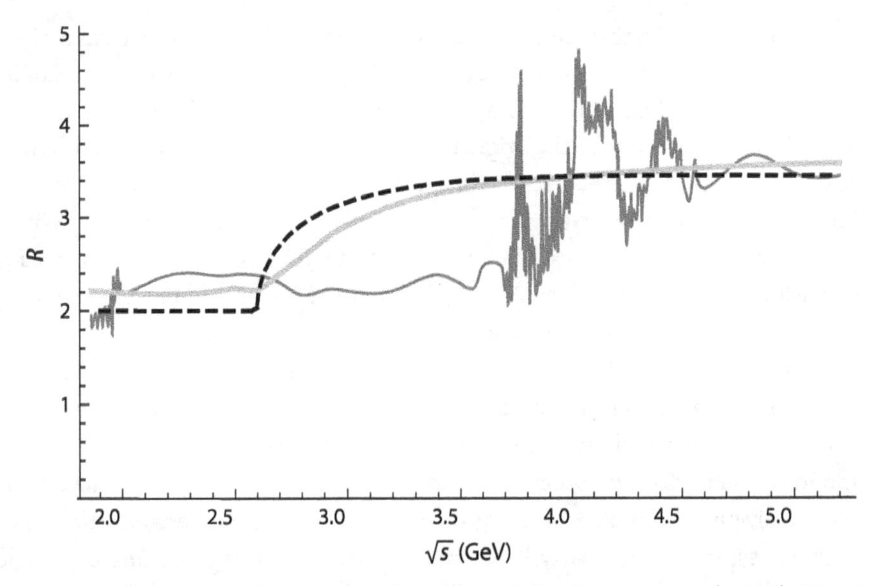

FIGURE 10.1. The ratios $R_h(s)$, $R_a(s)$, and $R_q(s)$, as defined in equation (10.38). The thin solid curve provides a schematic description of the data for R_h, where we do not present the error bars and omit, for the sake of demonstration, the J/ψ peak. The thick solid curve is R_a, where, for the smearing, we use a Gaussian function with a width of 0.35 GeV. The dashed curve is for R_q.

Here, 3 is the number of colors, Q_q is the electric charge of quark q, and the sum goes over all quarks q with masses $2m_q < \sqrt{s}$. Equation (10.39) applies at leading order in perturbation theory. Specifically, we neglect higher-order QCD and QED effects, set the phase-space factor to 1, and do not consider weak interaction effects.

The idea is to measure $R_h(s)$ and then calculate $R_a(s)$ and compare it to $R_q(s)$. Quark-hadron duality tells us that $R_a \approx R_q$, so long as the smoothing in R_a is done on a region that is large enough (i.e., larger than $\Lambda_{\rm QCD}$). Figure 10.1 demonstrates the validity of this prediction. We plot R_h, R_a, and R_q both below and above the charm threshold. It can be seen that while R_h is very different from R_q, the agreement between the smoothed R_a and R_q is impressive. The deviations between R_a and R_q are attributed to higher-order effects, experimental errors, and the fact that the duality is not exact.

10.5.2 Jets

When particles collide at high energy, the hard-scattering process can be described by the UV part of QCD. Consider, for example, the hard-scattering process of $e^+e^- \to q\bar{q}$. Confinement tells us that our detectors will not observe a quark-antiquark pair going back to back (unlike the case for muons in $e^+e^- \to \mu^+\mu^-$). Instead, the detectors observe a variety of hadrons. This is semiclassically understood as having $q\bar{q}$ pairs popping up from the vacuum to combine with the original partons to form color-singlet final states.

The number of $q\bar{q}$ pairs that are created depends on the relative momentum of the original partons. The higher this momentum, the more $q\bar{q}$ pairs are likely to be created. The point to emphasize is that, so long as many hadrons are produced, most of them travel roughly in the same direction as the original partons. The collection of hadrons related to a single original parton comes under the name *jet*.

While there is no practical first-principle way to relate jets to the UV properties of QCD, the fact that jets are formed gives us ways to probe QCD at short distances. To a good approximation, the $e^+e^- \to q\bar{q}$ transition appears as $e^+e^- \to 2j$ (here, j represents a jet), while the $e^+e^- \to q\bar{q}g$ transition appears as $e^+e^- \to 3j$.

The main challenge is to relate the jets to the underlying parton process. The source of the problem is that a jet constitutes a finite-sized object, and there is inherent ambiguity in defining it. For every definition, there is some probability that the jets do not directly correspond to the underlying process that we assume. For example, if we have two partons that are produced almost collinearly, such that the separation between them is much smaller than a typical jet size, we will interpret the set of resulting hadrons as a single jet. There is no simple solution to this problem. What is done in practice is to use simulations, determine the probability that such a situation occurs, and correct for it. There is, then, an intrinsic theoretical error in translating an experimental measurement of, say, $e^+e^- \to 2j$, to the calculated $e^+e^- \to q\bar{q}$.

Beyond the identification of jets, we can learn more about the short-distance physics by the internal properties of the jet. Some of the properties depend on the parton that generates the jet. For example, one can, statistically, tell a quark jet from a gluon jet. Other properties are universal. For example, the most likely mesons to be produced in the jet are pions, since they are the lightest. Baryons are less likely than mesons to be created within a jet. We can understand this fact by the need to pop up two $q\bar{q}$ pairs from the vacuum, and the combinatorics is such that the three $q\bar{q}$ pairs are more likely to end up in three mesons rather than in two baryons.

10.5.3 Parton Distribution Functions

High-energy collider experiments can be divided into three main classes, distinguished by the initial state: lepton-lepton, lepton-hadron, and hadron-hadron. When there are no initial state hadrons, perturbative QCD is simplest to apply. See, for example, the discussion of $e^+e^- \to$ hadrons in section 10.5.1. For deep inelastic scattering (DIS), $e + p \to e + X$ (and also for νp scattering); and for proton-proton collisions, $p + p \to X$ (and also for $p\bar{p}$ collisions), a more complicated formalism is required. This formalism allows one to combine first-principle calculations of the perturbative aspects with data-based input for the nonperturbative aspects of the relevant processes.

The key ingredient in this formalism is factorization. One factorizes the scattering process to a high-energy part, where the cross section can be calculated perturbatively, and a low-energy QCD part, where the nonperturbative aspects of the initial hadron(s) are parameterized. This formalism for scattering is analogous to the treatment of hadron decays discussed in section 10.3, where nonperturbative aspects were parameterized in terms of decay constants and form factors.

In the quark model, the proton is made of three valence quarks (uud). However, the actual structure of the proton is much more complicated. In the limit where the energy of the probe (e.g., the photon exchanged in electron-proton scattering) in the proton rest frame goes to infinity, due to the asymptotic freedom of QCD, the probe "sees" pointlike particles, which we call *partons*. The partons include the three valence quarks, as well as gluons and quark-antiquark pairs of various flavors. The spectrum of the partons is the nonperturbative part that we need to parameterize. This spectrum is called the *Parton Distribution Function* (PDF).

In the limit where the proton energy is much larger than its mass, one can neglect the transverse momenta of the partons, and thus each parton w_i has a momentum, p_i, that is in the same direction as the momentum of the proton p_p:

$$p_i^\mu = \xi_i \times p_p^\mu, \tag{10.40}$$

where ξ_i is a number that satisfies $0 < \xi_i < 1$. We denote the PDF of the parton w_i as $f_i(\xi)$. It gives the probability density to find the parton w_i with a fraction of the proton momentum ξ. All the parton momenta must add up to the proton momentum as follows:

$$\int_0^1 d\xi \sum_i f_i(\xi)\xi = 1. \tag{10.41}$$

For example, take DIS, where an electron of momentum p_e interacts with a proton of momentum p_p by exchanging a virtual photon of momentum p_γ. The cross section in the parton model is given by

$$\sigma(e^- p \to e^- X) = \int d\xi \sum_i f_i(\xi)\sigma(e^- w_i(\xi p_p) \to e^- X). \tag{10.42}$$

Here, $\sigma(e^- w_i(\xi p_p) \to e^- X)$ can be calculated perturbatively, while the $f_i(\xi)$'s are nonperturbative in nature and currently can be determined from data. The PDF depends on the momentum exchange in the process, $Q^2 = -p_\gamma^2$. The higher Q^2, the finer the structure of the proton that is resolved by the probe.

As a second example, take the inclusive cross section for Z-boson production in $p\bar{p}$ collisions. The total cross section can be written as follows:

$$\sigma(p\bar{p} \to Z + X) = \sum_{i,j} \int d\xi_1 d\xi_2 f_{i/p}(\xi_1) f_{j/\bar{p}}(\xi_2) \times \sigma(ij \to Z + X)(\xi_1\xi_2 s), \tag{10.43}$$

where $s = (p_p + p_{\bar{p}})^2$ and $f_{i/p}$ ($f_{j/\bar{p}}$) represents the PDF of parton w_i (w_j) in the proton (antiproton).

A more advanced calculation employs Generalized Parton Distribution Functions (GPDFs). This is a generalization of the PDF, in that the GPDF parameterizes the parton distributions as functions of additional variables, such as transverse momenta and spin.

Appendix

10.A Names and Quantum Numbers for Hadrons

The naming scheme for all the particles is nicely summarized by the PDG. Here, we discuss only the hadrons that we mention in the book, which are usually the lightest of a given flavor composition.

We start with mesons. Each meson is described by its total spin (J), internal parity (P), and principal quantum number. Neutral mesons are also characterized by their internal charge conjugation parity (C). Mesons made of u and d quarks and antiquarks are also characterized by their isospin.

The lightest state of a given quark composition is the pseudoscalar meson ($J^P = 0^-$). Those who carry a flavor quantum number (like the π^+) are stable (in the sense explained in section 10.2.4). The stable, pseudoscalar mesons play an important role in the investigation of flavor physics, as they decay via the weak interaction. Their list is given in table 10.1. Note that the four neutral meson pairs with well-defined $U(1)_s \times U(1)_c \times U(1)_b$ quantum numbers are not mass eigenstates. In fact, each pair mixes into a pair of mesons of well-defined masses and decay widths. This phenomenon is discussed in detail in appendix 13.A of chapter 13. The unflavored pseudoscalar mesons are unstable: they decay via QCD or QED. The ones that involve only light quarks are π^0 and η.

Some of the weakly decaying pseudoscalar mesons are, in fact, not the lightest that are charged under a symmetry. Yet final states that would be allowed by symmetry considerations are excluded by phase-space considerations. For example, $D_s \to DK$ is excluded because $m_{D_s} < m_D + m_K$, and $B_c \to BD$ is excluded because $m_{B_c} < m_B + m_D$.

Table 10.1. List of weakly decaying $J^P = 0^-$ mesons with their quark decomposition, $U(1)_S \times U(1)_C \times U(1)_b$ charge, mass and lifetime.

Meson	$q\bar{q}'$	(S, C, B)	Mass (GeV)	τ (s)
$\pi^\pm$	$u\bar{d}, \bar{u}d$	$(0, 0, 0)$	0.140	2.6×10^{-8}
$K^\pm$	$u\bar{s}, \bar{u}s$	$(\pm1, 0, 0)$	0.494	1.2×10^{-8}
$K^0, \overline{K^0}$	$d\bar{s}, \bar{d}s$	$(\pm1, 0, 0)$	0.498	See table note
$D^\pm$	$c\bar{d}, \bar{c}d$	$(0, \pm1, 0)$	1.87	1.0×10^{-15}
$D^0, \overline{D^0}$	$c\bar{u}, \bar{c}u$	$(0, \pm1, 0)$	1.87	4.1×10^{-16}
$D_s^\pm$	$c\bar{s}, \bar{c}s$	$(\pm1, \pm1, 0)$	1.97	5.0×10^{-15}
$B^\pm$	$u\bar{b}, \bar{u}b$	$(0, 0, \pm1)$	5.28	1.6×10^{-12}
$B^0, \overline{B^0}$	$d\bar{b}, \bar{d}b$	$(0, 0, \pm1)$	5.28	1.5×10^{-12}
$B_s, \overline{B_s}$	$s\bar{b}, \bar{s}b$	$(\mp1, 0, \pm1)$	5.37	1.5×10^{-12}
$B_c^\pm$	$c\bar{b}, \bar{c}b$	$(0, \pm1, \pm1)$	6.28	5.1×10^{-13}

Note: For the neutral K-meson mass eigenstates, K_S and K_L, the respective lifetimes are $\tau_S = 9.0 \times 10^{-11}$ s and $\tau_L = 5.1 \times 10^{-8}$ s. See also appendix 13.A.

The next-to-lightest mesons are the vector mesons ($J^P = 1^-$). In the quark model, they have $L = 0$, and their spin comes from the spin of their constituents. Those with only light quark (u and d) constituents include the isospin triplet $\rho(770)$, and the isospin singlet $\omega(782)$. The flavored vector mesons are indicated by stars relative to the pseudoscalars: K^*, D^*, B^*, and B_s^*. The unflavored vectors that do not have light quarks are $\phi(1020)$ (made of $s\bar{s}$), J/ψ (made of $c\bar{c}$), and Υ (made of $b\bar{b}$). Their excited states are denoted in a way that is very similar to atomic physics. For example, $\Upsilon(4S)$ is the S-wave state with the principal quantum number $n = 4$.

Finally, we briefly consider the names of some of the stable baryons (i.e., those that cannot decay via QCD or QED). Those with three light quarks (u, d combinations) are the proton p and the neutron n. Baryons with two u and/or d quarks include Λ, Λ_c, Λ_b. The subindex is in correspondence with the single heavy quark, with no index implying a single strange quark.

10.B Extracting $|V_{ud}|$

In this section, we give an example of how we use the approximate symmetries of QCD to extract a weak interaction parameter. Specifically, we use the isospin symmetry, as discussed in section 10.4.1, to extract the value of $|V_{ud}|$. The key processes to measure are $d \to u$ transitions in β decay, where the quark level amplitude has a factor of $|V_{ud}|$ (see figure 10.2). Next, we briefly discuss one way to do it with nuclear β decay.

Consider the rate of nuclear beta decay:

$$\Gamma(C \to Be\bar{\nu}_e) = N_f \, |\mathcal{A}(C \to Be\bar{\nu}_e)|^2, \tag{10.44}$$

where C and B are nuclei, and the proportionaly factor, N_f, includes phase-space effects and other numerical factors, and it is known for each specific decay. The tree-level contributions are presented in figure 10.2. The amplitude for the nuclear beta decay can be written as

$$\mathcal{A}(C \to Be\bar{\nu}_e) \propto V_{ud}\, \mathcal{M}^\mu, \qquad \mathcal{M}^\mu = \langle B(p_B)|V^\mu - A^\mu|C(p_C)\rangle, \tag{10.45}$$

where $V^\mu = \bar{u}\gamma^\mu d$ and $A^\mu = \bar{u}\gamma^\mu\gamma_5 d$. We obtain

$$\Gamma(C \to Be\bar{\nu}_e) = N_f \, |V_{ud}|^2 \times |\mathcal{M}^\mu|^2. \tag{10.46}$$

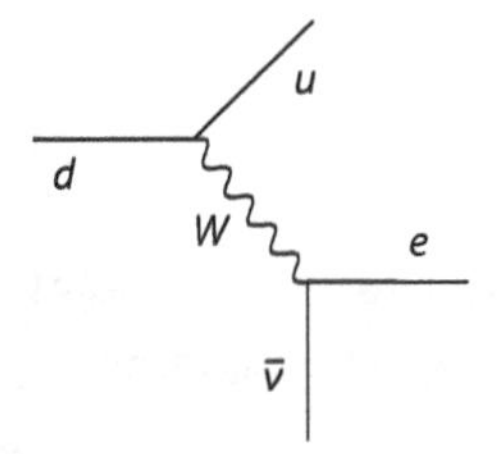

FIGURE 10.2. Beta decay at the quark level. At the upper vertex, we have a factor of V_{ud}.

We learn that, to extract $|V_{ud}|$ from the experimental measurement of the decay rate, we need to know the magnitude of the hadronic matrix element, $\mathcal{M}^\mu$.

Cases where the matrix element of either A_μ or V_μ vanishes are much simpler to handle. If, for example, both C and B are $J^P = 0^+$ nuclei, then $\langle B|A^\mu|C\rangle = 0$ and we have (see equation (10.22))

$$\mathcal{M}^\mu = \langle B(p_B)|V^\mu|C(p_C)\rangle = f_+(q^2)(p_C + p_B)^\mu + f_-(q^2)q^\mu, \qquad q \equiv p_C - p_B. \tag{10.47}$$

We now invoke considerations related to the isospin symmetry of QCD. We consider cases where C and B are members of the same isospin multiplet. Thus, in the isospin limit, $m_C = m_B$, and consequently $p_C^2 = p_B^2$. Let us invoke the Ward identity related to the isospin symmetry:

$$0 = q_\mu \mathcal{M}^\mu = f_+(q^2)(p_C^2 - p_B^2) + f_-(q^2)q^2 = f_-(q^2)q^2, \tag{10.48}$$

where, for the second equality, we used $p_C^2 = p_B^2$. We conclude that, for finite q^2 in the isospin limit,

$$f_-(q^2) = 0. \tag{10.49}$$

Thus, we need to find only the value of $f_+(q^2)$. In the isospin limit, $p_B = p_C = p$, and therefore $q = 0$. Since isospin breaking is a very small effect, the physical values of q^2 satisfy $0 < q^2/\Lambda_{QCD}^2 \ll 1$, and thus

$$f_+(q^2) \approx f_+(0). \tag{10.50}$$

The final step, then, is to find the value of $f_+(0)$. To do that, we note that in terms of the isospin $SU(2)$ algebra, the vector operator $\bar{u}\gamma^\mu d$ is the raising operator, which changes I_3 by 1 unit. Thus,

$$\langle B(p)|V^\mu|C(p)\rangle = 2p^\mu f_+(0) = 2p^\mu \langle j, m+1|J_+|j, m\rangle = 2p^\mu \sqrt{(j-m)(j+m+1)}, \tag{10.51}$$

where J_+ is the usual ladder operator of $SU(2)$ and m is the I_3 quantum number of C. Plugging the value of the matrix element (equation (10.51)) into equation (10.46), we obtain

$$\Gamma(C \to Be\bar{\nu}_e) = 4m_C^2(j-m)(j+m+1)\, N_f\, |V_{ud}|^2, \tag{10.52}$$

where we use $p^2 = m_C^2$.

We learn that by using Lorentz invariance and isospin, we can overcome the challenges of QCD. Concretely, we use the measured nuclear beta decay rate to extract $|V_{ud}|$, up to the small theoretical uncertainties that come from isospin-breaking effects.

10.C Extracting $|V_{cb}|$

In this section, we give another example of how we use the approximate symmetries of QCD to extract a weak interaction parameter. Specifically, we use heavy quark symmetry, as discussed in section 10.4.2, to extract $|V_{cb}|$.

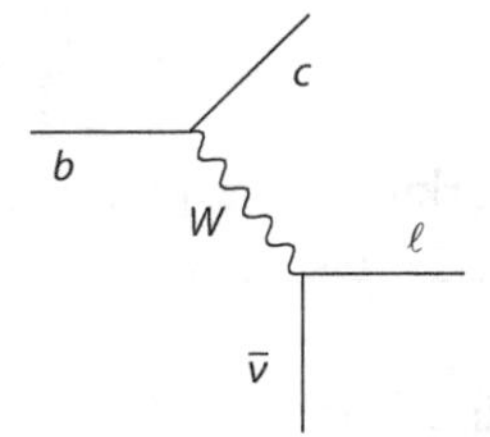

FIGURE 10.3. Semileptonic b decay at the quark level. At the upper vertex, we have a factor of V_{cb}.

Consider the semileptonic B-meson decay. The tree-level contributions are presented in figure 10.3. The discussion goes along similar lines to the discussion in section 10.B. The decay rate is given by

$$\Gamma(B^- \to D^0 \ell^- \bar{\nu}) = N_f \, |\mathcal{A}(B^- \to D^0 e \bar{\nu}_e)|^2, \tag{10.53}$$

where N_f is a known normalization factor. We learn that to extract $|V_{cb}|$ from the measured decay rates, we need to obtain the hadronic matrix elements.

The matrix element of $B \to D\ell\bar{\nu}$ depends on two form factors, just like the case of $K \to \pi\ell\bar{\nu}$ (see equation (10.22)):

$$\langle D^0(p_D)|V^\mu|B^-(p_B)\rangle = f_+(q^2)(p_B + p_D)^\mu + f_-(q^2)(p_B - p_D)^\mu, \qquad V^\mu = \bar{c}\gamma^\mu b, \tag{10.54}$$

The contribution of $f_-(q^2)$ to the decay rate is proportional to m_ℓ^2. Thus, this contribution is suppressed by m_ℓ^2/m_B^2 and can be safely neglected for the cases of $\ell = e, \mu$. Our task, then, is to find $f_+(q^2)$.

To make progress, we use heavy quark symmetry and consider the specific kinematic point of no D recoil. At this point, the outgoing D is at rest in the B rest frame, which corresponds to the maximal momentum transfer, $(q^2)_{\max}$. In the $m_Q/\Lambda_{\mathrm{QCD}} \to \infty$ limit (with $Q = b, c$), the velocity of the b quark is the same as the velocity of the B^- meson, and the velocity of the c quark is the same as the velocity of the D^0 meson. Thus, in the heavy quark limit, at the zero recoil point, the light DoF are insensitive to decay of the b quark since there is no change in the source of the color field. Thus, the overlap of the wave functions at that point is just 1:

$$f_+[(q^2)_{\max}] = 1. \tag{10.55}$$

One uses this result to extract $|V_{cb}|$ from the $B \to D$ decay rate.

Ideally, we would take the measurement at the zero recoil point, but the rate vanishes at that point. Thus, we must use a model for the q^2-dependence of the form factor and fit it to the data, which introduces another source of uncertainty, on top of the corrections to the heavy quark limit. The current total theoretical uncertainty is estimated to be at the level of only a few percentage points.

For Further Reading

- For general review of properties of hadrons, see the PDG in Workman et al. [1].
- More on the notion of quark-hadron duality can be found, for example, in Shifman [32].
- Reviews of heavy quark symmetry can be found, for example, in Manohar and Wise [33], Grinstein [34], and Neubert [35].
- For a review of jets and parton distribution functions see, for example, Sterman [36].

Problems

Question 10.1: Using the PDG

Find the PDG entries of B^+, D_s^+, and Λ. For each of them, write their quark composition, mass, lifetime, charge, and spin.

Question 10.2: Using the PDG even more

Using the PDG, answer the following questions:

1. Read the quark model review from the PDG and explain what P, C, J, I, and G stand for. For each of these quantum numbers, indicate if they are (*i*) exact in nature, (*ii*) exact in QCD and QED, or (*iii*) approximately conserved in QCD.
2. Find the mass, width, and the abovementioned quantum numbers of π^0, η, ρ^0, and ω.
3. Find the branching ratios of the η, ρ^0, and ω decays to two pions and to three pions.
4. From the answer to item 3, it is evident that η does not decay to two pions and the ω decay rate to two pions is highly suppressed, while the ρ decays dominantly to two pions. Based on the quantum numbers listed in item 2, explain these results.

Question 10.3: Leptonic pion decay

Consider the purely leptonic pion decays:

$$\pi^+ \to \ell^+ \nu_\ell. \tag{10.56}$$

1. Based on the masses of the relevant particles, what final-state leptons are allowed?
2. Draw the diagram for this decay. To take into account that the initial state is a pion, we need to write the amplitude using the relevant hadronic matrix element. Using equation (10.14), show that the amplitude is given by

$$\mathcal{A}_{\pi\to\ell\nu} = -\frac{g^2}{4m_W^2} V_{ud} f_\pi p_\pi^\mu \left[\bar{u}_\ell \gamma_\mu P_L v_\nu\right], \tag{10.57}$$

where u and v are the standard notations for the spinors and P_L is a projection operator defined in equation (1.3). Explain why you can approximate the W propagator as $1/m_W^2$.

3. Using the amplitude in equation (10.57), show that the decay width is given by equation (10.15):

$$\Gamma(\pi \to \mu\nu) = \frac{G_F^2 |V_{ud}|^2}{8\pi} m_\mu^2 m_\pi \left(1 - \frac{m_\mu^2}{m_\pi^2}\right)^2 f_\pi^2. \tag{10.58}$$

The decay rate of a particle of mass M and momentum p into two particles of masses m_1 and m_2 is given by

$$\Gamma = \frac{|\mathbf{p}|}{32\pi^2 M^2} \int d\Omega \sum |\mathcal{A}|^2,$$

$$|p| = \frac{\sqrt{[M^2 - (m_1+m_2)^2][M^2 - (m_1 - m_2)^2]}}{2M}, \tag{10.59}$$

where the sum goes over the outgoing particle polarizations.

4. For massless neutrinos,

$$R \equiv \frac{\Gamma(\pi^+ \to e^+ \nu_e)}{\Gamma(\pi^+ \to \mu^+ \nu_\mu)} = \left(\frac{m_e}{m_\mu}\right)^2 \left(\frac{m_\pi^2 - m_e^2}{m_\pi^2 - m_\mu^2}\right)^2. \tag{10.60}$$

Repeat the calculation of R for massive neutrinos. (We discuss ways that neutrinos can be massive in chapter 14.) Specifically, assume that ν_e and ν_μ are mass eigenstates with masses m_1 and m_2, respectively. Check that your result agrees with equation (10.60) in the case of massless neutrinos.

5. Using experimental data derive bounds on m_1 and m_2. To do so, assume that $m_1^2 \ll m_e^2$ and that $m_2^2 \ll m_\mu^2$, expand the ratio R to first-order in m_1^2/m_e^2 and in m_2^2/m_μ^2, and write

$$\frac{R}{R_0} = 1 + \left(\frac{m_1^2}{m_e^2}\right) a_e - \left(\frac{m_2^2}{m_\mu^2}\right) a_\mu, \tag{10.61}$$

where R_0 is the value of R for massless neutrinos. Find the expression for a_e and a_μ.

6. Find the numerical values of a_e and a_μ and show that they are both positive. Argue that there is a flat direction; that is, a curve in the (m_1, m_2) plane where $R = R_0$.

7. The experimental data give

$$R = (1.2327 \pm 0.0023) \times 10^4. \tag{10.62}$$

Using equation (10.60) would give the theory prediction of $R_0 = 1.2833 \times 10^{-4}$. However, to obtain the bound, we have to take into account higher-order corrections, which shift the result as follows:

$$R_0 = (1.2352 \pm 0.0001) \times 10^{-4}. \tag{10.63}$$

Working to 2σ, derive numerical bounds on m_1 for $m_2 = 0$, and on m_2 for $m_1 = 0$. Explain why the bound on m_1 is much stronger than the one on m_2. For each of the two cases, check whether the leading-order expansion that was used is valid. (Note: these bounds are not very strong compared to other bounds on neutrino masses; see appendix 14.B in chapter 14.)

Question 10.4: Form factors in B decays

Consider the semileptonic $B \to D^*\ell\nu$ decay. The differences between the D-meson case in equation (10.54) and the D^*-meson case that we discuss here is that the D-meson is a pseudoscalar and the D^*-meson is a vector. We define ϵ as the D^* polarization vector (which satisfies $\epsilon_\mu p^\mu_{D^*} = 0$); and p_B (p_{D^*}) as the B (D^*) momentum, $q \equiv (p_B - p_{D^*})$ and $P \equiv (p_B + p_{D^*})$.

The way that we construct the form factors here follows similar arguments used in section 10.3.3. The dynamical variables that we can use are the various momenta. Note that the amplitude must be proportional to the polarization of the outgoing particles. For an outgoing spin-1 particle, this results in a factor of ϵ^*. (For an outgoing spin-0 particle, this factor is a scalar, which is normalized to be just the number 1.)

1. Show that the vector form factor can be written as

$$\langle D^*(p_D, \epsilon)|V^\mu|\bar{B}(p_B)\rangle = g(q^2)\varepsilon^{\mu\nu\alpha\beta}\epsilon^*_\nu P_\alpha q_\beta, \tag{10.64}$$

where ε is the totally antisymmetric tensor.

2. Show that the axial vector form factor can be written as

$$\langle D^*(p_D, \epsilon)|A^\mu|\bar{B}(p_B)\rangle = f(q^2)\epsilon^{*\mu} + \epsilon^*_\nu p^\nu_B\left[a_+(q^2)P^\mu + a_-(q^2)q^\mu\right]. \tag{10.65}$$

Question 10.5: Pseudoscalar decay constant

Consider the matrix element of the pseudoscalar operator P:

$$P = \langle K(p)|\bar{s}\gamma_5 u|0\rangle. \tag{10.66}$$

We aim to relate this to the matrix element of the axial vector operator A^μ:

$$A^\mu = \langle K|\bar{s}\gamma^\mu\gamma_5 u|0\rangle = -if_K p^\mu. \tag{10.67}$$

1. Using the equation of motion, $i\not{\partial} q = m_q q$, show that

$$i\partial_\mu(\bar{s}\gamma_\mu\gamma_5 u) = -(m_s + m_u)\bar{s}\gamma_5 u. \tag{10.68}$$

2. Using equation (10.68) and $i\partial_\mu = p_\mu$, show that

$$P = if_K \frac{m_K^2}{m_u + m_s}. \tag{10.69}$$

11

Beyond the Standard Model

In spite of the enormous experimental success of the Standard Model, we know that it is not the complete theory of nature. In this chapter, we explain this statement and discuss the formalism and the experimental probes to be used in case the physics that extends the Standard Model takes place at a high energy scale.

11.1 Introduction

One obvious reason that we know that the Standard Model is not the full theory of nature is that it does not include gravity. There are, however, reasons to think that, beyond gravity and the Standard Model list of elementary particles and fundamental interactions, there must be degrees of freedom (DoF) that are as yet unknown to us. These reasons can be roughly divided into four classes:

1. Experiments: Measurements that are inconsistent with the Standard Model predictions
2. Cosmology and astrophysics: Observations that cannot be explained by the Standard Model
3. Fine-tuning: Parameters whose values can be explained in the Standard Model only with accidental fine-tuned cancellations among several contributions
4. Clues: Various nongeneric features that are just parameterized in the Standard Model, but not explained

We elaborate on this list with specific examples in sections 11.2 and 11.3.

Models that extend the Standard Model by adding DoF, and often by imposing larger symmetries, come under the general name of "Beyond the Standard Model," or BSM for short. The fact that experiments have not observed any particles related to such hypothetical new fields tells us that either these new particles are very heavy or that their couplings to the Standard Model particles are very weak. In light-BSM scenarios, where the new DoF are at or below the weak scale, the Standard Model is not a good low-energy effective theory. Each such feebly coupled BSM scenario requires a specific discussion of how to probe it. We do not discuss such theories any further in this book.

The situation is different for heavy-BSM scenarios, where the new DoF are well above the weak scale. There is a unified framework that allows one to understand the possible probes ofheavy-BSM scenarios while remaining agnostic about the details of the new DoF. We present this framework in section 11.5 and employ it in the next four chapters.

Direct searches for BSM physics aim to produce the new particles on-shell and study their properties. Numerous such searches have been conducted but as of now, no BSM particle has been discovered. Instead, these searches have set combination oflower bounds on the masses and upper bounds on the couplings of such states to Standard Model states. Roughly speaking, the lower bounds on the masses of particles with order 1 couplings to the Standard Model particles are of order 1 TeV. What sets this scale is the center of mass energy of the most powerful accelerator in action (the Large Hadron Collider [LHC]). Indirect searches for BSM physics aim to observe virtual effects of the new states at low energies. We discuss this method in chapters 12, 13, and 14.

11.2 Experimental and Observational Problems

There are several experimental results and observational data that cannot be explained within the Standard Model. They provide the most direct evidence that we need to extend the Standard Model. Here, we present the three points that are the most robust.

Neutrino masses. There are several related pieces of experimental evidence for BSM physics from the neutrino sector. All these measurements prove that the neutrinos are massive, in contrast to the Standard Model prediction that they are massless. First, measurements of the ratio of ν_μ to ν_e fluxes of atmospheric neutrinos and the directional dependence of the ν_μ flux are different from the Standard Model predictions. Both facts are beautifully explained by neutrino masses and mixing that lead to $\nu_\mu - \nu_\tau$ oscillations. Second, measurements of the solar neutrino flux find that, while the sun produces only electron-neutrinos, their flux on Earth is significantly smaller than the total flux of neutrinos. This puzzle is beautifully explained by $\nu_e - \nu_{\mu,\tau}$ mixing. Both the atmospheric neutrino result and the solar neutrino result are now confirmed by terrestrial accelerator and reactor neutrino experiments. We discuss neutrino physics in detail in chapter 14.

The baryon asymmetry of the universe (BAU). There is observational evidence for BSM physics from cosmology. The features of the Cosmic Microwave Background (CMB) radiation imply a certain baryon asymmetry of the universe. Similarly, the standard Big Bang Nucleosynthesis (BBN) scenario is consistent with the observed abundances of light elements only for a certain range of baryon asymmetry, consistent with the CMB constraint. Baryogenesis, the dynamical generation of a baryon asymmetry, requires *CP* violation. The *CP* violation in the Standard Model generates a baryon asymmetry that is smaller by at least 10 orders of magnitude than the observed asymmetry. This implies that there must be new sources of *CP* violation beyond the Standard Model. Furthermore, baryogenesis requires a departure from thermal equilibrium at a very short time after the Big Bang, and the one provided by the Standard Model is not of the right kind. We elaborate on this in section 15.3 in chapter 15.

Dark matter. The evidence for dark matter—particles that are electrically neutral and do not carry the color charge of strong interactions—comes from several observations: rotation curves in galaxies, gravitational lensing, the CMB, and the large-scale structure of the universe. The neutrinos of the Standard Model do constitute dark matter, but their abundance is too small to be all the dark matter abundance. Thus, there must be DoF beyond those of the Standard Model. We elaborate on this in section 15.2 in chapter 15.

11.3 Theoretical Considerations

Some Standard Model parameters are small. We distinguish between two classes of small parameters. "Technically natural" small parameters are those that, if set to zero, allow the theory to gain an extra symmetry. The small parameters that are not technically natural are those where the symmetry of the theory is not enlarged when they are set to zero. An equivalent way to distinguish the two classes is based on their renormalization properties: For technically natural parameters, the renormalization is multiplicative, while for nontechnically natural parameters, it is additive. For a technically natural parameter, loop corrections are proportional to the parameter itself, and if the parameter is set to be small at tree level, it remains small to all orders in perturbation theory. For a nontechnically natural parameter, the radiative corrections are not proportional to the tree-level parameter, and in cases where the radiative corrections are much larger than the measured value of the parameter, the smallness of the parameter can only be maintained by fine-tuned cancellation between the tree-level and loop-level contributions.

The existence in the Standard Model of small parameters that are not technically natural is suggestive of BSM frameworks, where the smallness of the parameters is protected against large radiative corrections by some symmetry. There are two parameters of this kind in the Standard Model: m_h^2 and $\theta_{\rm QCD}$.

The Higgs fine-tuning problem. Within the Standard Model, the mass term of the Higgs, μ^2, defined in equation (7.15) in chapter 7, gets additive, quadratically divergent, radiative corrections (see question 11.1 at the end of this chapter). Given that the Standard Model is an effective theory, the radiative corrections must be finite and proportional to the scale above which the Standard Model is no longer valid. The higher the cutoff scale above the weak scale, the stronger the fine-tuned cancellation between the tree-level mass term and the radiative corrections must be. In particular, if there is no BSM physics below $m_{\rm Pl}$, the bare mass term and the loop contributions have to cancel each other out to an accuracy of about 34 orders of magnitude. Among the theories that aim to solve the Higgs fine-tuning problem, supersymmetry and composite Higgs have been intensively studied and searched for.

The strong *CP* problem. The *CP*-violating $\theta_{\rm QCD}$ parameter contributes to the electric dipole moment of the neutron d_N. The experimental upper bound on d_N puts an upper bound on $\theta_{\rm QCD}$ of $\mathcal{O}(10^{-9})$ (see appendix 8.A.1 in chapter 8). The smallness of $\theta_{\rm QCD}$ is not technically natural. Among the theories that aim to solve the strong *CP* problem, the Peccei-Quinn mechanism, as well as its prediction of an ultra-light pseudoscalar, the axion, have been intensively studied and searched for.

Other features of the Standard Model parameters provide hints for the existence of BSM physics because they are nongeneric, but they are not related to nontechnically natural small parameters. We mention two of them next.

The flavor parameters. The Yukawa couplings are small (except for y_t) and hierarchical. For example, the Yukawa coupling of the electron is of $\mathcal{O}(10^{-5})$. These are technically natural small numbers, but their nongeneric structure—smallness and hierarchy—is suggestive of BSM physics. Among the theories that aim to solve this puzzle, the Froggatt-Nielsen mechanism, $U(2)$ flavor models, and models of extra dimensions have been intensively studied.

Grand unification. The three gauge couplings of the strong, weak, and electromagnetic interactions seem to converge to a unified value at a high energy scale (see question 11.2). The Standard Model cannot explain this fact, which is just accidental within this model. Yet it can be explained if the gauge group of the Standard Model is part of a larger simple group. This idea is called Grand Unified Theory (GUT), and among the relevant unifying groups, $SU(5)$ and $SO(10)$ have been intensively studied.

11.4 The BSM Scale

The Standard Model has a single mass scale that we call the *weak scale* and denote by $\Lambda_{\rm EW}$. It can be represented by the masses of the weak force carriers, m_W or m_Z, or by the vacuum expectation value (VEV) of the Higgs field, v. As an order of magnitude estimate, we take $\Lambda_{\rm EW} \sim 10^2$ GeV.

Some of the problems of the Standard Model presented here are suggestive of where the BSM scale lies. We present these well-motivated energy scales in decreasing order. Of course, there could be more than one scale for the BSM physics.

- **The Planck scale, $m_{\rm Pl} \sim 10^{19}$ GeV.** The Planck scale constitutes a cutoff scale of all quantum field theories (QFTs). At this scale, gravitational effects become as important as the known gauge interactions and cannot be neglected.
- **The GUT scale, $\Lambda_{\rm GUT} \sim 10^{16}$ GeV.** The GUT scale is the one where the three gauge couplings of the Standard Model roughly unify (see question 11.2). It is an indication that at that scale, the GUT symmetry group is broken into the Standard Model symmetry group. For example, in $SU(5)$ GUT models, $\Lambda_{\rm GUT}$ can be represented by the VEV of the scalar field that breaks $SU(5) \to SU(3) \times SU(2) \times U(1)$, or by the masses of the gauge bosons that correspond to the broken $SU(5)$ generators.
- **The seesaw scale, $\Lambda_\nu \sim 10^{15}$ GeV.** As will be explained in chapter 14, the value of the neutrino masses m_ν hints that new DoF appear at or below the so-called seesaw scale, $\Lambda_\nu \sim v^2/m_\nu$. This scale is intriguingly close to the GUT scale, and thus the two might be in fact related to the same BSM physics.

- **The Higgs fine-tuning scale $\Lambda_{\rm FT} \sim 1$ TeV.** No fine-tuning is necessary to explain the smallness of m_h^2 if radiative corrections are cut off at a scale $\Lambda_{\rm FT}$ of order $4\pi m_h/y_t \sim$ TeV.
- **The WIMP scale $\Lambda_{\rm DM} \sim 1$ TeV.** If the dark matter particles are weakly interacting massive particles (WIMPs) (see section 15.2.3 in chapter 15), the cross section of their annihilation that is required to explain the dark matter abundance is of order $1/(20 \text{ TeV})^2$. If the relevant coupling is of order α_W, the relevant scale is of order 1 TeV.

11.5 The SMEFT

As has been argued here, the Standard Model is not a full theory of nature. If the BSM DoF are at a scale $\Lambda \gg \Lambda_{\rm EW}$, then the Standard Model is a good low-energy, effective theory that is valid below Λ. In such a case, the Standard Model Lagrangian should be extended to include all nonrenormalizable terms, suppressed by powers of Λ:

$$\mathcal{L}_{\rm SMEFT} = \mathcal{L}_{\rm SM} + \frac{1}{\Lambda} O_{d=5} + \frac{1}{\Lambda^2} O_{d=6} + \cdots , \qquad (11.1)$$

where $\mathcal{L}_{\rm SM}$ is the renormalizable Standard Model Lagrangian and $O_{d=n}$ represents operators that are products of Standard Model fields, of overall dimension n in the fields, and transforming as singlets under the Standard Model gauge group. The Standard Model that has been extended to include such nonrenormalizable terms is called the Standard Model effective field theory (SMEFT). For physics at an energy scale E well below Λ, the effects of operators of dimension $n > 4$ are suppressed by $(E/\Lambda)^{n-4}$. Thus, in general, the higher the dimension of an operator, the smaller its effect at low energies.

Nonrenormalizable operators are generated by extensions of the Standard Model, which introduce new DoF that are much heavier than the electroweak scale. By studying nonrenormalizable operators, we allow the most general extension of the Standard Model and remain agnostic about its specific structure. At the same time, constraints on nonrenormalizable terms can be translated into constraints on specific BSM models.

The low-energy effects of nonrenormalizable operators are small. Thus, when we study them, we also have to consider loop effects in the Standard Model. In the previous chapters, we have studied the Standard Model at tree level, and with only renormalizable terms. From this point on, we extend our discussion to include the effects of loop corrections and nonrenormalizable terms.

We can classify the effects of including loop corrections and nonrenormalizable terms into three broad categories:

- *Forbidden processes:* Various processes are forbidden by the accidental symmetries of the Standard Model. Nonrenormalizable terms, but not loop corrections, can break these accidental symmetries and allow forbidden processes to occur. Examples include neutrino masses, discussed in chapter 14, and proton decay.

- *Rare processes:* Within the Standard Model, various processes that are not forbidden do not occur at tree level. Here, both loop corrections and nonrenormalizable terms can contribute. Examples include flavor-changing neutral current (FCNC) processes, discussed in chapter 13.
- *Tree-level processes:* Often, tree-level processes in a particular sector depend on a small subset of the Standard Model parameters. This situation leads to relations among the various processes within this sector. These relations are violated by both loop effects and nonrenormalizable terms. Here, precision measurements and precision theory calculations are needed to observe these small effects. Examples include electroweak precision measurements, discussed in chapter 12.

As for the last two types of effects, where loop corrections and nonrenormalizable terms may both contribute, their use in phenomenology can be divided into two eras. Before all the Standard Model particles have been directly discovered and all the Standard Model parameters measured, one could assume the validity of the renormalizable Standard Model and indirectly measure the properties of the yet-unobserved Standard Model particles. Indeed, the masses of the charm quark, the top quark, and the Higgs boson were first indirectly measured in this way. Once all the Standard Model particles have been observed and the parameters measured directly, the loop corrections can be quantitatively determined and the effects of nonrenormalizable terms in the SMEFT can be unambiguously probed. Thus, at present, all three classes of processes are suitable for searching for BSM physics.

11.6 Examples of SMEFT Operators

In this section, we give a few examples of nonrenormalizable terms in the SMEFT. Most of these operators and their implications are discussed in detail in the upcoming chapters.

At dimension-five, there is a single class of operators:

$$O_{ij}^{LL\phi\phi} = \phi^T \phi L_{Li}^T L_{Lj}. \tag{11.2}$$

These terms are discussed in detail in chapter 14. They break the accidental $U(1)_e \times U(1)_\mu \times U(1)_\tau$ symmetry of the Standard Model, and allow processes that are forbidden in the renormalizable Standard Model, such as neutrino oscillations and leptonic FCNC decays (e.g., $\mu \to e\gamma$). The observed values of neutrino masses, $m_\nu \sim 0.05$ eV, put an upper bound on the scale suppressing these terms, $\Lambda \lesssim 10^{15}$ GeV, where we used $m_\nu \sim v^2/\Lambda$.

At $d = 6$, there is a large number of four-fermion operators affecting flavor physics. Here are a few examples of such four-quark operators:

$$O_{(1)}^{QQ} = (\overline{Q_{Lp}} \gamma_\mu Q_{Lr})(\overline{Q_{Ls}} \gamma^\mu Q_{Lt}),$$

$$O_{(3)}^{QQ} = (\overline{Q_{Lp}} \gamma_\mu \tau^a Q_{Lr})(\overline{Q_{Ls}} \gamma^\mu \tau^a Q_{Lt}),$$

$$O^{DD} = (\overline{D_{Rp}}\gamma_\mu D_{Rr})(\overline{D_{Rs}}\gamma^\mu D_{Rt}),$$
$$O^{QD}_{(1)} = (\overline{Q_{Lp}}\gamma_\mu Q_{Lr})(\overline{D_{Rs}}\gamma^\mu D_{Rt}),$$
$$O^{QD}_{(8)} = (\overline{Q_{Lp}}\gamma_\mu T^A Q_{Lr})(\overline{D_{Rs}}\gamma^\mu T^A D_{Rt}), \quad (11.3)$$

where τ^a are the Pauli matrices, contracting two $SU(2)_L$-doublets into a triplet; and the T^A are the Gell-Mann matrices, contracting $SU(3)_C$ triplets and anti-triplets into an octet. These terms are discussed in detail in chapter 13. They contribute to various FCNC processes, such as neutral meson mixing. The consistency of observables related to $K^0 - \overline{K}^0$ mixing with the renormalizable Standard Model predictions (to cite one example) puts lower bounds on the scale suppressing these $d=6$ terms, $\Lambda \gtrsim 10^6 - 10^8$ GeV (depending on the Lorentz structure of the operator and on whether the observable is CP conserving or violating).

At $d=6$, there are four operators that modify the electroweak vector-boson propagators at tree level:

$$O_{WB} = (H^\dagger \tau^a H) W^a_{\mu\nu} B_{\mu\nu},$$
$$O_{HH} = |H^\dagger D_\mu H|^2,$$
$$O_{BB} = (\partial_\rho B_{\mu\nu})^2,$$
$$O_{WW} = (D_\rho W^a_{\mu\nu})^2. \quad (11.4)$$

These terms, discussed in detail in chapter 12, modify various electroweak precision measurements. The consistency of the relevant observables with the renormalizable Standard Model predictions puts lower bounds on the scale suppressing the O_{WB} and O_{HH} terms, $\Lambda \gtrsim 10^4$ GeV.

11.6.1 Baryon Number Violation

Equation (11.2) implies that the accidental symmetry of the SMEFT truncated at $d=5$ is just $U(1)_B$. This symmetry is, however, broken by four classes of $d=6$ terms:

$$O^{QQQL} = \varepsilon_{\alpha\beta\gamma}\varepsilon_{jk}\varepsilon_{il} Q^{i\alpha}_{Lp} Q^{j\beta}_{Lr} Q^{k\gamma}_{Ls} L^l_{Lt},$$
$$O^{UUDE} = \varepsilon_{\alpha\beta\gamma} D^\alpha_{Rp} U^\beta_{Rr} U^\gamma_{Rs} E_{Rt},$$
$$O^{QQUE} = \varepsilon_{\alpha\beta\gamma}\varepsilon_{ij} Q^{i\alpha}_{Lp} Q^{j\beta}_{Lr} U^\gamma_{Rs} E_{Rt},$$
$$O^{DUQL} = \varepsilon_{\alpha\beta\gamma}\varepsilon_{ij} D^\alpha_{Rp} U^\beta_{Rr} Q^{i\gamma}_{Ls} L^j_{Lt}, \quad (11.5)$$

where α, β, and γ are $SU(3)_C$ indexes, i, j, k, l are $SU(2)_L$ indexes, and p, r, s, t are flavor indexes.

These terms result in proton decay. The decay rate is given by

$$\Gamma_p \sim \frac{m_p^5}{\Lambda^4}. \tag{11.6}$$

The lower bound on the proton lifetime,

$$\tau_p \gtrsim 10^{34} \text{ years}, \tag{11.7}$$

puts a lower bound on the scale suppressing these $d=6$ terms, $\Lambda \gtrsim 10^{16}$ GeV.

All four classes of B-violating operators respect $B-L$. Thus, they lead to a proton decay into $L=-1$ final states, such as $p \to e^+\pi^0$. The $B-L$ symmetry is not a symmetry of the SMEFT truncated at $d=6$, however, since it already has been broken by the $d=5$ terms.

11.6.2 Higgs Decays

The following classes of $d=6$ operators affect the Yukawa interactions:

$$\begin{aligned}
O_{QUH} &= \phi^\dagger\phi\, \overline{Q_{Lp}}\tilde{\phi}U_{Rr}, \\
O_{QDH} &= \phi^\dagger\phi\, \overline{Q_{Lp}}\phi D_{Rr}, \\
O_{LEH} &= \phi^\dagger\phi\, \overline{L_{Lp}}\phi E_{Rr}.
\end{aligned} \tag{11.8}$$

Replacing the three ϕ-fields with their VEVs, these terms contribute to the mass matrices M^f. Replacing two of the ϕ fields with their VEVs, these terms contribute to the effective Yukawa matrices Y^f.

The Standard Model predicts four generic features of the Yukawa interactions that apply at tree level:

- Proportionality: $y_i/y_j = m_i/m_j$
- Factor of proportionality: $y_i/m_i = \sqrt{2}/v$
- Diagonality: $Y_{ij} = 0$ for $i \neq j$
- CP conservation: $\mathcal{I}m(y_i/m_i) = 0$

All four relations are violated by both Standard Model loop corrections and the SMEFT operators of equation (11.8).

To demonstrate how such effects arise in the SMEFT, it is convenient to define the following parameters:

$$T_R^f \equiv \frac{v^2}{2\Lambda^2}\frac{\mathcal{R}e(C^f)}{y_f}, \quad T_I^f \equiv \frac{v^2}{2\Lambda^2}\frac{\mathcal{I}m(C^f)}{y_f}, \tag{11.9}$$

where C^f/Λ^2 is the coefficient of the corresponding $d=6$ term $\left(\text{e.g., } C^t = C_{tt}^{QUH}\right)$. In the presence of these terms, the mass and the effective Yukawa coupling of a fermion f are given by

$$m_f = \frac{y_f v}{\sqrt{2}} \sqrt{\left(1 + T_R^f\right)^2 + T_I^{f2}},$$

$$y_f^{\text{eff}} = y_f \frac{1 + 4T_R^f + 3T_R^{f2} + 3T_I^{f2} + 2iT_I^f}{\sqrt{\left(1 + T_R^f\right)^2 + T_I^{f2}}}, \tag{11.10}$$

leading to

$$\left|\frac{y_f^{\text{eff}}}{m_f}\right|^2 = \frac{2}{v^2} \times \frac{\left(1 + 3T_R^f\right)^2 + 9T_I^{f2}}{\left(1 + T_R^f\right)^2 + T_I^2},$$

$$\mathcal{I}m\left(m_f y_f^{\text{eff}*}\right) = -\sqrt{2} v y_f^2 T_I^f. \tag{11.11}$$

The consistency of various Higgs production and decay processes with the renormalizable Standard Model predictions, as well as the upper bound on the electric dipole moment of the electron, put lower bounds on the scale suppressing the C^t, C^b, C^τ, and C^μ operators, $\Lambda \gtrsim 10^4$ GeV.

For Further Reading

- For a general review of BSM, see, for example, Csaki, Lombardo, and Telem [37].
- For a general review of the Higgs fine-tuning problem, see, for example, chapter 1 of Terning [38].

Problems

Question 11.1: Radiative corrections to the Higgs mass

In this question, we discuss the Higgs fine-tuning problem. We explore the quadratic divergences in the Higgs mass and show an explicit example of a BSM model that can make the radiative corrections small.

For simplicity, we consider only the correction to m_h^2 due to a top loop. The relevant term in the Lagrangian is

$$\mathcal{L} = y_t \left(\frac{h + v}{\sqrt{2}}\right) \overline{t_L} t_R + \text{h.c.} \tag{11.12}$$

The top mass is given by $m_t = y_t v/\sqrt{2}$.

1. Draw the one-loop correction to the two-point function of the Higgs boson due to the top quark.
2. This diagram corresponds to correction to the Higgs mass, Δ_f, where we put the subindex f to remind you that it comes from an internal fermion loop. Evaluate the diagram and show that

$$-i\Delta_f = -2y_t^2 N_C \int \frac{d^4k}{(2\pi)^4}\left[\frac{1}{k^2-m_t^2}+\frac{2m_t^2}{(k^2-m_t^2)^2}\right], \tag{11.13}$$

where $N_C = 3$ is the number of colors. Remember to include the minus sign due to the fermion loop. (The factor of $-i$ is just a convention that is cancelled after a Wick rotation.)

3. Using the general formula

$$\int \frac{d^4k}{(2\pi)^4}\frac{1}{(k^2-A)^n} = \frac{i(-1)^n}{16\pi^2} I_n(A), \qquad I_n(A) = \int_0^\infty \frac{xdx}{(x+A)^n}, \tag{11.14}$$

write Δ_f in terms of I_1 and I_2, and show that

$$\Delta_f = \frac{2N_C|y_t|^2}{16\pi^2}\left[2m_t^2 I_2(m_t^2) - I_1(m_t^2)\right]. \tag{11.15}$$

4. Both I_1 and I_2 diverge. To regulate the divergencies, we cut off the integrals inside I_1 and I_2 at Λ^2; that is, we write

$$I_n(A, \Lambda^2) = \int_0^{\Lambda^2} \frac{xdx}{(x+A)^n}. \tag{11.16}$$

Evaluate $I_1(m_t^2, \Lambda^2)$ and $I_2(m_t^2, \Lambda^2)$. What are the degrees of divergence in Λ (logarithmic, linear, quadratic, etc.) for I_1 and I_2?

5. Find Δ_f as a function of Λ. Denote the answer as $\Delta_f(\Lambda^2)$. What is the degrees of divergence in Λ (logarithmic, linear, quadratic, etc.) for $\Delta_f(\Lambda^2)$?
6. Assuming that the cutoff is the Planck scale, $\Lambda = m_{\rm Pl} \sim 10^{19}$ GeV, find $\Delta_f(m_{\rm Pl}^2)$ and the ratio $\Delta_f(m_{\rm Pl}^2)/m_h^2$. The fact that the numerical value of the ratio is so large constitutes the Higgs fine-tuning problem.
7. We now move to find an explicit model that effectively generates the cutoff. We add a scalar field to the model, $\tilde{t}$, with the following coupling to the Higgs field:

$$-\mathcal{L}_{\tilde{t}h} = \frac{\lambda_{th}}{2}\,|\tilde{t}|^2\, h^2 + \mu\,|\tilde{t}|^2\, h, \tag{11.17}$$

where the coupling λ_{th} is dimensionless and μ has a mass dimension of 1. Draw the two $\tilde{t}$-mediated, one-loop diagrams that contribute to the Higgs mass.

8. Calculate each of the diagrams and denote them as $\Delta_s^{(1)}$ (for the one proportional to λ_{th}) and $\Delta_s^{(2)}$ (for the one proportional to μ^2). Denote the sum of these two diagrams as Δ_s (s stands for scalar) and show that

$$\Delta_s = \frac{1}{16\pi^2}\left[\lambda_{th} I_1(m_{\tilde{t}}^2) - \mu^2 I_2(m_{\tilde{t}}^2)\right]. \tag{11.18}$$

9. Consider a system with N_s such scalars. Show that, for $N_s = 2N_C$, $m_t = m_{\tilde{t}}$, $\lambda_{th} = |y_t|^2$, and $\mu^2 = 2m_t^2|y_t|^2$, the sum of the fermion and scalar loops vanishes; that is, show that in this case,
$$\Delta_f + 2N_C\Delta_s = 0. \tag{11.19}$$
10. Keep $N_s = 2N_C$ and $\lambda_{th} = |y_t|^2$, but allow arbitrary values for μ^2 and $m_{\tilde{t}}$. Show that for $\mu, m_{\tilde{t}} \ll \Lambda$, the quadratic divergence cancels.

What we learn is that adding to the Standard Model scalars with masses that are not much above the weak scale and with specific couplings results in a theory where the Higgs mass does not have quadratic divergence. This is important as logarithmic divergencies are fundamentally different from the quadratic ones, as they are renormalizable, and the exact low-energy properties are not sensitive to the details of the UV physics. In particular, models of supersymmetry provide such a mechanism.

Question 11.2: Coupling unification

The three gauge couplings of the Standard Model roughly meet (i.e., assume the same value) at a high scale. This fact could be a hint that the three gauge groups unify into a GUT. In this question, you are asked to investigate this aspect of GUT. We assume that the GUT group is $SU(5)$. It is spontaneously broken into $SU(3)_C \times SU(2)_L \times U(1)_Y$ at the GUT scale, Λ_{GUT}. Thus, at Λ_{GUT},

$$g_{\text{GUT}} = g_1 = g_2 = g_3, \tag{11.20}$$

where

$$g_1 = \sqrt{\frac{5}{3}}\,g', \qquad g_2 = g, \qquad g_3 = g_s. \tag{11.21}$$

(The factor of $\sqrt{5/3}$ arises from embedding the $U(1)_Y$ group into the $SU(5)$ group.)

We discuss running in section 4.4 in chapter 4 and question 5.3 in chapter 5. We assume that there are no fields with masses between Λ_{EW} and Λ_{GUT}. Then the solution for the Renormalization Group Equation (RGE) is given by equation (5.35), which in this case can be written as

$$\frac{1}{\alpha(\mu_{\text{GUT}})} = \frac{1}{\alpha_i(\mu_{\text{EW}})} + \frac{B_i}{2\pi}\log\left(\frac{\mu_{\text{EW}}}{\mu_{\text{GUT}}}\right), \tag{11.22}$$

where $\alpha_i = g_i^2/(4\pi)$. The difference between the running of the various couplings is encoded in the B_is. For the running of g', B is given by a generalization of equation (4.25) in chapter 4:

$$B_1 = \frac{1}{3}\left(\sum_f 2n_f Y_f^2 + \sum_b n_b Y_b^2\right), \tag{11.23}$$

where the sum over f (b) is a sum over all Weyl fermions (complex scalars), Y is the hypercharge, and n_f indicates the number of internal DoF (e.g., $n_Q = 6$ and $n_E = 1$). For the running of the $SU(N)$ gauge couplings (g_2 and g_3), B_N is given by a generalization of equation (4.26) in chapter 4:

$$B_N = \frac{1}{3}\left(\sum_f n_f + \frac{1}{2}\sum_b n_b - 11N\right). \tag{11.24}$$

1. Calculate the B_i coefficients for α_i $(i = 1, 2, 3)$. Remember to take into account the normalization of g_1 with respect to g'.
2. Using $\alpha^{-1}(m_Z) \approx 128$, and equations (7.60) and (7.73) in chapter 7 and equation (5.16) in chapter 5, find the central values of $\alpha_i(m_Z)$ for $i = 1, 2, 3$.
3. Using one-loop RGE running and the central values of $\alpha_i(m_Z)$, calculate the scale where each pair of couplings meet, M_{GUT}^{ij}, with $i \neq j$.
4. We can define a measure of missing a grand unification by

$$r_m = \frac{\Delta M_{\text{GUT}}}{M_{\text{GUT}}}, \tag{11.25}$$

   where ΔM_{GUT} is the difference between the largest and smallest M_{GUT}^{ij}, and M_{GUT} is their average. Find r_m.
5. We used one-loop RGE and central values of the couplings. What do you expect to be the relative effects on r_m due to higher-order corrections and due to the experimental uncertainties of α_i? Taking into account these uncertainties, is it possible that $r_m = 0$, in which case the Standard Model is consistent with GUT, with no additional DoF between Λ_{EW} and M_{GUT}?
6. A smaller value for r_m is obtained once an extra $(1, 2)_{+1/2}$ scalar field with a weak scale mass is added to the Standard Model. Calculate r_m in this case. Is $r_m = 0$ consistent within this model?

What we learn is that GUT is an attractive idea based on coupling unification. There are more aspects of GUT that make it a well-motivated extension to the Standard Model. Yet, to date, there is no direct experimental evidence for it.

12

Electroweak Precision Measurements

In this chapter, we discuss electroweak precision measurements (EWPM). These measurements provide a test of the Standard Model beyond tree level and a probe of Beyond the Standard Model (BSM) physics in the gauge and scalar sectors of the theory.

12.1 Introduction

Consider a situation in which a class of processes is described at tree level by only one sector of a theory, and where this sector depends on only a small number of parameters. If the number of observables is larger than the number of parameters, relations among the observables are predicted. At the quantum level, however, the tree-level relations are violated and the processes depend on all parameters of the theory. In some cases, the tree-level predictions follow from a symmetry that is respected by the relevant sector, but not by other sectors of the theory. Violations of such symmetry-based relations are particularly sensitive to loop effects. The EWPM program takes advantage of these features of quantum theories.

At tree level, all flavor-diagonal electroweak processes depend on only three parameters of the renormalizable Standard Model Lagrangian. In the language of the Standard Model Lagrangian, the three parameters are g, g', and v. Alternatively, one can work with the combinations of these parameters that are measured most precisely: α, m_Z, and G_F. The number of relevant observables is much larger than three. Thus, at tree level, a large number of relations among these observables are predicted. However, these predictions are violated by Standard Model loop effects, and possibly by nonrenormalizable operators that are generated by BSM physics.

The full Standard Model has 18 parameters. The EWPM allow us to probe some of the additional 15 parameters through their modification of the tree-level relations. A total of 11 of the 15 parameters (8 of the Yukawa couplings and the 3 Cabibbo-Kobayashi-Maskawa [CKM] mixing angles) are small, and consequently they have negligible effects on deviations from the tree-level relations. The four large parameters are the Kobayashi-Maskawa (KM) phase δ; the strong coupling constant g_s; the Higgs self–coupling λ (or equivalently, the Higgs mass, m_h); and the top Yukawa coupling y_t (or, equivalently, the top mass, m_t). The KM phase has negligible effects on flavor-diagonal processes. As for the strong coupling constant, its universality and the absence of direct couplings of the electroweak vector bosons to gluons combine to make its effect on the relevant parameters very small. Thus, in practice, there are only two additional Standard Model parameters that have significant effects on the EWPM: m_t and m_h. In the past, when these masses had not yet been directly measured, the EWPM were used to determine their values. Now that the top quark and the Higgs boson have been discovered and their masses are known from direct measurements, EWPM are used to probe nonrenormalizable operators (i.e., BSM physics).

As for the experimental aspects of the EWPM program, the relevant processes can be divided into two classes: low energy and high energy. The low-energy observables involve processes with a characteristic energy scale well below m_W and m_Z, such that the intermediate W-boson or Z-boson are far off-shell. The high-energy observables are measured in processes where the W-boson or the Z-boson are on-shell. The low-energy EWPM include measurements of G_F and α, as well as data from neutrino scattering, deep inelastic scattering (DIS), atomic parity violation (APV), and low-energy e^+e^- scattering. The high-energy EWPM include measurements of the masses, total widths, and partial decay widths of the W- and Z-bosons.

12.2 The Weak Mixing Angle

12.2.1 The Weak Mixing Angle at One Loop

To illustrate the way that EWPM are used, we consider three definitions of the weak angle. Each definition involves a different set of observables. At tree level, all three definitions are equivalent and correspond to (see equation (7.29) in chapter 7)

$$\tan\theta \equiv \frac{g'}{g}. \tag{12.1}$$

At the one-loop level, however, they differ, as follows:

- Definition in terms of α, G_F and m_Z:

$$\sin^2 2\theta_0 \equiv \frac{4\pi\alpha}{\sqrt{2}G_F m_Z^2}. \tag{12.2}$$

  Quantitatively, θ_0 is defined in terms of the best measured observables and thus has the smallest experimental uncertainties.
- Definition in terms of m_W and m_Z:

$$\sin^2\theta_W \equiv 1 - \frac{m_W^2}{m_Z^2}. \tag{12.3}$$

  This definition is based on the tree-level relation $\rho = 1$ discussed in section 7.3.2; see equation (7.35) in chapter 7.

- Definition in terms of g_V^f, the vectorial couplings of the Z-boson to fermion f:

$$\sin^2\theta_*^f \equiv \frac{T_3^{f,L} - g_V^f}{2Q^f}, \tag{12.4}$$

  where g_V^f is defined in equation (7.77) and we use the fact that all Standard Model fermions have $T_3^R = 0$. In principle, we have nine definitions of θ_*, one for each charged fermion type. See question 12.3 at the end of this chapter for one of the ways to obtain experimental sensitivity to g_V^f.

For any given model (i.e., the Standard Model or BSM), one can compute all the relevant corrections to the tree-level Standard Model predictions as functions of the model parameters. For pedagogical and practical purposes, however, we work in a simple framework that represents a large class of models. To do this, we make the following working assumption:

- The only significant effects are in the electroweak gauge boson propagators. These effects are called *oblique corrections.*

This assumption implies that all θ_*^f are equal and the relevant one-loop effects are flavor universal. Thus, from now on, we will not write the f super-index of θ_*. In some models, the assumption that all corrections are oblique does not hold, and yet the corrections are flavor universal.

We denote the oblique corrections to the electroweak gauge boson propagators by $\Pi_{AB}(q^2)$. The propagators P_{AB} are defined as follows:

$$P_{AB}(q^2) = \frac{-i}{q^2 - m_A^2}\left[\delta_{AB} + \frac{-i\Pi_{AB}(q^2)}{q^2 - m_B^2}\right]. \tag{12.5}$$

Taking charge conservation into account, we learn that there are four Π_{AB}s that do not vanish. We can choose the set Π_{WW}, Π_{ZZ}, $\Pi_{\gamma\gamma}$, and $\Pi_{\gamma Z}(=\Pi_{Z\gamma})$, corresponding to the mass eigenstates. Alternatively, we can use Π_{+-}, Π_{33}, Π_{00}, and Π_{30}, where the sub-indices $+, -, 3$, and 0 refer to the interaction eigenstates W^+, W^-, W_3, and B, respectively. The transition between the two sets is straightforward.

At $q^2 = m_A^2$, the relevant Π_{AB} can be identified as a correction to the mass of the corresponding gauge boson. Thus, gauge invariance (specifically, $m_\gamma^2 = 0$) guarantees that

$$\Pi_{\gamma\gamma}(q^2 = 0) = \Pi_{\gamma Z}(q^2 = 0) = 0. \tag{12.6}$$

To proceed further, we note that the following two points, which lead to considerable simplifications:

- High-energy EWPM depend on $\Pi_{WW}(m_W^2)$, $\Pi_{ZZ}(m_Z^2)$, and $\Pi_{\gamma Z}(m_Z^2)$.

- For low-energy EWPM, we can expand $\Pi_{AB}(q^2 \ll m_W^2)$ in q^2 and assume that terms of order $(q^2)^2$ and higher can be neglected:

$$\Pi_{AB}(q^2 \ll m_W^2) = \Pi_{AB}(0) + q^2\Pi'_{AB}(0), \qquad \Pi'(q^2) \equiv \frac{d\Pi(q^2)}{dq^2}. \tag{12.7}$$

Thus, the four functions $\Pi_{AB}(q^2)$ are replaced by nine numbers.

We now present the effects of the various corrections to the propagators on the three differently defined θs. Instead of considering each of the three definitions separately, we consider the two independent differences between them. This has two advantages: First, these differences depend only on the loop corrections and are independent of the tree-level contribution. Second, a logarithmic divergence that appears in the Standard Model loop corrections to each of the $\sin^2\theta$s cancels out in the difference. (Naively, one may expect quadratic divergences, but these are prevented by the Ward identities.) We have

$$\sin^2\theta_0 - \sin^2\theta_* = \frac{\sin^2 2\theta}{4\cos 2\theta}\left[\Pi'_{\gamma\gamma}(0) + \frac{\Pi_{WW}(0)}{m_W^2} - \frac{\Pi_{ZZ}(m_Z^2)}{m_Z^2}\right] + \sin\theta\cos\theta\frac{\Pi_{\gamma Z}(m_Z^2)}{m_Z^2}, \tag{12.8}$$

$$\sin^2\theta_W - \sin^2\theta_* = -\frac{1}{m_Z^2}\left[\Pi_{WW}(m_W^2) - \frac{m_W^2}{m_Z^2}\Pi_{ZZ}(m_Z^2)\right] + \sin\theta\cos\theta\frac{\Pi_{\gamma Z}(m_Z^2)}{m_Z^2}. \tag{12.9}$$

On the right-hand side of these equations, we do not specify which of the three θs we use, since the difference accounts for a higher-order correction. In equation (12.8), the terms in parentheses on the right-hand side correspond to corrections to α, G_F, and m_Z, respectively. In equation (12.9), the terms in parentheses on the right-hand side correspond to corrections to the tree-level contributions to m_W^2 and m_Z^2:

$$\Delta m_W^2 = \Pi_{WW}(m_W^2), \qquad \Delta m_Z^2 = \Pi_{ZZ}(m_Z^2). \tag{12.10}$$

The corrections to the g_V^f coupling arise from mixing between the off-shell photon and the Z-boson. We learn that, as anticipated, the loop effects are different for the three definitions. Once we extract the values of $\sin^2\theta$ from each of the three sets of observables, we can probe these effects.

12.2.2 The Weak Mixing Angle within the Standard Model

Our analysis so far has applied to all models with only oblique corrections. In this section, we discuss the specific case of the Standard Model. We make the following approximation:

- We consider only one-loop diagrams involving the top and bottom quarks.

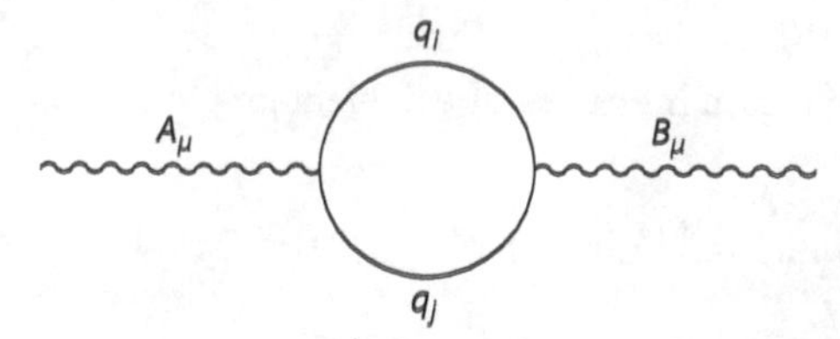

FIGURE 12.1. A general form of an oblique correction diagram.

Within the Standard Model, these diagrams provide the leading oblique corrections. We thus need to calculate diagrams of the general form that can be seen in figure 12.1.

We do not reproduce the details of the calculation here, but only give the final results:

$$\sin^2\theta_0 - \sin^2\theta_* = \frac{3\alpha}{16\pi\cos^2 2\theta}\frac{m_t^2 - m_b^2}{m_Z^2},$$
$$\sin^2\theta_W - \sin^2\theta_* = \frac{-3\alpha}{16\pi\sin^2\theta}\frac{m_t^2 - m_b^2}{m_Z^2}. \tag{12.11}$$

Equations (12.11) demonstrate the main point of this section: One can use the observables of the EWPM to determine parameters outside the pure electroweak sector. Before the discovery of the top quark, the measured values of the observables on the left-hand side of these equations were used to determine m_t (see question 12.3). Moreover, the Standard Model was tested by examining whether the ranges allowed for m_t from the two sets of observables overlap.

Let us explain the various factors in equation (12.11). The factor of $\alpha/(16\pi)$ is typical of electroweak one-loop effects. The factor of 3 is the color factor of the quarks in the loop. The θ dependence is different in the two cases, reflecting the specific combination of observables involved in each case. The factor of $(m_t^2 - m_b^2)/m_Z^2$ deserves a more detailed discussion, which is provided in section 12.3.

An interesting aspect of equation (12.11) has to do with the quadratic dependence on the top mass, which seems to violate the *decoupling theorem*. This theorem states that the effect of heavy states on low-energy observables must go to zero as their mass goes to infinity. The intuition behind this theorem is straightforward. The heavier a state is, the smaller its effects (when off-shell) become. This can be understood based on the uncertainty principle, on second-order perturbation theory, or simply by considering the form of propagators in quantum field theory (QFT). Yet the theorem does not apply to the top contributions to EWPM, which become larger as m_t becomes heavier. The solution to the puzzle lies in the fact that the Standard Model quarks acquire their masses from the Higgs mechanism. Consequently, their Yukawa couplings are proportional to their masses. The heavier the top, the stronger its Yukawa coupling becomes. Indeed, the top-related loop corrections to EWPM depend on the top coupling to the longitudinal W and Z, which are the degrees of freedom (DoF) that were eaten by the spontaneous symmetry breaking (SSB). The quadratic dependence on the top mass reflects the proportionality of the loop corrections to the top Yukawa coupling, not to its mass. (In fact, the $m_t \to \infty$ limit cannot be taken because perturbation theory no longer holds.)

Additional oblique corrections of interest come from loop diagrams with internal Higgs bosons. At the one-loop level, the m_h-dependence is logarithmic, and thus the EWPM are less sensitive to m_h than they are to m_t. This result is known as the *screening theorem*. We do not discuss that in detail here. Two loop contributions are proportional to m_h^2, but they are small because of the extra loop factor. We note, however, that the precision of the EWPM was good enough to have sensitivity to the Higgs-related oblique corrections and provided an allowed range for the Higgs mass well before it was actually measured. The present allowed range from EWPM (removing all direct measurements of the Higgs mass, production, and decay rates) is 60 GeV $\leq m_h \leq$ 127 GeV at the 90 percent confidence level. Indeed, the directly measured value of the Higgs mass (see equation (7.25) in chapter 7) is within this range.

12.3 Custodial Symmetry

The Standard Model Higgs potential has an accidental symmetry. This so-called custodial symmetry predicts tree-level relations among various observables. In this section, we focus on the Standard Model: the fact that the custodial symmetry is not a symmetry of the full Standard Model makes EWPM sensitive to Standard Model parameters that break the custodial symmetry. In the next section, we focus on BSM physics: The fact that the symmetry is accidental makes EWPM sensitive to nonrenormalizable terms that violate the symmetry.

Consider the Standard Model Higgs potential (see equation (7.15) in chapter 7):

$$V = \mu^2(\phi^\dagger\phi) + \lambda(\phi^\dagger\phi)^2. \tag{12.12}$$

Since ϕ is a complex, $SU(2)_L$-doublet scalar field, it has 4 DoF:

$$\phi = \begin{pmatrix} \phi_1 + i\phi_2 \\ \phi_3 + i\phi_4 \end{pmatrix}. \tag{12.13}$$

The scalar potential, when written in terms of these four components, depends only on the combination $\phi_1^2 + \phi_2^2 + \phi_3^2 + \phi_4^2$, and thus it has manifestly an $SO(4)$ symmetry. The 4 DoF form an $SO(4)$-quartet (4), which is a vector in the space. The potential depends only on the length of the vector, which is the singlet combination of two 4s. At the algebra level, $SO(4) \sim SU(2) \times SU(2)$. Of the six generators, four are also generators of the gauge group $SU(2)_L \times U(1)_Y$. The two extra generators are then related to an accidental symmetry of the scalar sector of the Standard Model.

The vacuum expectation value (VEV) of the Higgs field breaks three of the generators, leaving (within the pure Higgs sector) an unbroken $SU(2)$ symmetry. This is the symmetry that is called *custodial symmetry*. It would have applied to the full Standard Model Lagrangian if two conditions were met: $g' = 0$ and $Y^d = Y^u$. In this limit, the (W_1, W_2, W_3) DoF transform as a triplet under the custodial symmetry. Consequently, the mass terms induced by the $SU(2)_L \times U(1)_Y$ spontaneous symmetry breaking are equal for these 3 DoF.

The most general mass matrix in the (W_1, W_2, W_3, B) basis, which is consistent with $U(1)_{\rm EM}$ gauge invariance, is given by

$$\begin{pmatrix} m_W^2 & 0 & 0 & 0 \\ 0 & m_W^2 & 0 & 0 \\ 0 & 0 & m_Z^2 c_W^2 & m_Z^2 c_W s_W \\ 0 & 0 & m_Z^2 c_W s_W & m_Z^2 s_W^2 \end{pmatrix}, \tag{12.14}$$

where $s_W \equiv \sin\theta_W$ and $c_W \equiv \cos\theta_W$. The custodial symmetry requires that the top three diagonal terms be equal and thus that $m_Z^2 c_W^2 = m_W^2$ (namely, the $\rho = 1$ relation).

The custodial symmetry holds at tree level for models with any number of scalar doublets and singlets. It is, however, not a symmetry of the full Standard Model. In particular, it is broken by the Yukawa coupling since $Y^d \neq Y^u$ (see question 12.4). The largest breaking parameter, then, is $y_t^2 - y_b^2$. This breaking is communicated to the Higgs sector by loop effects, resulting in violation of the predictions that follow the custodial symmetry. This is the reason that the leading correction to the $\rho = 1$ relation is proportional to $m_t^2 - m_b^2$ (see equation (12.11)).

12.4 Probing BSM

Within the Standard Model, the EWPM program is sensitive at tree level to three input parameters and at the loop level to a few more. Since we have more observables than relevant Standard Model parameters, the EWPM can be used to test the Standard Model. So far, no significant deviation from the Standard Model was found. Furthermore, as already mentioned, all the relevant Standard Model parameters are now directly measured, and therefore the data can be used to constrain BSM physics.

As for the probing of BSM physics with EWPM, one can decide to do one of two options. First, one can consider a specific model, calculate the new contributions to the observables, and constrain the BSM parameters by comparing to the experimental results (just as we did for the Standard Model itself). In section 12.4.3, we will demonstrate this by a brief discussion of the four-generation extension of the Standard Model. Second, one can consider the effects of nonrenormalizable terms without committing to a specific model. Under reasonable assumptions, which will be specified next, there is only a small number of dimension-six terms in the Standard Model effective field theory (SMEFT) that affect the EWPM. In section 12.4.1, we discuss these operators and how they are constrained by EWPM.

12.4.1 *Nonrenormalizable Operators and the q^2 Expansion*

In general, there are many operators that affect the EWPM. However, the number of operators with potentially significant effects is much smaller within a large class of models that fulfill the following three conditions:

- The scale of the BSM physics is much higher than the electroweak-breaking scale, $\Lambda \gg m_W$. (This condition holds, by definition, for BSM physics whose effects can be represented by nonrenormalizable terms.)
- The BSM physics generates only oblique corrections to the relevant observables.
- There are no contributions from BSM vector bosons.

(The analysis given here applies to a broader class of models than implied by the third condition, but we assume that stronger condition for the sake of simplicity.)

The contributions to oblique corrections from nonrenormalizable terms can come from tree-level diagrams or loop diagrams. We consider here only tree-level BSM contributions. Consequently, our analysis concerns only operators that involve exactly two electroweak gauge fields and no fermions. In addition to the two electroweak gauge fields, the operators can have derivatives and Higgs fields. Tree-level contributions come from replacing the Higgs fields in these operators with their VEVs. There are no such dimension-five operators and thus the leading contributions (for $\Lambda \gg v$) come from dimension-six operators.

With these considerations, we learn that there are only four dimension-six terms that contribute to the oblique corrections:

$$\mathcal{L}_{\text{o.c.}} = \frac{1}{\Lambda^2}\left(c_{WB}\mathcal{O}_{WB} + c_{HH}\mathcal{O}_{HH} + c_{BB}\mathcal{O}_{BB} + c_{WW}\mathcal{O}_{WW}\right), \tag{12.15}$$

where C_{WB}, C_{HH}, C_{BB}, and C_{WW} are dimensionless coefficients, and the operators are defined as follows (note that in the research literature, a variety of normalizations are used):

$$\begin{aligned}
\mathcal{O}_{WB} &= (H^\dagger \tau_a H) W_a^{\mu\nu} B_{\mu\nu} \to \frac{v^2}{2} W_3^{\mu\nu} B_{\mu\nu}, \\
\mathcal{O}_{HH} &= |H^\dagger D^\mu H|^2 \to \frac{v^4}{16}(g W_3^\mu - g' B^\mu)^2, \\
\mathcal{O}_{BB} &= (\partial^\rho B^{\mu\nu})^2, \\
\mathcal{O}_{WW} &= (D^\rho W_a^{\mu\nu})^2,
\end{aligned} \tag{12.16}$$

where τ_a are the Pauli matrices, and the arrows correspond to replacing the Higgs field by its VEV.

In addition to the expansion in inverse powers of Λ, which led us to neglect SMEFT terms of dimension $d > 6$, we can expand in powers of q^2. This expansion leads to considerable simplification of the analysis. Since we deal with oblique corrections, we expand the vacuum polarization amplitudes $\Pi_{AB}(q^2)$ with $AB = \{+-, 33, 00, 30\}$:

$$\Pi_{AB}(q^2 \ll \Lambda^2) = \Pi_{AB}(0) + q^2 \Pi'_{AB}(0) + \frac{(q^2)^2}{2}\Pi''_{AB}(0) + \cdots. \tag{12.17}$$

Within the SMEFT, there are only two dimensionful parameters: v and Λ. The dependence of the various $\Pi_{AB}^{(n)} \equiv d^n \Pi_{AB}/(dq^2)^n$ on v comes from replacing the Higgs field with its VEV, and thus they can depend only on positive powers of v. On the other hand, the various $\Pi_{AB}^{(n)}$ depend only on inverse powers of Λ. For example, the contributions of dimension-six terms are proportional to $1/\Lambda^2$. Since Π_{AB} is dimension-two (and $\Pi_{AB}^{(n)}$ is dimension $2-2n$) in mass, we can have only the following dependencies on v and Λ of the various terms in the q^2 expansion:

$$\begin{aligned}
&\Pi_{AB}(0) \propto v^2,\ v^4/\Lambda^2, \cdots, \\
&\Pi'_{AB}(0) \propto 1,\ v^2/\Lambda^2, \cdots, \\
&\Pi''_{AB}(0) \propto 1/\Lambda^2, \cdots, \\
&\Pi_{AB}^{(n)}(0) \propto 1/\Lambda^{2n-2},\ v^2/\Lambda^{2n}, \cdots .
\end{aligned} \tag{12.18}$$

Since we neglect terms with a dimension higher than six, we can truncate the q^2 expansion at order $(q^2)^2$: Terms of order $(q^2)^3$ and higher are suppressed by at least $1/\Lambda^4$. One more observation from these findings is that the dimension-six operators affect $\Pi(0)$, $\Pi'(0)$, and $\Pi''(0)$. Specifically, examining equation (12.16), we learn that $\mathcal{O}_{WB}$ contributes to $\Pi'_{30}(0)$; $\mathcal{O}_{HH}$ contributes to $\Pi_{33}(0)$, $\Pi_{00}(0)$, and $\Pi_{30}(0)$; $\mathcal{O}_{BB}$ contributes to $\Pi''_{00}(0)$; and $\mathcal{O}_{WW}$ contributes to $\Pi''_{33}(0)$ and $\Pi''_{+-}(0)$.

We note that the q^2 expansion of equation (12.7) applies to any model, but only to low-energy observables, $q^2 \ll m_W^2$. The q^2 expansion of equation (12.18) applies to all observables, but only to BSM physics models that are characterized by a scale that is much higher than the electroweak scale, $\Lambda \gg v$.

12.4.2 The S, T, and U Parameters

To study the contributions of dimension-six terms to oblique corrections, we need to keep terms up to Π'' in the q^2 expansion. Thus, the four functions, $\Pi_{AB}(q^2)$, are replaced by 12 numbers, $\Pi_{AB}(0)$, $\Pi'_{AB}(0)$, and $\Pi''_{AB}(0)$. Out of these numbers, two combinations vanish due to the $U(1)_{\rm EM}$ gauge invariance, $\Pi_{\gamma\gamma}(0)=\Pi_{\gamma Z}(0)=0$. Three other combinations are fixed by the three tree-level Standard Model parameters. The remaining seven parameters are then related to observables and can be fitted. The seven parameters are usually called S, T, U, V, W, X, and Y. From the theoretical side, one can obtain the contributions of the four dimension-six terms to these parameters and finally constrain their coefficients in the Lagrangian.

To provide a simple demonstration of the relation to observables, we truncate the expansion at order q^2. This leaves only three parameters that neither vanish nor are fixed by the tree-level parameters. These are the S, T, and U parameters, defined as follows:

$$\begin{aligned}
\alpha T &= \frac{\Pi_{WW}(0)}{m_W^2} - \frac{\Pi_{ZZ}(0)}{m_Z^2}, \\
\frac{\alpha S}{\sin^2 2\theta} &= \Pi'_{ZZ}(0) - \frac{2\cos 2\theta}{\sin 2\theta}\Pi'_{Z\gamma}(0) - \Pi'_{\gamma\gamma}(0),
\end{aligned}$$

$$\frac{\alpha U}{4\sin^2\theta} = \Pi'_{WW}(0) - \cos^2\theta\, \Pi'_{ZZ}(0) - \sin 2\theta\, \Pi'_{\gamma Z}(0) - \sin^2\theta\, \Pi'_{\gamma\gamma}(0). \quad (12.19)$$

They are related to the differences between $\sin^2\theta_0$, $\sin^2\theta_W$, and $\sin^2\theta_*$, presented in equation (12.9), as follows:

$$\sin^2\theta_* - \sin^2\theta_0 = \frac{\alpha}{4\cos 2\theta}\left(S - \sin^2 2\theta\, T\right),$$
$$\sin^2\theta_0 - \sin^2\theta_W = \frac{\alpha\cos^2\theta}{\cos 2\theta}\left(-\frac{1}{2}S - \cos^2\theta\, T + \frac{\cos 2\theta}{4\sin^2\theta}U\right). \quad (12.20)$$

A few comments are in order at this point:

- The S, T, and U parameters receive one-loop contributions in the Standard Model, and possibly BSM contributions as well. One can subtract the Standard Model values and redefine them such that within the Standard Model, they vanish. With this new definition, a nonzero value would be a sign of BSM physics. The values presented next correspond to this definition.
- S, T, and U are pure numbers. Their scaling by α^{-1} relative to Π_{AB} (or Π'_{AB}) means that they roughly reflect the allowed size of BSM physics contributions compared to the Standard Model one-loop contributions.
- T represents custodial symmetry–breaking contributions. The deviation from unity of the low-energy ratio of charged- to neutral-current amplitudes, denoted in the literature by $\rho_*(0)$, is a purely custodial symmetry–breaking observable and is proportional to T.
- S parameterizes new physics contributions of order q^2/Λ^2 and is custodial symmetry invariant. The deviation of the wave function renormalization constant of the Z-boson (denoted in the literature by Z_{Z*}) from unity is proportional to S.
- U represents custodial symmetry–breaking contributions of order q^2/Λ^2 and is very small in many BSM models. In particular, within the SMEFT, U arises from a dimension-eight operator, while S and T arise from dimension-six operators. Thus, the U parameter rarely provides a significant constraint.

The EWPM determine the allowed ranges for these parameters. The strongest bounds arise from S and T, but there are experimental correlations among them that must be taken into account (see figure 12.2). Setting $U=0$, motivated by the fact that in the SMEFT, U is suppressed by an extra factor of v^2/Λ^2 compared to T and S, the current experimental bounds read as

$$S = +0.00 \pm 0.07, \qquad T = +0.05 \pm 0.06. \quad (12.21)$$

The contributions of the dimension-six operators to S and T are given by

$$S = \frac{2\sin 2\theta}{\alpha}\frac{v^2}{\Lambda^2}c_{WB}, \qquad T = -\frac{1}{2\alpha}\frac{v^2}{\Lambda^2}c_{HH}, \quad (12.22)$$

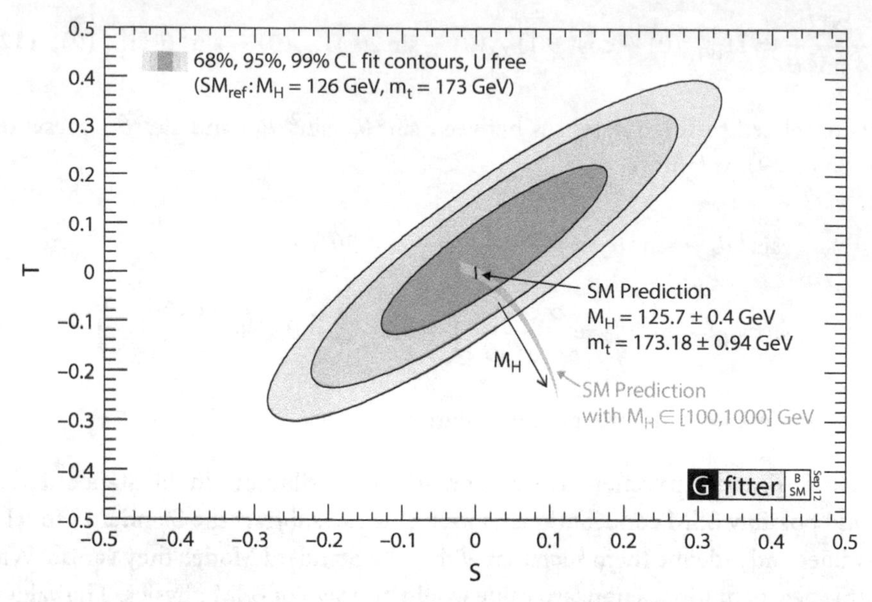

FIGURE 12.2. The current bounds on S and T. Taken from the Gfitter Group [39].

where c_{WB} and c_{HH} are defined in equation (12.15). We note that T is proportional to c_{HH}, which is the only custodial symmetry–breaking parameter among the four couplings of equation (12.15). Using the experimental upper bounds of equation (12.21) as reference points gives

$$\frac{\Lambda}{\sqrt{|c_{WB}|}} \gtrsim 14 \text{ TeV} \sqrt{\frac{0.07}{|S|}}, \qquad \frac{\Lambda}{\sqrt{|c_{HH}|}} \gtrsim 6 \text{ TeV} \sqrt{\frac{0.11}{|T|}}. \tag{12.23}$$

The lesson to draw from this discussion is that EWPM currently have an accuracy of an order of 10 percent of the Standard Model one-loop corrections, and consequently, they probe BSM up to a scale that is roughly two orders of magnitude above the weak scale.

12.4.3 The Four-Generation Standard Model

The four-generation Standard Model is excluded by several measurements, such as by the invisible width of the Z-boson (see question 8.3 in chapter 8). Yet it provides a simple example of constraining a specific BSM physics model by the EWPM.

Here, we consider a four-generation model where the fourth neutrino is massive, without providing an explicit mechanism for generating its mass. We further assume that all the fourth-generation fermions are heavier than the top quark and the splitting within each multiplet is small. We further assume no significant flavor mixing between the fourth generation and the three known ones. Under these assumptions, we have

$$S = \frac{1}{2\pi}\left[1 - \frac{1}{3}\log\left(\frac{m_{u_4}^2}{m_{d_4}^2}\right)\right] + \frac{1}{6\pi}\left[1 + \log\left(\frac{m_{\nu_4}^2}{m_{e_4}^2}\right)\right],$$

$$T = \frac{1}{2\pi \sin^2(2\theta_W)} \left(\frac{m_{u4}^2 - m_{d4}^2}{m_Z^2} \right) + \frac{1}{6\pi \sin^2(2\theta_W)} \left(\frac{m_{\nu 4}^2 - m_{e4}^2}{m_Z^2} \right), \quad (12.24)$$

where m_{f_4} are the masses of the fourth-generation fermions. We learn that T is related to the mass splitting between t_4 and b_4 and between ν_4 and e_4, which are the custodial symmetry–breaking parameters in this model. In the case of a degenerate quark doublet and a degenerate lepton doublet, we have $T=0$ and $S=2/(3\pi)$. In a way, the S parameter "counts" the number of extra generations. The current measurement of S by itself excludes an extra degenerate generation (so that $T=U=0$) at the 7σ level. Yet, for nondegenerate fermions, when the correlation with T is taken into account, the bounds on extra generations are less severe.

For Further Reading

- A summary of the relevant EWPM observables and the relevant data can be found in the Particle Data Group (PDG), Workman et al. [1].
- Reviews on EWPM can be found, for example, in the PDG review of the Standard Model, Workman et al. [1], and in Skiba [40].
- Read more about custodial symmetry in Willenbrock [41] and Logan [42]. For a review of the screening theorem see, for example, Gunion, Haber, Kane, and Dawson [43].

Problems

Question 12.1: Algebra

The oblique corrections to various observables are given by the following expressions:

$$\begin{aligned}
m_Z^2 &= (g^2 + g'^2)(v^2/4) + \Pi_{ZZ}(m_Z^2),\\
m_W^2 &= g^2(v^2/4) + \Pi_{WW}(m_W^2),\\
4\pi\alpha &= [g^2 g'^2/(g^2 + g'^2)][1 + \Pi'_{\gamma\gamma}(0)],\\
G_F/\sqrt{2} &= [1/(2v^2)][1 - \Pi_{WW}(0)/m_W^2],\\
g_V^f &= T_3^{L,f} - 2Q_f\{[g'^2/(g^2 + g'^2)] - (e/\sqrt{g^2 + g'^2})[\Pi_{\gamma Z}(m_Z^2)/m_Z^2]\}.
\end{aligned} \quad (12.25)$$

Derive equation (12.9) from this set of equations and the definitions in equations (12.2), (12.3), and (12.4).

Question 12.2: Extracting $\sin^2\theta_*$

To extract θ_*, we need to measure processes that depend on the couplings of the Z-boson to fermions. There are many such processes. Here, we consider two of these, the ratio of decay rates,

$$R_\ell \equiv \frac{\Gamma(Z \to \ell^+\ell^-)}{\Gamma(Z \to \nu\bar{\nu})}, \tag{12.26}$$

where ℓ is a charged fermion, and the forward-backward asymmetry in $e^+e^- \to \mu^+\mu^-$ scattering close to the Z-pole:

$$A^\mu_{\rm FB} \equiv \frac{\sigma^\mu_F - \sigma^\mu_B}{\sigma^\mu_F + \sigma^\mu_B}, \qquad \sigma^\mu_F = 2\pi \int_0^1 d(\cos\theta)\,\frac{d\sigma^\mu}{d\cos\theta}, \qquad \sigma^f_B = 2\pi \int_{-1}^0 d(\cos\theta)\,\frac{d\sigma^\mu}{d\cos\theta}, \tag{12.27}$$

where θ is the angle between the incoming e^- and the outgoing μ. You are asked to show that a combination of the two can be used to extract θ_*, as defined in equation (12.4).

1. Show that, neglecting the small phase-space effects and working at tree level,

$$R_\ell = 2\left[(g_V^\ell)^2 + (g_A^\ell)^2\right]. \tag{12.28}$$

2. Show that, neglecting the photon-mediated contribution and considering only the Z-mediated contribution at tree level,

$$A^\mu_{\rm FB} = \frac{3}{4} A_e A_\mu, \qquad A_\ell = \frac{2 g_V^\ell g_A^\ell}{(g_V^\ell)^2 + (g_A^\ell)^2}. \tag{12.29}$$

3. Find $g_V^\ell - g_A^\ell$ in terms of R_ℓ and $A^\mu_{\rm FB}$, and then use equation (12.4) to obtain $\sin^2\theta_*$.

Question 12.3: Predicting m_t

Here, we demonstrate how m_t had been predicted using EWPM data before the top quark was discovered.

1. Using equation (12.11), find the expression for $\sin^2\theta_0 - \sin^2\theta_W$.
2. Use the central values of G_F, α, m_Z, m_W, and m_b to determine $\sin^2\theta_0 - \sin^2\theta_W$ and from that, find m_t.
3. Compare your result of m_t to the direct measurement. Comment on any differences you find between the two.

Question 12.4: Custodial symmetry

Consider the Higgs sector of the Standard Model in the $g' \to 0$ limit:

$$\mathcal{L}_\phi = (D_\mu\phi)^\dagger (D^\mu\phi) - V_\phi, \tag{12.30}$$

where

$$D_\mu\phi = \left(\partial_\mu + i\frac{g}{2}\sigma^a W^a_\mu\right)\phi, \qquad V_\phi = \mu^2(\phi^\dagger\phi) + \lambda(\phi^\dagger\phi)^2. \tag{12.31}$$

1. Show that V_ϕ is invariant under an $SO(4)$ symmetry, where the four components of the Higgs field as defined in equation (12.13) transform as 4.

At the algebra level, $SO(4) \sim SU(2)_L \times SU(2)_R$. Here, $SU(2)_L$ is the global version of the Standard Model gauge group with the same name. Under $SU(2)_L \times SU(2)_R$, the Higgs field transforms as $(2, 2)$. Then we write it as a matrix:

$$\Phi = \frac{1}{\sqrt{2}}\left(\widetilde{\phi} \quad \phi\right) = \frac{1}{\sqrt{2}}\begin{pmatrix} \phi_3 - i\phi_4 & \phi_1 + i\phi_2 \\ -\phi_1 + i\phi_2 & \phi_3 + i\phi_4 \end{pmatrix}, \tag{12.32}$$

where $\widetilde{\phi}$ is defined in equation (8.14).

2. Show that $\widetilde{\phi}^\dagger \widetilde{\phi} = \phi^\dagger \phi$, and use that result to show that

$$\mathrm{Tr}\left[\Phi^\dagger \Phi\right] = \phi^\dagger \phi. \tag{12.33}$$

3. Show that $\mathcal{L}_\phi$, defined in equation (12.30), can be written as

$$\mathcal{L}_\phi = \mathrm{Tr}\left[(D_\mu \Phi)^\dagger (D^\mu \Phi)\right] - V_\Phi, \tag{12.34}$$

where

$$D_\mu \Phi = \partial_\mu \Phi + i\frac{g}{2}\sigma \cdot W_\mu \Phi, \qquad V_\Phi = \mu^2\left(\mathrm{Tr}\left[\Phi^\dagger \Phi\right]\right) + \lambda\left(\mathrm{Tr}\left[\Phi^\dagger \Phi\right]\right)^2. \tag{12.35}$$

4. Under $SU(2)_L \times SU(2)_R$, the W-boson field transforms as $(3, 1)$: $W^\mu \to LW^\mu L^\dagger)$, while the Φ field transforms as $\Phi \to L\Phi R^\dagger$, where L and R are group elements of $SU(2)_L$ and $SU(2)_R$, respectively. Using these transformation properties, show that $\mathcal{L}_\phi$ is invariant under $SU(2)_L \times SU(2)_R$.
5. We now consider SSB. We assign the VEV to ϕ_3. Show that the $SO(4)$ symmetry of V_ϕ is broken down to $SO(3)$, with (ϕ_1, ϕ_2, ϕ_4) transforming as a triplet (3).
6. Using

$$\langle\Phi\rangle = \frac{1}{2}\begin{pmatrix} v & 0 \\ 0 & v \end{pmatrix}, \tag{12.36}$$

show that

$$L\langle\Phi\rangle \neq \langle\Phi\rangle \qquad \langle\Phi\rangle R^\dagger \neq \langle\Phi\rangle, \tag{12.37}$$

so that both $SU(2)_L$ and $SU(2)_R$ are broken.

7. Show that

$$L\langle\Phi\rangle L^\dagger = \langle\Phi\rangle. \tag{12.38}$$

Thus, $\mathcal{L}_\phi$ is invariant under simultaneous $SU(2)_L$ and $SU(2)_R$ transformations with $L = R$, such that an $SU(2)_D$ subgroup remains unbroken. This $SU(2)_D$ is identified with the unbroken $SO(3)$ of the Higgs potential discussed in item 5.

8. Show that (W_1, W_2, W_3) forms a triplet under $SU(2)_D$. This implies that they have the same mass, and then it leads to the $\rho = 1$ relation.

There are two sources of explicit breaking of the $SO(4)$ symmetry: the hypercharge coupling and the Yukawa couplings. We discuss them in turn next.

9. Show that $U(1)_Y$ is a subgroup of $SU(2)_R$. To do this, show that under $U(1)_Y$,

$$\Phi \to \Phi e^{-\frac{i}{2}\sigma_3\theta}, \qquad (12.39)$$

and compare it to the transformation of Φ under $SU(2)_R$.

10. Argue that, because $U(1)_Y$ is gauged while the rest of $SU(2)_R$ is not, $g' \neq 0$ implies the breaking of $SU(2)_R$ and of $SU(2)_D$.
11. Assign Q_L to $(2,1)$ and $Q_R = (U_R, D_R)$ to $(1,2)$ of $SU(2)_L \times SU(2)_R$. Show that

$$\mathcal{L}_{\text{Yuk}} = Y\overline{Q_L}\Phi\, Q_R \qquad (12.40)$$

is invariant under $SU(2)_L \times SU(2)_R$.

12. Show that equation (12.40) is the same as the quark terms in $\mathcal{L}_{\text{Yuk}}$ of the Standard Model (equation (8.16) in chapter 8) when $Y^u = Y^d = Y$.
13. Argue that, since $Y^u \neq Y^d$, $SU(2)_R$ and $SU(2)_D$ are explicitly broken.
14. Argue that any loop correction to the $\rho = 1$ relation must be proportional to either g' or to the mass splitting between the up- and down-sector fermions.

13

Flavor-Changing Neutral Currents

In this chapter, we discuss flavor-changing neutral currents (FCNCs) and *CP* violation in the quark sector. These processes are sensitive to loop effects and nonrenormalizable operators. The structure of FCNCs is affected by the hierarchies in the quark masses and in the Kobayashi-Maskawa (CKM) mixing angles, and thus it is far from generic.

13.1 Introduction

Electroweak precision measurements (EWPM), discussed in chapter 12, test the Standard Model and probe Beyond the Standard Model (BSM) physics. To achieve these goals, one goes through the following steps:

1. Calculating loop corrections to relevant processes
2. Checking the consistency of the various observables within the Standard Model
3. Obtaining the allowed ranges for nonrenormalizable terms

Similarly, FCNCs test the Standard Model and probe BSM physics. The two classes of processes—EWPM and FCNCs—are sensitive to different sectors of the Standard Model. Thus, they probe different classes of dimension six operators, and the information extracted from them is complementary. In both cases, the experimental sensitivity is to BSM contributions that are of a comparable size to the corresponding Standard Model one-loop contributions. Yet the FCNC processes can probe new physics at higher scales because, in addition to the loop suppression (see section 9.4 in chapter 9), the Standard Model contributions are suppressed by small flavor parameters, as discussed in section 13.2.

There are additional interesting differences between EWPM and FCNCs. First, while for EWPM the leading contributions are at tree level and loop contributions provide small corrections, for FCNC there are no tree-level contributions and the loop diagrams constitute the leading contributions. Second, while for EWPM the Quantum ChromoDynamics (QCD) corrections are small and the calculations very precise, for FCNCs the QCD effects are significant, leading to large hadronic uncertainties. Thus, for EWPM, precise calculations are necessary and possible. For FCNCs, precise calculations are desirable but less crucial, and at the same time harder to achieve.

The fact that FCNCs can probe physics at scales that are much higher than the energy scale of the relevant experiments has been demonstrated several times in the history of particle physics. FCNC processes have played an important role in predicting the existence of Standard Model particles before they were directly discovered, as well as in predicting their masses:

- The smallness of $\Gamma(K_L \to \mu^+\mu^-)$ led to predicting the charm quark (see question 9.6 in chapter 9).
- The size of the mass difference in the neutral kaon system, Δm_K, led to a successful prediction of the charm mass (see question 13.2 at the end of this chapter).
- The measurement of *CP* violation in kaon decays led to predicting the third generation (see appendix 13.C.3).
- The size of the mass difference in the neutral *B* system, Δm_B, led to a successful prediction of the top mass (see question 13.2).

As mentioned here and discussed in section 9.4 in chapter 9, FCNC processes cannot be mediated at tree level in the Standard Model. Yet since there is no symmetry that forbids them, they are mediated at the loop level. Concretely, W-mediated interactions lead to FCNCs at the one-loop level. Since the W-boson couplings are charged current flavor changing, an even number of insertions of W-boson couplings are needed to generate an FCNC process. We consider two classes of FCNCs based on the change in F (the charge under the global $[U(1)]^6$ flavor symmetry of the QCD Lagrangian):

- FCNC decays ($\Delta F = 1$ processes) have two insertions of W-couplings.
- Neutral meson mixings ($\Delta F = 2$ processes) have four insertions of W-couplings.

13.2 CKM and GIM Suppression in FCNC Decays

In this section, we discuss FCNC meson decays, which are $\Delta F = 1$ processes. To demonstrate the generic features of one-loop FCNCs, we consider the example of $s \to d$ transitions. Since the change of flavor quantum numbers is $\Delta s = -\Delta d = 1$, this transition belongs to the class of $\Delta F = 1$ processes. The FCNC part of any process that involves $s \to d$ transition is plotted in figure 13.1(a). In figure 13.1(b), for example, we show a full diagram that contributes to the decay $K \to \pi\nu\bar{\nu}$ and includes figure 13.1(a) as a subdiagram. (Diagrams with such a topology are usually called *penguin diagrams*.)

By inspecting the diagram in figure 13.1(a), we learn that its flavor structure is given by

$$\mathcal{A}_{s\to d} \sim \sum_{i=u,c,t} (V_{is}V_{id}^*) f(x_i), \qquad x_i = \frac{m_i^2}{m_W^2}, \tag{13.1}$$

where $f(x_i)$ depends on the specific decay. CKM unitarity implies

$$V_{ud}^* V_{us} + V_{cd}^* V_{cs} + V_{td}^* V_{ts} = 0. \tag{13.2}$$

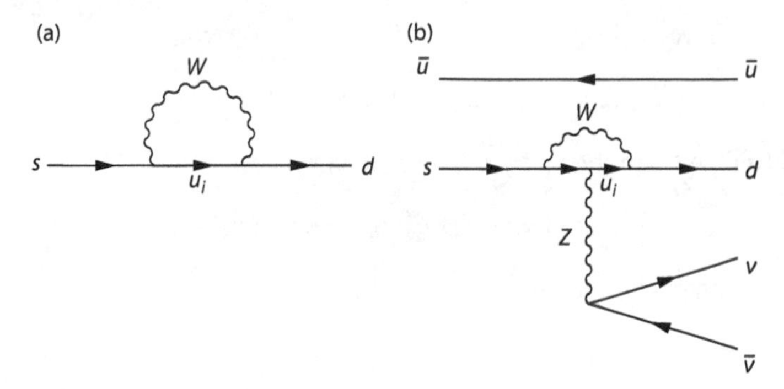

FIGURE 13.1. (a) One-Loop diagrams for a $\Delta F = 1$ $s \to d$ FCNC. (b) A one-loop diagram that contributes to $K^- \to \pi^- \nu\bar{\nu}$.

We can then use this unitarity condition to eliminate one of the three CKM terms in the sum in equation (13.1). We choose to eliminate the u-term and write

$$\mathcal{A}_{s\to d} \sim \sum_{i=c,t} [f(x_i) - f(x_u)]\, V_{is} V_{id}^*. \tag{13.3}$$

We draw the following lessons from these findings:

- The contribution of the m_i-independent terms in $f(x_i)$ to $\mathcal{A}_{s\to d}$ vanishes when summed over all internal quarks.
- $\mathcal{A}_{s\to d}$ would vanish if the up-type quarks were all degenerate, and therefore, it must depend on the mass-splittings among the up-type quarks.

The explicit dependence on the mass-splittings among the quarks depends on the process. In many cases, for small x_i we have

$$f(x_i) \sim x_i. \tag{13.4}$$

Using this crude approximation, we can write

$$\mathcal{A}_{s\to d} \sim \left[(x_t - x_u) V_{ts} V_{td}^* + (x_c - x_u) V_{cs} V_{cd}^*\right]. \tag{13.5}$$

Inspecting equation (13.5), we identify two suppression factors:

- *CKM suppression:* The amplitude is proportional to at least one off-diagonal CKM matrix element. Given the specific structure of the CKM matrix, off-diagonal elements are small. Specifically, $|V_{ts} V_{td}^*| \sim \lambda^5$ and $|V_{cs} V_{cd}^*| \sim \lambda$, where λ is the Wolfenstein parameter defined in equation (9.7).
- *GIM suppression:* The amplitude is proportional to mass differences between the up-type quarks. In particular, $(x_c - x_u) \sim (m_c/m_W)^2$. The suppression by factors of small quark masses is called the Glashow-Iliopoulos-Maiani (GIM) mechanism.

While we derive the results based on one specific example, the CKM and GIM suppressions play a role in all FCNC processes. For the other FCNCs in the down sector, $b \to q$

with $q = d, s$, we have

$$\mathcal{A}_{b\to q} \sim (x_t - x_u)V_{tb}V_{tq}^* + (x_c - x_u)V_{cb}V_{cq}^*. \tag{13.6}$$

With regard to FCNCs in the up sector, for $c \to u$, we have

$$\mathcal{A}_{c\to u} \sim (x_b - x_d)V_{cb}^*V_{ub} + (x_s - x_d)V_{cs}^*V_{us}, \tag{13.7}$$

and, for $t \to q$ with $q = u, c$, we have

$$\mathcal{A}_{t\to q} \sim (x_b - x_d)V_{tb}^*V_{qb} + (x_s - x_d)V_{ts}^*V_{qs}. \tag{13.8}$$

The CKM suppression applies to FCNC decay rates. It does not necessarily apply, however, to the corresponding branching ratios. The reason is that branching ratios depend on the ratio between the FCNC decay rates and the full decay width which, in the down sector, is CKM suppressed, and thus the ratio of CKM factors is not necessarily small. In particular, the leading flavor-changing charged current (FCCC) K decay rate is suppressed by $|V_{us}| \sim |V_{cs}V_{cd}|$ and the leading (FCCC) B decay rate is suppressed by $|V_{cb}| \sim |V_{tb}V_{ts}|$.

Several remarks are in order:

- While the $f(x_i) \sim x_i$ approximation is not valid for the top quark, it gives a reasonable order of magnitude estimate, and we use it for the purpose of demonstration. For example, while $x_t/x_c \sim 10^4$, we have for $f(x)$ defined in equation (13.11), $f(x_t)/f(x_c) \approx 10^3$.
- The exact form of the dependence on the mass splitting is process dependent, but in all cases, the amplitude vanishes when the internal quarks are degenerate. We refer to quadratic dependence $[x_i - x_j]$ as *hard GIM*, and to logarithmic dependence $[\log(x_i/x_j)]$ as *soft GIM*.
- The size of FCNC amplitudes increases with the mass of the internal quark. The reason that this does not violate the decoupling theorem is the same as discussed in the EWPM case (see section 12.2.2 in chapter 12).

13.2.1 Examples: $K \to \pi\nu\bar{\nu}$ and $B \to \pi\nu\bar{\nu}$

As examples of $\Delta F = 1$ processes, we consider the semileptonic decays:

$$K^+ \to \pi^+\nu\bar{\nu}, \qquad B^+ \to \pi^+\nu\bar{\nu}, \tag{13.9}$$

which proceed via $\bar{s} \to \bar{d}\nu\bar{\nu}$ and $\bar{b} \to \bar{d}\nu\bar{\nu}$ transitions.

Consider the following ratios of FCNC-to-FCCC semileptonic decay rates:

$$R_{K\pi} = \frac{\Gamma(K^+ \to \pi^+\nu\bar{\nu})}{\Gamma(K^+ \to \pi^0 e^+\nu)} = \frac{3}{2}\frac{g^4}{16\pi^2}\left|\frac{V_{ts}^*V_{td}[f(x_t) - f(x_u)] + V_{cs}^*V_{cd}[f(x_c) - f(x_u)]}{V_{us}}\right|^2, \tag{13.10}$$

$$R_{B\pi} = \frac{\Gamma(B^+ \to \pi^+\nu\bar{\nu})}{\Gamma(B^+ \to \pi^0 e^+\nu)} = \frac{3}{2}\frac{g^4}{16\pi^2}\left|\frac{V_{tb}^*V_{td}[f(x_t) - f(x_u)] + V_{cb}^*V_{cd}[f(x_c) - f(x_u)]}{V_{ub}}\right|^2,$$

where

$$f(x) = \frac{x}{8}\left[\frac{2+x}{1-x} - \frac{3x-6}{(1-x)^2}\log x\right]. \tag{13.11}$$

For small x, we have

$$f(x \ll 1) \approx \frac{x(3\log x + 1)}{4}. \tag{13.12}$$

Since $f(x_u) \ll 1$, we have to a very good approximation $f(x_i) - f(x_u) \approx f(x_i)$ for $i = c, t$.

A few comments are in order with regard to equation (13.10):

- The factor of 3 comes from summing over the neutrino flavors, while the factor of $1/2$ is an isospin factor between the $P^+ \to \pi^+$ and $P^+ \to \pi^0$ ($P = K, B$) transitions.
- The $g^4/16\pi^2$ is the loop suppression factor.
- For $R_{K\pi}$, the t-term is CKM-suppressed, $|V_{ts}V_{td}/V_{us}| \sim \lambda^4$, but not GIM-suppressed, $f(x_t) = \mathcal{O}(1)$. The c-term is GIM-suppressed, $f(x_c) = \mathcal{O}(m_c^2/m_W^2)$, but not CKM-suppressed, $|V_{cs}V_{cd}/V_{us}| \simeq 1$. The two terms contribute comparably.
- For $R_{B\pi}$, the t-term is neither CKM-suppressed, $|V_{tb}V_{td}/V_{ub}| = \mathcal{O}(1)$, nor GIM-suppressed, $f(x_t) = \mathcal{O}(1)$. The c-term is not CKM-suppressed, $|V_{cb}V_{cd}/V_{ub}| = \mathcal{O}(1)$, but it is GIM-suppressed, $f(x_c) = \mathcal{O}(m_c^2/m_W^2)$. Thus, the contribution of the c-term is negligible.
- As a result of the different CKM and GIM suppression factors, we obtain numerically very different predictions:

  $$R_{K\pi} \sim 10^{-9}, \qquad R_{B\pi} \sim 10^{-4}. \tag{13.13}$$

  These predictions have not been fully tested yet, as we have only experimental upper bounds, $R_{K\pi} \lesssim 10^{-8}$ and $R_{B\pi} \lesssim 0.18$.

The comparison of $R_{K\pi}$ and $R_{B\pi}$ demonstrates how the CKM and GIM suppression factors depend crucially on the specific quarks involved and how they come into play in determining the various FCNC rates. While for the two specific examples presented here, there are only experimental upper bounds, many FCNC decays have been observed and their rates measured. To date, all measured FCNC decay rates in the quark sector agree with the Standard Model predictions.

13.3 CKM and GIM Suppression in Neutral Meson Mixing

In this section, we discuss neutral meson mixing, which is a $\Delta F = 2$ process. This phenomenon is observed and measured via meson oscillations, as discussed in Appendix 13.A. In Appendix 13.A.4, we present the explicit Standard Model calculation. In this section, we show that the general lessons learned from $\Delta F = 1$ processes about the loop, CKM, and GIM suppression factors of FCNC mostly carry over to $\Delta F = 2$ processes.

To demonstrate the features of $\Delta F = 2$ FCNC processes, we consider the example of $K^0 - \overline{K}^0$ mixing. It is generated by the $s\bar{d} \to d\bar{s}$ transition, which is a $\Delta s = -\Delta d = 2$

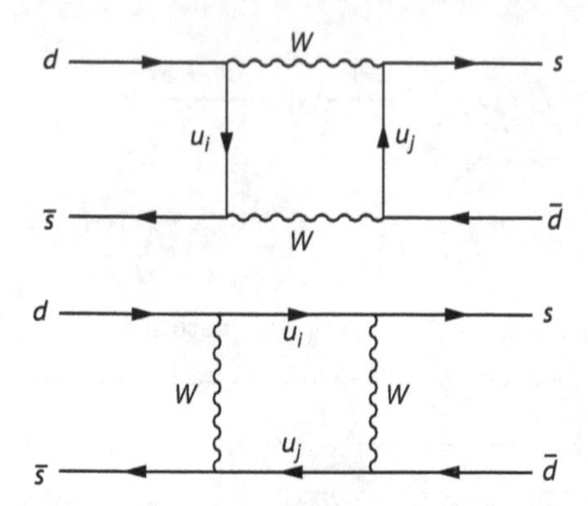

FIGURE 13.2. The one-loop diagrams for $\Delta F = 2$ FCNC.

process. The leading diagrams for this transition are plotted in figure 13.2. By inspecting this diagram, we learn that its flavor structure is given by

$$\mathcal{A}_{s\bar{d}\to d\bar{s}} \sim \sum_{i,j=u,c,t} (V_{is}V_{id}^* V_{js}V_{jd}^*)\, S(x_i, x_j), \tag{13.14}$$

where $S(x_j, x_i)$ $[x_i \equiv m_i^2/m_W^2]$ is given explicitly in equation (13.68). We draw the following lessons from these findings:

- The contribution of the m_i-independent terms in $S(x_i, x_j)$ to $\mathcal{A}_{s\bar{d}\to d\bar{s}}$ vanishes when summed over all internal quarks.
- $\mathcal{A}_{s\bar{d}\to d\bar{s}}$ would vanish if the up-type quarks were all degenerate, and therefore, it must depend on the mass-splittings among the up-type quarks.

To proceed, we use the unitarity condition of equation (13.2) and approximate $x_u = 0$ to eliminate the u-terms in the sum. We obtain

$$\mathcal{A}_{s\bar{d}\to d\bar{s}} \sim (V_{cs}V_{cd}^*)^2 S(x_c, x_c) + 2V_{cs}V_{cd}^* V_{ts}V_{td}^* S(x_c, x_t) + (V_{ts}V_{td}^*)^2 S(x_t, x_t). \tag{13.15}$$

We conclude that $\Delta F = 2$ amplitudes have, in addition to the loop suppression factor, the following suppression factors:

- *CKM suppression:* The amplitude is proportional to a least two off-diagonal CKM matrix elements.
- *GIM suppression:* The amplitude depends on the mass differences between the up-type quarks.

While we derive the results based on one example, the CKM and GIM suppressions play a role in the mixing of all neutral mesons. In fact, there are three more $\Delta F = 2$ amplitudes that we should consider:

$$\mathcal{A}_{b\bar{d}\to d\bar{b}} \sim (V_{cb}V_{cd}^*)^2 S(x_c, x_c) + 2V_{cb}V_{cd}^* V_{tb}V_{td}^* S(x_c, x_t) + (V_{tb}V_{td}^*)^2 S(x_t, x_t),$$

$$\mathcal{A}_{b\bar{s}\to s\bar{b}} \sim (V_{cb}V_{cs}^*)^2 S(x_c, x_c) + 2V_{cb}V_{cs}^*V_{tb}V_{ts}^* S(x_c, x_t) + (V_{tb}V_{ts}^*)^2 S(x_t, x_t),$$

$$\mathcal{A}_{c\bar{u}\to u\bar{c}} \sim (V_{cs}^*V_{us})^2 S(x_s, x_s) + 2V_{cs}^*V_{us}V_{cb}^*V_{ub} S(x_s, x_b) + (V_{cb}^*V_{ub})^2 S(x_b, x_b), \tag{13.16}$$

which correspond to $B^0 - \overline{B}^0$, $B_s^0 - \overline{B_s}^0$, and $D^0 - \overline{D}^0$ mixing, respectively.

13.3.1 Examples: Δm_K, Δm_B, and Δm_{B_s}

The hadronic process of $K^0 - \overline{K}^0$ mixing proceeds via the $s\bar{d} \to d\bar{s}$ quark transition and leads to the mass-splitting Δm_K between the two neutral kaon mass eigenstates. The Standard Model calculation gives (see equation (13.73))

$$\frac{|\Delta m_K|}{m_K} = \frac{g^4}{96\pi^2}\frac{m_K^2}{m_W^2}\frac{B_K f_K^2}{m_K^2}\Big|(V_{cs}^*V_{cd})^2 S(x_c, x_c) + 2V_{cs}^*V_{cd}V_{ts}^*V_{td} S(x_c, x_t)$$
$$+(V_{ts}^*V_{td})^2 S(x_t, x_t)\Big|. \tag{13.17}$$

To estimate the relative size of the three terms, we note that

$$\frac{|V_{ts}^*V_{td}|}{|V_{cs}^*V_{cd}|} \sim 10^{-3}, \qquad \frac{S(x_c, x_t)}{S(x_c, x_c)} \sim 10, \qquad \frac{S(x_t, x_t)}{S(x_c, x_c)} \sim 10^4. \tag{13.18}$$

We conclude that the contributions of the terms proportional to $S(x_t, x_t)$ are smaller by a factor of $\mathcal{O}(100)$ than the contribution of the $S(x_c, x_c)$ term and can thus be neglected:

$$\frac{\Delta m_K}{m_K} \approx \frac{B_K f_K^2}{m_K^2} \times \frac{g^4}{96\pi^2} \times \frac{m_K^2}{m_W^2} \times |V_{cs}V_{cd}|^2 \times \frac{m_c^2}{m_W^2}. \tag{13.19}$$

The $B_K f_K^2/m_K^2$ factor encodes the QCD hadronic matrix element. The m_K^2/m_W^2 factor is related to the fact that the flavor-changing processes are W-mediated. This factor also exists in FCCC tree-level processes. The other three factors are the following:

- The $g^4/(96\pi^2)$ factor represents one-loop suppression.
- The $|V_{cs}V_{cd}|^2$ factor represents CKM suppression.
- The m_c^2/m_W^2 factor represents GIM suppression.

The $B^0 - \overline{B}^0$ and $B_s - \overline{B}_s$ mixing amplitudes are given in equations (13.70) and (13.71), respectively. In both cases, the $S(x_t, x_t)$ is the largest of the S-functions, while the CKM factors are of the same order in all three terms. We thus have

$$\frac{\Delta m_B}{m_B} \propto \frac{g^4}{96\pi^2}\left(\frac{m_t^2}{m_W^2}\right)|V_{tb}V_{td}|^2, \qquad \frac{\Delta m_{B_s}}{m_{B_s}} \propto \frac{g^4}{96\pi^2}\left(\frac{m_t^2}{m_W^2}\right)|V_{tb}V_{ts}|^2. \tag{13.20}$$

We learn that the GIM- and CKM-suppression factors are different among the various neutral meson systems of the down sector, as follows:

- $B_s^0 - \overline{B}_s^0$ mixing: CKM suppression by $|V_{tb}V_{ts}|^2 \sim 2 \times 10^{-3}$ and no GIM suppression
- $B^0 - \overline{B}^0$ mixing: CKM suppression by $|V_{tb}V_{td}|^2 \sim 10^{-4}$ and no GIM suppression
- $K^0 - \overline{K}^0$ mixing: CKM and GIM suppression by $|V_{cs}V_{cd}|^2(m_c^2/m_W^2) \sim 10^{-5}$

We learn that the Standard Model predicts hierarchy among the $\Delta F = 2$ processes:

$$\frac{\Delta m_K}{m_K} \ll \frac{\Delta m_B}{m_B} \ll \frac{\Delta m_{B_s}}{m_{B_s}}. \tag{13.21}$$

The experimental results,

$$\frac{\Delta m_K}{m_K} = 7.0 \times 10^{-15}, \qquad \frac{\Delta m_B}{m_B} = 6.3 \times 10^{-14}, \qquad \frac{\Delta m_{B_s}}{m_{B_s}} = 2.1 \times 10^{-12}, \tag{13.22}$$

show that this pattern is indeed realized in nature.

13.3.2 CP Suppression

In some cases, *CP*-violating observables are CKM suppressed beyond their *CP*-conserving counterparts. Whether this is the case can be understood by examining the relevant unitarity triangle: The *CP*-violating observables depend on the area of it, while *CP*-conserving observables depend on the length-squared of one side. Thus, in cases where the unitarity triangle is squashed (such as the *sd* and *bs* triangles), we can have a situation where the area of the triangle, $|J_{\text{CKM}}|/2 \sim \lambda^6$, is much smaller than the length-squared of one of its sides, resulting in an extra suppression for *CP*-violating observables. Explicitly, for FCNCs in the down sector, we have

$$\begin{aligned}
sd:&\ J_{\text{CKM}}/|V_{us}V_{ud}|^2 = \mathcal{O}(\lambda^4),\\
bs:&\ J_{\text{CKM}}/|V_{tb}V_{ts}|^2 = \mathcal{O}(\lambda^2),\\
bd:&\ J_{\text{CKM}}/|V_{tb}V_{td}|^2 = \mathcal{O}(1).
\end{aligned} \tag{13.23}$$

CP asymmetries measure the ratios between the *CP*-violating difference between two *CP*-conjugate rates and the *CP*-conserving sum of these rates, as follows:

- *CP* violation in $K^0 - \overline{K}^0$ mixing is the source of δ_L, with the *CP* asymmetry in $K_L \to \pi\ell\nu$ as defined in equation (13.86).
- *CP* violation in the interference of $B_s - \overline{B_s}$ mixing with $b \to c\bar{c}s$ decay is the source of $\mathcal{Im}(\lambda_{\psi\phi})$, with the *CP* asymmetry in $B_s \to \psi\phi$ as defined similarly to equation (13.88).
- *CP* violation in the interference of $B^0 - \overline{B}^0$ mixing with $b \to c\bar{c}d$ decay is the source of $\mathcal{Im}(\lambda_{D^+D^-})$, with the *CP* asymmetry in $B \to D^+D^-$ as defined in equation (13.88).

The pattern of a possible significant *CP* suppression in the *sd* sector, possible intermediate *CP* suppression in the *bs* sector, and no *CP* suppression in the *bd* sector is manifest in the Standard Model predictions:

$$\begin{aligned}
\delta_L &\propto J_{\text{CKM}}/|V_{us}V_{ud}|^2 \sim 10^{-3},\\
\mathcal{I}m(\lambda_{\psi\phi}) &\propto J_{\text{CKM}}/|V_{tb}V_{ts}|^2 \sim 10^{-2},\\
\mathcal{I}m(\lambda_{D^+D^-}) &\propto J_{\text{CKM}}/|V_{tb}V_{td}|^2 \sim 1.
\end{aligned} \tag{13.24}$$

Experiments confirm this pattern:

$$\begin{aligned}
\delta_L &= (3.34 \pm 0.07) \times 10^{-3},\\
\mathcal{I}m(\lambda_{\psi\phi}) &= (5.0 \pm 2.0) \times 10^{-2},\\
\mathcal{I}m(\lambda_{D^+D^-}) &= -0.76^{+0.15}_{-0.13}.
\end{aligned} \tag{13.25}$$

13.3.3 Summary

Within the Standard Model, we identify four possible suppression factors of FCNC processes relative to FCCC ones:

1. Loop suppression
2. CKM suppression
3. GIM suppression in processes that are not dominated by the top quark contribution
4. *CP* suppression in some of the processes related to squashed unitarity triangles

Harking back to the distinction that we made in section 9.3 in chapter 9 between *a* Standard Model and *the* Standard Model, the loop suppression of FCNC is a feature of *a* Standard Model, as it is the outcome of the symmetry and particle content. On the other hand, the CKM, GIM, and *CP* suppressions are features of *the* Standard Model, as they are the outcome of the specific values of the Standard Model parameters—namely, the smallness of CKM mixing angles and of quark masses.

13.4 Testing the CKM Sector

Within the Standard Model, the CKM matrix is the only source of flavor-changing processes and *CP* violation. In section 9.3 of chapter 9, we use only tree-level processes to extract the values of CKM parameters. Here, we add FCNC to the set of CKM measurements to form a global test of the Standard Model. The primary question is whether the long list of measurements can be fitted by the four CKM parameters.

The present status of our knowledge of the absolute values of the various entries in the CKM matrix is given in equation (9.5) in chapter 9. The values there take into account all the relevant tree-level and loop processes. Yet, as explained previously, the test of the Standard Model is stronger when we reduce all this to the four CKM parameters.

Indeed, the following ranges for the four Wolfenstein parameters are consistent with all measurements:

$$\lambda \approx 0.2250 \pm 0.0007, \qquad A \approx 0.83 \pm 0.02, \qquad \rho \approx 0.16 \pm 0.01, \qquad \eta \approx 0.35 \pm 0.01. \tag{13.26}$$

For the purpose of demonstration, it is useful to project the individual constraints onto the (ρ, η) plane:

- Charmless semileptonic B decays can be used to extract (see question 9.3 in chapter 9)
$$\left|\frac{V_{ub}}{V_{cb}}\right|^2 = \lambda^2(\rho^2 + \eta^2). \tag{13.27}$$
- $B \to DK$ decays can be used to extract (see appendix 13.C.1)
$$\tan \gamma = \left(\frac{\eta}{\rho}\right). \tag{13.28}$$
- $S_{\psi K_S}$, the CP asymmetry in $B \to \psi K_S$, is used to extract (see appendix 13.C.2)
$$\sin 2\beta = \frac{2\eta(1-\rho)}{(1-\rho)^2 + \eta^2}. \tag{13.29}$$
- The CP asymmetries of various $B \to \pi\pi$, $B \to \rho\pi$, and $B \to \rho\rho$ decays depend on the phase (see question 13.4)
$$\alpha = \pi - \beta - \gamma. \tag{13.30}$$
- The ratio between the mass splittings in the B and B_s systems depends on (see appendix 13.A.4)
$$\left|\frac{V_{td}}{V_{ts}}\right|^2 = \lambda^2[(1-\rho)^2 + \eta^2]. \tag{13.31}$$
- The CP violation in $K \to \pi\pi$ decays, ϵ_K, depends in a complicated way on ρ and η (see equation (13.111)).

The resulting constraints are shown in figure 13.3. The consistency of the various constraints is impressive. This is a triumph of the Standard Model, in that such a variety of measurements, with different sources of uncertainties, all agree to a high precision. We conclude that the flavor structure of the Standard Model passes a highly nontrivial test.

13.5 Probing BSM

In this section, we go beyond testing the self-consistency of the CKM picture of flavor physics and CP violation. The aim is to quantify how much room is left for BSM physics in the flavor sector and to translate these constraints into lower bounds on the scale of

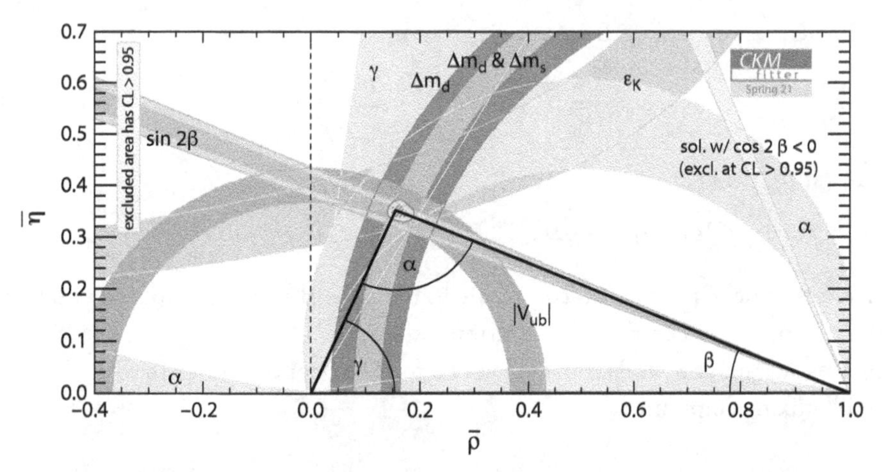

FIGURE 13.3. The allowed region in the ρ, η plane. Superimposed are the individual constraints from charmless semileptonic B decays ($|V_{ub}|$), mass differences in the B^0 (Δm_d) and B_s (Δm_s) neutral meson systems, and CP violation in $K \to \pi\pi$ (ε_K), $B \to \psi K$ ($\sin 2\beta$), $B \to \pi\pi, \rho\pi, \rho\rho$ (α), and $B \to DK$ (γ). Taken from CKMfitter Group collaboration [28].

higher-dimension, flavor-violating operators. We make the working assumption that

- The contribution of BSM physics to FCCCs, that are at tree-level, can be neglected.

On the other hand, we allow BSM physics of arbitrary size and phase to contribute to FCNC processes.

13.5.1 New Physics Contributions to $B^0 - \overline{B}^0$ Mixing

We consider BSM effects in the FCNC process of $B^0 - \overline{B}^0$ mixing, which plays a role in the mass-splitting Δm_B and in the CP asymmetry $S_{\psi K_S}$. The Standard Model amplitude is given by equation (13.70). The modification of the mixing amplitude by general BSM physics can be parameterized as follows:

$$M_{B\overline{B}} = M_{B\overline{B}}^{\rm SM} \Delta_d, \tag{13.32}$$

where Δ_d is a dimensionless complex parameter. BSM physics will be signaled by $\Delta_d \neq 1$. Our aim is to find the phenomenological constraints on Δ_d.

Our first step is to use all relevant tree-level processes which, under our assumption here, can be used to determine the CKM parameters. This was done in section 9.3, and the results of this fit were shown in figure 9.2. Our second step is to use $\Delta B = 2$ processes to determine Δ_d:

- The mass splitting between the two neutral B-mesons is given by

$$\Delta m_B = 2|M_{B\overline{B}}^{\rm SM}(\rho, \eta)| \times |\Delta_d|. \tag{13.33}$$

- The CP asymmetry in $B \to \psi K_S$ is given by

$$S_{\psi K_S} = \sin\left[2\arctan\left(\frac{\eta}{1-\rho}\right) + \arg(\Delta_d)\right]. \tag{13.34}$$

The results of the fit are

$$\mathcal{Re}\,(\Delta_d) = +0.94^{+0.18}_{-0.15}, \qquad \mathcal{Im}\,(\Delta_d) = -0.11^{+0.11}_{-0.05}. \tag{13.35}$$

We learn that BSM physics can contribute to the $B^0 - \overline{B}^0$ mixing amplitude up to about 20 percent of the Standard Model contribution.

Analogous upper bounds can be obtained for BSM contributions to the $K^0 - \overline{K}^0$ and $B_s^0 - \overline{B}_s^0$ mixing amplitudes.

13.5.2 Probing the SMEFT

Assuming that new degrees of freedom (DoF) that have flavor-changing couplings to quarks are much heavier than the electroweak-breaking scale, their effects on low-energy processes, such as neutral meson mixing, can be presented as higher-dimension operators. Then, bounds such as equation (13.35) constrain the coefficients of such operators.

Consider a simple example, where we have a single dimension-six operator that contributes to Δ_d:

$$\frac{z_{bd}}{\Lambda^2}(\overline{Q_{Ld}}\gamma_\mu Q_{Lb})(\overline{Q_{Ld}}\gamma^\mu Q_{Lb}), \tag{13.36}$$

where Q_{Ld} and Q_{Lb} are the $SU(2)$-doublet quark fields whose $T_3 = -1/2$ members are the d_L and b_L fields (see equation (8.30)), and where we separated the coefficient into a dimensionless complex coupling, z_{bd}, and a high energy scale, Λ. We further define $\tilde{\Lambda} = \Lambda/\sqrt{|z_{bd}|}$. We consider the bound that can be obtained from Δm_B. Comparing equations (13.63) and (13.36), we obtain

$$|\Delta_d| - 1 = \left|\frac{1}{\tilde{\Lambda}^2 C_{\rm SM}}\right| \approx \frac{1}{\tilde{\Lambda}^2} \times \frac{2\pi^2}{|V_{td}^* V_{tb}|^2 S(x_t, x_t) G_F^2 m_W^2}. \tag{13.37}$$

The bound of equation (13.35) translates into a lower bound on $\tilde{\Lambda}$:

$$\tilde{\Lambda} \gtrsim 10^3 \text{ TeV}. \tag{13.38}$$

Using $S_{\psi K_S}$, one can obtain analogous bounds for various operators that contribute to CP violation in $B^0 - \overline{B}^0$ mixing. We can also obtain bounds for operators that affect other $\Delta F = 2$ processes, some of which are given in table 13.1.

The following points are worth emphasizing:

- BSM physics can contribute to FCNC at a level comparable to the Standard Model contributions, even if it takes place at a scale that is five orders of magnitude above the electroweak scale.
- If the BSM physics has a generic flavor structure (i.e., $z_{ij} = \mathcal{O}(1)$), then its scale must be above 10^4 TeV.

Table 13.1. Lower bounds from *CP* conserving and *CP* violating $\Delta F = 2$ processes on the scale of new physics $\tilde{\Lambda}$ for $(\overline{Q}_{Li}\gamma_\mu Q_{Lj})(\overline{Q}_{Li}\gamma^\mu Q_{Lj})$ operators

i,j	$\tilde{\Lambda}$ [TeV] *CP* conserving	$\tilde{\Lambda}$ [TeV] *CP* violating	Observables
s,d	9.8×10^2	1.6×10^4	Δm_K; ϵ_K
b,d	6.6×10^2	9.3×10^2	Δm_B; $S_{\psi K}$
b,s	1.4×10^2	2.5×10^2	Δm_{B_s}; $S_{\psi\phi}$

- It could be that there are BSM particles with $m \sim$ TeV, but then the flavor structure of their couplings must be far from generic, $|z_{ij}| \ll 1$.
- The pattern of the bounds—namely, those from K^0 are stronger than those from B^0, which in turn are stronger than those from B_s—is directly related to the strength of the flavor (CKM and GIM) suppression we have in the Standard Model, as discussed in section 13.3.1. The reason for that is that the experimental accuracy and the QCD uncertainties are similar for the three cases.

It is instructive to compare the FCNC and EWPM bounds. The EWPM bounds are of order 10 TeV (see equation (12.23)), while FCNC bounds, as can be seen from table 13.1, are as high as 10^4 TeV. In both cases, the Standard Model explains the data well and the strength of the bound depends on the theoretical and experimental errors, which are roughly at the same level. The difference in the strength of the FCNC and the EWPM bounds stems from the difference in the suppression factors. In both cases, the relevant Standard Model effects are at the one-loop level. Yet, for the FCNC, as previously discussed, there are additional suppression factors: CKM, GIM, and *CP*. These factors result in much smaller Standard Model effects, and thus stronger bounds.

Appendix

In this appendix, we discuss the phenomenology of meson mixing and *CP* asymmetries, both in general (sections 13.A and 13.B) and in the Standard Model (sections 13.A.4 and 13.C). We include the general discussion here since the formalism is rarely discussed in quantum field theory (QFT) courses, and yet these processes play a major role in probing and testing the Standard Model.

13.A Neutral Meson Mixing and Oscillation

13.A.1 Introduction

Our discussion refers to the *CP*-conjugate pair of neutral pseudoscalar mesons, $P^0 - \overline{P}^0$, for each of the four cases, $P = K, D, B$, and B_s. Each of these has well-defined flavor quantum numbers, and is the lightest with these quantum numbers. Thus, within QCD and QED, they are stable and do not mix with their antiparticle. The weak interaction, however, does not respect these flavor symmetries, and thus P^0 and $\overline{P}^0$ decay. Moreover, since these states are neutral under the unbroken symmetries of the Standard Model, the weak interaction leads to mixing among them. This mixing is generated by $P^0 \leftrightarrow \overline{P}^0$ transition amplitudes. The mixing lifts the degeneracy between the masses m_P and $m_{\overline{P}}$, resulting in two mass eigenstates that are superpositions of P^0 and $\overline{P}^0$. The mass eigenstates have different masses ($\Delta m \neq 0$), and different widths ($\Delta\Gamma \neq 0$).

In all four cases ($P = K, D, B, B_s$), the mass difference Δm is experimentally measurable and theoretically informative:

- In general, the mesons that are produced by QCD and quantum electro dynamics (QED) interactions are those with well-defined flavor quantum numbers. The fact that, for the neutral K, D, B, and B_s mesons, the flavor eigenstates are not mass eigenstates implies that the time evolution of these states is nontrivial. In particular, they undergo flavor oscillations with measurable oscillation frequencies set by Δm.
- Neutral meson mixing is an FCNC process. Thus, as discussed in the main text, Δm provides, within the Standard Model, indirect measurements of CKM parameters, and probes BSM up to very high energy scales.

Note that flavor mixing and flavor oscillations, while related, are not the same. Flavor mixing, or flavor-changing transitions, refer to amplitudes or processes where the initial and final states carry different flavor quantum numbers. The term *flavor oscillation* refers to a phenomenon where the time evolution of a state exhibits sinusoidal change of flavor quantum numbers.

Consider the case where initially, at $t = 0$, the neutral meson state is some specific combination of P^0 and $\overline{P}^0$:

$$|\psi_P(0)\rangle = a(0)|P^0\rangle + b(0)|\overline{P}^0\rangle\,. \tag{13.39}$$

For example, an initial pure P^0 state corresponds to $a(0) = 1$ and $b(0) = 0$. The state evolves in time and also acquires components that correspond to all possible decay final states $\{f_1, f_2, \ldots\}$:

$$|\psi_P(t)\rangle = a(t)|P^0\rangle + b(t)|\overline{P}^0\rangle + c_1(t)|f_1\rangle + c_2(t)|f_2\rangle + \cdots\,. \tag{13.40}$$

As amplitude such as $\langle\overline{P}^0|H|P^0\rangle$ represents flavor mixing. The resulting sinusoidal time dependence of $a(t)$ and $b(t)$ represents flavor oscillations.

13.A.2 Flavor Mixing

To obtain the time-dependent coefficients $a(t)$, $b(t)$, and $c_i(t)$ of equation (13.40), we need to find the matrix elements of the Hermitian Hamiltonian H between the initial state and all possible final states—a formidable task. Since our interest lies in obtaining only $a(t)$ and $b(t)$, however, we can use a simplified formalism, where we deal with a 2×2 matrix, with the matrix elements of an effective non-Hermitian Hamiltonian $\mathcal{H}$ between P^0 and

$\overline{P}^0$ states only. The nonhermiticity is related to the possibility of decays, which makes the $\{P^0, \overline{P}^0\}$ system an open one.

Before we proceed, let us clarify a semantic issue. The effective Hamiltonian $\mathcal{H}$ is a combination of operators. What we need for our purposes is its matrix elements between specific meson states. We denote the matrix element by $\mathcal{H}_{\alpha\beta}$ with $\alpha, \beta = P, \overline{P}$.

The complex matrix $\mathcal{H}$ can be written in terms of the Hermitian matrices M and Γ:

$$\mathcal{H} = M - \frac{i}{2}\Gamma. \tag{13.41}$$

The matrices M and Γ are associated with transitions via off-shell (dispersive) and on-shell (absorptive) intermediate states, respectively. Diagonal elements of M and Γ are associated with the flavor-conserving transitions $P^0 \to P^0$ and $\overline{P}^0 \to \overline{P}^0$. The CPT symmetry implies that the diagonal elements are equal: $M_{PP} = M_{\overline{P}\overline{P}}$ and $\Gamma_{PP} = \Gamma_{\overline{P}\overline{P}}$. The off-diagonal elements are associated with the flavor-changing transitions $P^0 \leftrightarrow \overline{P}^0$, and they are of significant interest to us. The phase

$$\theta_P = \arg(M_{P\overline{P}}\Gamma^*_{P\overline{P}}), \tag{13.42}$$

if different from zero ($\theta_P \neq 0$), implies CP violation.

Since $\mathcal{H}$ is not a diagonal matrix, the states that have well-defined masses and decay widths are not P^0 and $\overline{P}^0$, but rather the eigenvectors of $\mathcal{H}$. We denote the light and heavy eigenstates by P_L and P_H, with masses $m_H > m_L$. Another possible choice, which is standard for K-mesons, is to define the mass eigenstates according to their lifetimes. We denote the short- and long-lived eigenstates by K_S and K_L, with decay widths $\Gamma_S > \Gamma_L$. (The K_L-meson is experimentally found to be the heavier state.)

Diagonalizing $\mathcal{H}$, we find that the eigenstates of $\mathcal{H}$ are given by

$$|P_{L,H}\rangle = p|P^0\rangle \pm q|\overline{P}^0\rangle, \tag{13.43}$$

where

$$\left(\frac{q}{p}\right)^2 = \frac{M^*_{P\overline{P}} - (i/2)\Gamma^*_{P\overline{P}}}{M_{P\overline{P}} - (i/2)\Gamma_{P\overline{P}}}, \qquad |p|^2 + |q|^2 = 1. \tag{13.44}$$

Since $\mathcal{H}$ is not Hermitian, the eigenstates need not be orthogonal to each other, and $\langle P_H|P_L\rangle = |p|^2 - |q|^2$ can be different from zero. The eigenvalues of $\mathcal{H}$ can be written as

$$\mu_{H,L} = m_{H,L} + \frac{i}{2}\Gamma_{H,L}, \tag{13.45}$$

such that the masses and decay widths of the eigenstate are given by the real and imaginary parts of the eigenvalues, respectively. The average mass, m, and the average width, Γ, are given by

$$m \equiv \frac{m_H + m_L}{2}, \qquad \Gamma \equiv \frac{\Gamma_H + \Gamma_L}{2}. \tag{13.46}$$

The mass difference Δm and the width difference $\Delta\Gamma$ are defined as follows:

$$\Delta m \equiv m_H - m_L, \qquad \Delta\Gamma \equiv \Gamma_H - \Gamma_L, \tag{13.47}$$

where, Δm is positive by definition, while the sign of $\Delta\Gamma$ is to be determined experimentally. (Alternatively, one can use the states defined by their lifetimes to have $\Delta\Gamma \equiv \Gamma_S - \Gamma_L$ be positive by definition, in which case the sign of Δm has to be determined experimentally.) Solving the eigenvalue equation gives

$$(\Delta m)^2 - \frac{1}{4}(\Delta\Gamma)^2 = 4|M_{P\overline{P}}|^2 - |\Gamma_{P\overline{P}}|^2, \qquad \Delta m\Delta\Gamma = 4(M_{P\overline{P}}\Gamma^*_{P\overline{P}}). \tag{13.48}$$

Two limits are of particular interest:

- Consider the case where $\mathcal{H}$ is CP symmetric. Then, $\theta_P = 0$. As for the eigenvectors, equation (13.44) gives

$$\left|\frac{q}{p}\right| = 1. \tag{13.49}$$

  It follows that the mass eigenstates are also CP eigenstates and are orthogonal to each other, $\langle P_H|P_L\rangle = 0$. As for the eigenvalues, equation (13.48) gives

$$\Delta m = 2|M_{P\overline{P}}|, \qquad |\Delta\Gamma| = 2|\Gamma_{P\overline{P}}|. \tag{13.50}$$

- Another limit of interest is when we allow CP violation but consider $|\Gamma_{P\overline{P}}| \ll |M_{P\overline{P}}|$. Working to leading order in $|\Gamma_{P\overline{P}}/M_{P\overline{P}}|$, we have

$$\Delta m = 2|M_{P\overline{P}}|, \qquad |\Delta\Gamma| = 2|\Gamma_{P\overline{P}}| \cos\theta_P. \tag{13.51}$$

It is interesting to note that within the Standard Model, $\Delta m = 2|M_{P\overline{P}}|$ to a very good approximation for the four mesons that we consider. This is because $|\Gamma_{P\overline{P}}| \ll |M_{P\overline{P}}|$ for the B and B_s system and $|\theta_P| \ll 1$ for the K, D, and B_s systems.

13.A.3 Flavor Oscillation

To study the time evolution of the neutral mesons, it is convenient to define the dimensionless ratios:

$$x \equiv \frac{\Delta m}{\Gamma}, \qquad y \equiv \frac{\Delta\Gamma}{2\Gamma}, \tag{13.52}$$

By definition, $x \geq 0$ and $|y| \leq 1$. We further define the decay amplitudes of P^0 and $\overline{P}^0$ into a final state f:

$$A_f = \langle f|H|P^0\rangle, \qquad \overline{A}_f = \langle f|H|\overline{P}^0\rangle, \tag{13.53}$$

and parameter λ_f:

$$\lambda_f \equiv \frac{q}{p}\frac{\overline{A}_f}{A_f}. \tag{13.54}$$

Our normalization is such that

$$\Gamma(P^0 \to f) = |A_f|^2, \qquad \Gamma(\overline{P}^0 \to f) = |\overline{A}_f|^2. \tag{13.55}$$

We denote the time-evolved state of an initial state $|P^0\rangle$ by $|P^0(t)\rangle$, and that of an initial state $|\overline{P}^0\rangle$ by $|\overline{P}^0(t)\rangle$, where t is defined in the meson rest frame. For the mass eigenstates, the time evolution is simple:

$$|P_L(t)\rangle = e^{-im_L t - \frac{1}{2}\Gamma_L t}|P_L\rangle, \qquad |P_H(t)\rangle = e^{-im_H t - \frac{1}{2}\Gamma_H t}|P_H\rangle. \tag{13.56}$$

For the flavor eigenstates, the time evolution is more complicated:

$$|P^0(t)\rangle = g_+(t)|P^0\rangle - (q/p)g_-(t)|\overline{P}^0\rangle, \qquad |\overline{P}^0(t)\rangle = g_+(t)|\overline{P}^0\rangle - (p/q)g_-(t)|P^0\rangle, \tag{13.57}$$

where

$$g_\pm(t) = \frac{1}{2}\left(e^{-im_H t - \frac{1}{2}\Gamma_H t} \pm e^{-im_L t - \frac{1}{2}\Gamma_L t}\right). \tag{13.58}$$

We define the rescaled time dependent decay rates as

$$\hat{\Gamma}[P^0(t) \to f] \equiv \frac{d\Gamma[P^0(t) \to f]}{dt} \times \frac{1}{e^{-\Gamma t}|A_f|^2}. \tag{13.59}$$

They are given by

$$\begin{aligned} 2\hat{\Gamma}[P^0(t) \to f] &= (1 + |\lambda_f|^2)\cosh(y\Gamma t) + (1 - |\lambda_f|^2)\cos(x\Gamma t) \\ &\quad + 2\mathcal{R}e(\lambda_f)\sinh(y\Gamma t) - 2\mathcal{I}m(\lambda_f)\sin(x\Gamma t), \\ 2\hat{\Gamma}[\overline{P}^0(t) \to f] &= (1 + |\lambda_f|^{-2})\cosh(y\Gamma t) + (1 - |\lambda_f|^{-2})\cos(x\Gamma t) \\ &\quad + 2\mathcal{R}e(\lambda_f^{-1})\sinh(y\Gamma t) - 2\mathcal{I}m(\lambda_f^{-1})\sin(x\Gamma t). \end{aligned} \tag{13.60}$$

We next introduce the notion of *flavor tagging*. The flavor eigenstates P^0 and $\overline{P}^0$ have a well-defined flavor content. For example, B^0 has the quantum number of a $\bar{b}d$ state. The term *flavor tagging* refers to the experimental determination of whether a neutral P meson is in a P^0 or a $\overline{P}^0$ state. In some cases, the final state of the decay informs us about the flavor of the state at the time of the decay. This is the case when the decay is into a flavor-specific final state—namely, a state f that can come from either the P^0 or $\overline{P}^0$ state, but not from both. In other words, flavor-specific decays are cases where either $\overline{A}_f = 0$ or $A_f = 0$. Semileptonic decays provide very good flavor tags. Take, for example, the quark transitions $b \to c\mu^-\bar{\nu}$ and $\bar{b} \to \bar{c}\mu^+\nu$, which correspond to the following semileptonic B-meson decays:

$$\overline{B}^0 \to X_c\mu^-\bar{\nu}, \qquad B^0 \to X_{\bar{c}}\mu^+\nu, \tag{13.61}$$

where X_c ($X_{\bar{c}}$) corresponds to a hadronic system with a charm number of $+1$ (-1). The charge of the final lepton is correlated with the flavor of the initial meson: μ^+ comes from a B^0 decay, while μ^- comes from a $\overline{B}^0$ decay. Of course, before the meson decays, it could be in a superposition of B^0 and $\overline{B}^0$. The decay acts as a quantum measurement. In the case of semileptonic decays, it acts as a measurement of flavor.

The oscillation formalism is simplified in the case of flavor-tagged decay. Take the case of $\overline{A}_f = 0$, and therefore, $\lambda_f = 0$. We further simplify the discussion by assuming that $|q/p| = 1$ and $y = 0$. We obtain

$$\hat{\Gamma}[P^0(t) \to f] = \frac{1+\cos(\Delta m t)}{2}, \qquad \hat{\Gamma}[\overline{P}^0(t) \to f] = \frac{1-\cos(\Delta m t)}{2}. \tag{13.62}$$

We see that the flavor oscillates with a frequency of Δm in the rest frame. The general case of equation (13.60) involves deviations from pure exponential decay, which depend on both Δm and $\Delta\Gamma$. Moreover, for a flavor-specific final state, we can interpret $\hat{\Gamma}[P^0(t) \to f]$ as the time-dependent probability of $P^0 \to P^0$, and $\hat{\Gamma}[\overline{P}^0(t) \to f]$ as the time-dependent probability of $P^0 \to \overline{P}^0$.

13.A.4 Standard Model Calculations of the Mixing Amplitude

We present the Standard Model calculation of the mixing amplitude $M_{B\overline{B}}$ and its generalization to the other meson systems. The leading diagrams that contribute to $M_{B\overline{B}}$ are one-loop diagrams that are called *box diagrams* and are displayed in figure 13.2. We can write the transition amplitude as

$$\mathcal{A}_{B^0 \to \overline{B}^0} = C_{\rm SM}(\bar{d}_L \gamma_\mu b_L)(\bar{d}_L \gamma^\mu b_L). \tag{13.63}$$

The normalized matrix element is related to the amplitude via

$$M_{B\overline{B}} = \frac{1}{2m_B}\langle B^0|\mathcal{A}_{B^0 \to \overline{B}^0}|\overline{B}^0\rangle, \tag{13.64}$$

and thus

$$M_{B\overline{B}} = \frac{C_{\rm SM}}{2m_B}\langle B^0|(\bar{d}_L \gamma_\mu b_L)(\bar{d}_L \gamma^\mu b_L)|\overline{B}^0\rangle. \tag{13.65}$$

The nonperturbative QCD effects are encoded in the hadronic matrix element, which we parameterize as follows:

$$\langle B^0|(\bar{d}_L \gamma_\mu b_L)(\bar{d}_L \gamma^\mu b_L)|\overline{B}^0\rangle = -\frac{1}{3} m_B^2 B_B f_B^2, \tag{13.66}$$

where B_B is a number and f_B is the B-meson decay constant. Lattice calculations give $\sqrt{B_B} f_B \approx 0.22$ GeV. This is where the hadronic uncertainties lie.

The weak interaction effects are encoded in $C_{\rm SM}$, which is calculated from the box diagrams:

$$C_{\rm SM} = \frac{G_F^2 m_W^2}{2\pi^2} \times \left[(V_{cb}V_{cd}^*)^2 S(x_c, x_c) + (V_{tb}V_{td}^*)^2 S(x_t, x_t) + (V_{cb}V_{cd}^*)(V_{tb}V_{td}^*) S(x_t, x_c)\right], \tag{13.67}$$

where $x_i = m_i^2/m_W^2$, and we approximate $x_u = 0$ and S is the loop function:

$$S(x_i, x_j) = x_i x_j \left[-\frac{3}{4(1-x_i)(1-x_j)} + \frac{\log x_i}{(x_i - x_j)(1-x_i)^2}\left(1 - 2x_i + \frac{x_i^2}{4}\right) + \frac{\log x_j}{(x_j - x_i)(1-x_j)^2}\left(1 - 2x_j + \frac{x_j^2}{4}\right)\right]. \tag{13.68}$$

For $x_i, x_j \ll 1$, we have

$$S(x_i, x_j) \approx \begin{cases} x_i \log(x_j/x_i) & \text{for } x_i \ll x_j \ll 1, \\ x_i & \text{for } x_i = x_j \ll 1. \end{cases} \tag{13.69}$$

Note that $S(0, x) = 0$. Taking into account the values of the quark masses and CKM elements, we conclude that the term proportional to $S(x_t, x_t)$ dominates those proportional to $S(x_t, x_c)$ and $S(x_c, x_c)$, and thus

$$M_{B\overline{B}} \approx \frac{G_F^2}{12\pi^2} m_B m_W^2 (B_B f_B^2)(V_{tb} V_{td}^*)^2 S(x_t, x_t). \tag{13.70}$$

This result is subject to known radiative corrections that are of $O(1)$.

Equation (13.70) can be straightforwardly generalized to other systems. For $M_{B_s\overline{B}_s}$, we replace $d \to s$:

$$M_{B_s\overline{B}_s} \approx \frac{G_F^2}{12\pi^2} m_{B_s} m_W^2 (B_{B_s} f_{B_s}^2)(V_{tb} V_{ts}^*)^2 S(x_t, x_t)\,. \tag{13.71}$$

The ratio $\Delta m_B/\Delta m_{B_s}$ is particularly interesting:

$$\frac{\Delta m_B}{\Delta m_{B_s}} = \frac{m_B B_B f_B^2}{m_{B_s} B_{B_s} f_{B_s}^2} \left|\frac{V_{td}}{V_{ts}}\right|^2. \tag{13.72}$$

In the $SU(3)_F$ limit, the hadronic matrix elements of B and B_s are the same. Consequently, in the ratio of equation (13.72), the hadronic uncertainty is only in the correction to the $SU(3)_F$ limit, and therefore, it is small. Thus, the ratio $\Delta m_B/\Delta m_{B_s}$ provides an excellent measurement of $|V_{td}/V_{ts}|$.

For $M_{K\overline{K}}$, we replace $b \to s$:

$$M_{K\overline{K}} = \frac{G_F^2}{12\pi^2} m_K m_W^2 (B_K f_K^2) \left[(V_{cs} V_{cd}^*)^2 S(x_c, x_c) + (V_{ts} V_{td}^*)^2 S(x_t, x_t) + (V_{cs} V_{cd}^*)(V_{ts} V_{td}^*) S(x_t, x_c) \right]. \tag{13.73}$$

Lattice results gives $B_K = 0.86 \pm 0.24$.

For the four systems, $P = B, B_s, D, K$, the calculation of $M_{P\overline{P}}$ translates into the calculation of the mass-splitting $\Delta M_P = 2|M_{P\overline{P}}|$. (In the D system, however, the calculation of $M_{D\overline{D}}$ is complicated, so we do not discuss it here.)

The numerical values of the mixing parameters are presented in table 13.2. The Standard Model calculations outlined here agree well with the data.

Table 13.2. The experimental values of the neutral meson–mixing parameters.

P	m (GeV)	Γ (GeV)	x	y
K^0	0.498	3.68×10^{-15}	0.945 ± 0.001	-0.997
D^0	1.86	1.60×10^{-13}	0.0041 ± 0.0005	$+0.0063 \pm 0.0006$
B^0	5.28	4.33×10^{-13}	0.769 ± 0.004	-0.0005 ± 0.0050
B_s	5.37	4.34×10^{-13}	27.01 ± 0.10	-0.064 ± 0.004

Note: In all cases (including the K-meson system), we define x and y as in equations (13.47) and (13.52). For the K^0 system, the error on y is well below 10^{-3}, and thus we do not quote an error. For the B^0 system, there is only an upper bound on $|y|$.

13.B *CP* Violation

CP asymmetries arise when two processes related by *CP* conjugation differ in their rates. Given the fact that *CP* violation is related to a phase in the Lagrangian, all *CP* asymmetries must arise from interference effects.

To date, *CP* violation has been observed (at a level higher than 5σ) in about 30 hadron decay modes involving b, c, or s decays. It has not been established in other quark decays, or in the leptonic sector, or in flavor diagonal processes. Here, we present the formalism relevant to measuring *CP* asymmetries in meson decays.

13.B.1 Notations and Formalism

We discuss here the specific case of B-meson decays, but the discussion applies to all meson decays. Our starting point is equation (13.60), which gives the time-dependent decay rates of B^0 and $\overline{B}^0$. We also use the parameter λ_f, as defined in equation (13.54).

Consider A_f, the $B \to f$ decay amplitude, and $\overline{A}_{\overline{f}}$, the amplitude of the *CP* conjugate process, $\overline{B} \to \overline{f}$. There are two types of phases that may appear in these decay amplitudes:

- *CP*-odd phases, also known as *weak phases*. They are complex parameters in any Lagrangian term that contributes to A_f, and they appear in a complex conjugate form in $\overline{A}_{\overline{f}}$. In other words, *CP*-violating phases change sign between A_f and $\overline{A}_{\overline{f}}$. In the Standard Model, these phases appear only in the couplings of the $W^{\pm}$-bosons, and hence the term *weak phases*.
- *CP*-even phases, also known as *strong phases*. Phases can appear in decay amplitudes even when the Lagrangian parameters are all real. They arise from contributions of intermediate on-shell states and can be identified with the e^{-iHt} term in the time evolution Schrödinger equation. These *CP*-conserving phases appear with the same sign in A_f and $\overline{A}_{\overline{f}}$. In meson decays, the intermediate states are typically hadronic states with the same flavor quantum number as the final state, and their dynamics is driven by strong interactions, and hence the term *strong phases*.

It is useful to factorize an amplitude into three parts: the magnitude a_i, the weak phase ϕ_i, and the strong phase δ_i. If there are two such contributions, we write

$$A_f = a_1 e^{i(\delta_1+\phi_1)} + a_2 e^{i(\delta_2+\phi_2)}, \qquad \overline{A}_{\bar{f}} = a_1 e^{i(\delta_1-\phi_1)} + a_2 e^{i(\delta_2-\phi_2)}, \tag{13.74}$$

where we always can choose $a_1 \geq a_2$. It is also useful to define

$$\phi_f \equiv \phi_2 - \phi_1, \qquad \delta_f \equiv \delta_2 - \delta_1, \qquad r_f \equiv \frac{a_2}{a_1}. \tag{13.75}$$

For neutral meson mixing, it is useful to write

$$M_{B\overline{B}} = |M_{B\overline{B}}|e^{i\phi_M}, \qquad \Gamma_{B\overline{B}} = |\Gamma_{B\overline{B}}|e^{i\phi_\Gamma}, \tag{13.76}$$

such that for θ_B, defined in equation (13.42), we have

$$\theta_B = \phi_M - \phi_\Gamma. \tag{13.77}$$

Note that each of the phases appearing in equations (13.74) and (13.76) is convention dependent, but combinations such as $\delta_1 - \delta_2$, $\phi_1 - \phi_2$, $\phi_M - \phi_\Gamma$ are physical.

In neutral meson decays, the phenomenology of *CP* violation is particularly rich, thanks to the fact that meson mixing, as described in appendix 13.A, can contribute to the *CP* violating interference effects. One distinguishes three types of *CP* violation in meson decays, depending on which amplitudes interfere:

1. In decay: The interference is between two decay amplitudes.
2. In mixing: The interference is between the absorptive and dispersive mixing amplitudes.
3. In interference of decays with and without mixing: The interference is between the direct decay amplitude and a first-mix-then-decay amplitude.

In figure 13.4, we give a schematic description of these three types. *CP* violation in decay corresponds to interference between a_1 and a_2. *CP* violation in mixing corresponds to interference between $M_{B\overline{B}}$ and $\Gamma_{B\overline{B}}$. *CP* violation in interference of decays with and without mixing to interference between $\overline{A}_f$ and $M_{B\overline{B}}A_f$. We discuss these three types next.

For the discussion of *CP* violation in the $K^0 - \overline{K}^0$ system, we use a somewhat different notation. The reason is that, since the lifetimes of K_S and K_L are so different, experiments often identify these mass eigenstates rather than the flavor-tagged decays, as is done in

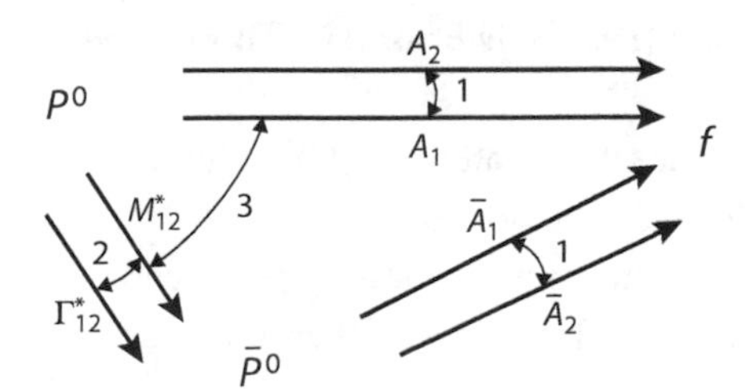

FIGURE 13.4. The three types of *CP* violation. 1: In decay. 2: In mixing. 3: In interference between mixing and decay.

most measurements of *CP* violation in the $B^0 - \overline{B}^0$ system. Thus, for *K*-mesons, we define

$$\epsilon_f \equiv \frac{1-\lambda_f}{1+\lambda_f}. \tag{13.78}$$

The converse relation reads as

$$\lambda_f \equiv \frac{1-\epsilon_f}{1+\epsilon_f}. \tag{13.79}$$

Historically, *CP* violation was first observed in the $K_L \to \pi^+\pi^-$ decay, and thus we denote $\epsilon_{\pi^+\pi^-} = \epsilon_K$. For modes with $|\overline{A}_f/A_f| - 1 \ll |q/p| - 1$, as is the case for $f = \pi^+\pi^-$, we can set $|\overline{A}_f/A_f| = 1$ and then we have $|q/p| = |\lambda_f|$.

13.B.2 CP Violation in Decay

CP violation in decay corresponds to

$$|\overline{A}_{\bar{f}}/A_f| \neq 1. \tag{13.80}$$

In charged particle decays, this is the only possible contribution to the *CP* asymmetry:

$$\mathcal{A}_f \equiv \frac{\Gamma(B^- \to f^-) - \Gamma(B^+ \to f^+)}{\Gamma(B^- \to f^-) + \Gamma(B^+ \to f^+)} = \frac{|\overline{A}_{f^-}/A_{f^+}|^2 - 1}{|\overline{A}_{f^-}/A_{f^+}|^2 + 1}. \tag{13.81}$$

Using equation (13.74), we obtain, for $r_f \ll 1$,

$$\mathcal{A}_f = 2r_f \sin\phi_f \sin\delta_f. \tag{13.82}$$

This result shows explicitly that we need two decay amplitudes (i.e., $r_f \neq 0$), with different weak phases ($\phi_f \neq 0, \pi$), and different strong phases ($\delta_f \neq 0, \pi$).

A few comments are in order:

- To have a large *CP* asymmetry, we need each of the three factors in equation (13.82) not to be small.
- A similar expression holds for the contribution of *CP* violation in decay in neutral meson decays. In this case, however, there are additional contributions from mixing, as discussed next.
- Another complication with regard to neutral meson decays is that it is not always possible to tell the flavor of the decaying meson (i.e., if it is B^0 or $\overline{B}^0$). This can be a problem or an advantage.
- In general, the strong phase is not calculable since it is related to QCD. This is not a problem if the aim is just to demonstrate *CP* violation, but it is if we want to extract the weak parameter, ϕ_f. In some cases, however, the strong phase can be independently measured, eliminating this particular source of theoretical uncertainty.

13.B.3 CP Violation in Mixing

CP violation in mixing corresponds to

$$|q/p| \neq 1. \qquad (13.83)$$

In decays of neutral mesons into flavor-specific final states ($\overline{A}_f = 0$, and consequently, $\lambda_f = 0$), and, in particular, semileptonic neutral meson decays, this is the only source of *CP* violation:

$$\mathcal{A}_{\rm SL}(t) \equiv \frac{\hat{\Gamma}[\overline{B}^0(t) \to \ell^+ X] - \hat{\Gamma}[B^0(t) \to \ell^- X]}{\hat{\Gamma}[\overline{B}^0(t) \to \ell^+ X] + \hat{\Gamma}[B^0(t) \to \ell^- X]} = \frac{1 - |q/p|^4}{1 + |q/p|^4}. \qquad (13.84)$$

Using equation (13.44), we obtain, for $|\Gamma_{B\overline{B}}/M_{B\overline{B}}| \ll 1$,

$$\mathcal{A}_{\rm SL} = -\left|\Gamma_{B\overline{B}}/M_{B\overline{B}}\right| \sin(\phi_M - \phi_\Gamma). \qquad (13.85)$$

Two comments are in order:

- Equation (13.84) implies that $\mathcal{A}_{\rm SL}(t)$, which is an asymmetry of time-dependent decay rates, is actually time independent.
- The calculation of $|\Gamma_{B\overline{B}}/M_{B\overline{B}}|$ is difficult since it depends on low-energy QCD effects. Hence, the extraction of the value of the *CP* violating phase $\phi_M - \phi_\Gamma$ from a measurement of $\mathcal{A}_{\rm SL}$ involves, in general, large hadronic uncertainties.

CP violation in $K^0 - \overline{K}^0$ mixing is measured via semileptonic asymmetry, which is defined as follows:

$$\delta_L \equiv \frac{\Gamma(K_L \to \ell^+ \nu_\ell \pi^-) - \Gamma(K_L \to \ell^- \nu_\ell \pi^+)}{\Gamma(K_L \to \ell^+ \nu_\ell \pi^-) + \Gamma(K_L \to \ell^- \nu_\ell \pi^+)} = \frac{1 - |q/p|^2}{1 + |q/p|^2} \approx 2\mathcal{R}e(\epsilon_K), \qquad (13.86)$$

where we use equation (13.79) and the fact that $|\epsilon_K| \ll 1$. This asymmetry is different from the one defined in equation (13.84), in that the decaying meson is the neutral mass eigenstate rather than the flavor eigenstate, and hence the different dependence on $|q/p|$.

13.B.4 CP Violation in Interference of Decays with and without Mixing

CP violation in interference of decays with and without mixing corresponds to

$$\mathcal{I}m(\lambda_f) \neq 0. \qquad (13.87)$$

A particular simple case is the *CP* asymmetry in decays into final *CP* eigenstates. Moreover, a situation that is relevant in many cases is when one can neglect the effects of *CP* violation in decay and in mixing (i.e., when $|\overline{A}_{f_{CP}}/A_{f_{CP}}| \approx 1$ and $|q/p| \approx 1$). In this case, $\lambda_{f_{CP}}$ is, to a good approximation, a pure phase, $|\lambda_{f_{CP}}| = 1$. We further consider the case where we can neglect y ($|y| \ll 1$). Then,

$$\mathcal{A}_{f_{CP}}(t) \equiv \frac{\Gamma[\overline{B}^0(t) \to f_{CP}] - \Gamma[B^0(t) \to f_{CP}]}{\Gamma[\overline{B}^0(t) \to f_{CP}] + \Gamma[B^0(t) \to f_{CP}]} = \mathcal{I}m(\lambda_{f_{CP}}) \sin(\Delta m_B t). \qquad (13.88)$$

(Experimentally, the coefficient of the $\sin(\Delta m_B t)$ term is denoted by S_f.) The approximations made here are valid in cases where $|\Gamma_{B\overline{B}}/M_{B\overline{B}}| \ll 1$ and $a_2 \ll a_1$, which lead to

$$\frac{q}{p} = \frac{M^*_{B\overline{B}}}{|M_{B\overline{B}}|} = e^{-i\phi_M}, \qquad \frac{\overline{A}_{f_{CP}}}{A_{f_{CP}}} = e^{-2i\phi_A}, \qquad (13.89)$$

where ϕ_M is defined in equation (13.76) and $\phi_A = \phi_1$ is defined in equation (13.74). We then get

$$\mathcal{I}m(\lambda_{f_{CP}}) = \mathcal{I}m\left(\frac{M^*_{B\overline{B}}}{|M_{B\overline{B}}|}\frac{\overline{A}_{f_{CP}}}{A_{f_{CP}}}\right) = -\sin(\phi_M + 2\phi_A). \qquad (13.90)$$

We learn that a measurement of a *CP* asymmetry in a process where these approximations are valid provides a direct probe of the weak phase between the mixing and the decay amplitudes.

For the case where we measure decays of the K_L and K_S mass eigenstates into final *CP*-even eigenstates, one obtains (see question 13.1)

$$\mathcal{A}^{\text{mass}}_{f_{CP}} \equiv \frac{\Gamma(K_L \to f_{CP})}{\Gamma(K_S \to f_{CP})} = \left|\frac{1 - \lambda_{f_{CP}}}{1 + \lambda_{f_{CP}}}\right|^2 = |\epsilon_{f_{CP}}|^2. \qquad (13.91)$$

In particular, for $f_{CP} = \pi^+\pi^-$, we have

$$\mathcal{A}^{\text{mass}}_{\pi^+\pi^-} = |\epsilon_K|^2. \qquad (13.92)$$

13.C Standard Model Calculations of *CP* Violation

In this section, we give several examples of *CP* violating observables that are used to test the CKM mechanism of flavor mixing and that are mentioned in the main text.

13.C.1 Extracting γ from $B \to DK$

The angle γ, defined in equation (9.20) in chapter 9, is extracted with small theoretical uncertainties from a combination of the following three $B \to DK$ decay modes:

$$B^+ \to D^0 K^+, \qquad B^+ \to \overline{D}^0 K^+, \qquad B^+ \to D_{CP} K^+, \qquad (13.93)$$

where D_{CP} is a D state that decays into a *CP* eigenstate. For the sake of concreteness, we take $D_{CP} = (K^+K^-)_D$–namely, the state that decays into K^+K^-. We neglect $D^0 - \overline{D}^0$

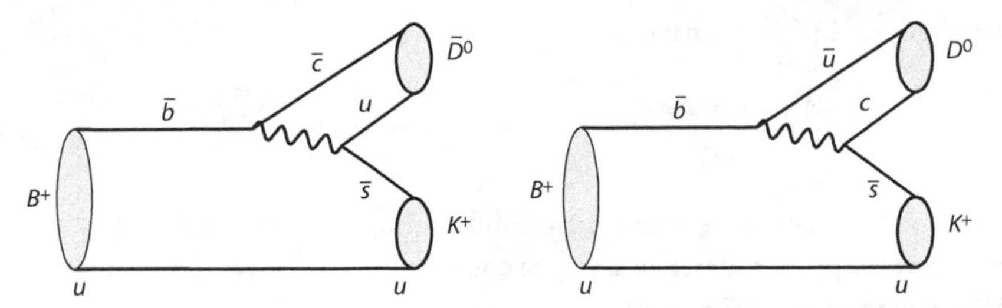

FIGURE 13.5. Diagrams for $B^+ \to \overline{D}^0 K^+$ (left) and (b) $B^+ \to D^0 K^+$ (right). There is one more diagram where the spectator u quark ends up in the $\overline{D}^0$ meson, but that is not shown.

mixing ($x_D = y_D = 0$) and CP violation in D decays. Then,

$$D_{CP} = \frac{1}{\sqrt{2}}(D^0 + \overline{D}^0). \tag{13.94}$$

The $B^+ \to D_{CP} K^+$ decay can proceed in two ways (see figure 13.5):

- $B^+ \to \overline{D}^0 K^+$, followed by $\overline{D}^0 \to K^+ K^-$. The corresponding quark transitions are $\bar{b} \to \bar{c} u \bar{s}$ followed by $\bar{c} \to \bar{s} s \bar{u}$. The CKM factors are

$$(V_{cb}^* V_{us})(V_{cs} V_{us}^*). \tag{13.95}$$

- $B^+ \to D^0 K^+$, followed by $D^0 \to K^+ K^-$. The corresponding quark transitions are $\bar{b} \to \bar{u} c \bar{s}$, followed by $c \to s \bar{s} u$. The CKM factors are

$$(V_{ub}^* V_{cs})(V_{cs}^* V_{us}). \tag{13.96}$$

CP violation comes from interference between these decay amplitudes. To a very good approximation, it is proportional to the relative weak phase:

$$\arg\left(\frac{V_{ub}^* V_{us}}{V_{cb}^* V_{cs}}\right) \approx -\gamma. \tag{13.97}$$

Let us define the following pairs of CP conjugate amplitudes:

$$\begin{aligned} A_{D^0 K^+} &\equiv A(B^+ \to D^0 K^+), & \overline{A}_{\overline{D}^0 K^-} &\equiv A(B^- \to \overline{D}^0 K^-), \\ A_{\overline{D}^0 K^+} &\equiv A(B^+ \to \overline{D}^0 K^+), & \overline{A}_{D^0 K^-} &\equiv A(B^- \to D^0 K^-), \\ A_{D_{CP} K^+} &\equiv A(B^+ \to D_{CP} K^+), & \overline{A}_{D_{CP} K^-} &\equiv A(B^- \to D_{CP} K^-). \end{aligned} \tag{13.98}$$

Since the $B^\pm$ decays into D^0 and $\overline{D}^0$ are tree level, with a single CKM phase for each amplitude, we have

$$|A_{D^0 K^+}| = |\overline{A}_{\overline{D}^0 K^-}|, \qquad |A_{\overline{D}^0 K^+}| = |\overline{A}_{D^0 K^-}|. \tag{13.99}$$

Given equation (13.94), we have

$$A_{D_{CP}K^+} = \frac{A_{D^0K^+} + A_{\overline{D}^0K^+}}{\sqrt{2}}, \qquad \overline{A}_{D_{CP}K^-} = \frac{\overline{A}_{\overline{D}^0K^-} + \overline{A}_{D^0K^-}}{\sqrt{2}}. \tag{13.100}$$

The decay amplitudes $A_{D^0K^+}$ and $A_{\overline{D}^0K^+}$ differ in magnitude, weak phase, and strong phase. The weak-phase difference is γ (see equation (13.97)). We define the ratio of size r and the strong-phase difference δ via

$$\frac{A_{\overline{D}^0K^+}}{A_{D^0K^+}} = re^{i(\delta+\gamma)}, \qquad \frac{\overline{A}_{D^0K^-}}{\overline{A}_{\overline{D}^0K^-}} = re^{i(\delta-\gamma)}. \tag{13.101}$$

We obtain the four decay rates as follows:

$$\begin{aligned}
\Gamma(B^+ \to D^0K^+) &= \Gamma(B^- \to \overline{D}^0K^-) = |A_{D^0K^+}|^2, \\
\Gamma(B^+ \to \overline{D}^0K^+) &= \Gamma(B^- \to D^0K^-) = |A_{D^0K^+}|^2 r^2, \\
\Gamma(B^+ \to D_{CP}K^+) &= |A_{D_{CP}K^+}|^2 = \frac{|A_{D^0K^+}|^2}{2}\left[1 + r^2 + 2r\cos(\delta+\gamma)\right], \\
\Gamma(B^- \to D_{CP}K^-) &= |A_{D_{CP}K^+}|^2 = \frac{|A_{D^0K^+}|^2}{2}\left[1 + r^2 + 2r\cos(\delta-\gamma)\right]
\end{aligned} \tag{13.102}$$

We have four measurements that depend on four parameters: γ, $|A_{D^0K^+}|$, r, and δ. Thus, we can extract γ up to a discrete ambiguity. The crucial point is that we do not need to calculate the magnitude of the amplitudes and the strong phase. Instead, they are extracted from the measurements. Thus, $B \to DK$ decays provide a theoretically clean determination of the *CP* violating phase γ.

13.C.2 Extracting β from $B \to D^+D^-$

The *CP* asymmetry in the $B \to D^+D^-$ decays can be translated into a value of $\sin 2\beta$ (β is defined in equation (9.20)), with small theoretical uncertainties. In practice, the best measurement of $\sin 2\beta$ is provided by $B \to \psi K_S$ decay, but it involves a theoretical subtlety, and therefore we leave its analysis to question 13.5.

Assuming that the approximations that led to equation (13.88) are valid, we have

$$\mathcal{A}_{D^+D^-}(t) = \mathcal{I}m(\lambda_{D^+D^-})\sin(\Delta m_B t). \tag{13.103}$$

We thus need to calculate $\lambda_{D^+D^-}$ as follows:

- Given that $|\Gamma_{B\overline{B}}/M_{B\overline{B}}| \ll 1$, equation (13.70) leads to

$$\frac{q}{p} = \frac{M^*_{B\overline{B}}}{|M_{B\overline{B}}|} = \frac{V^*_{tb}V_{td}}{V_{tb}V^*_{td}}. \tag{13.104}$$

- The decay is dominated by the tree-level $b \to c\bar{c}d$ decay, so we neglect loop corrections and obtain

$$\frac{\overline{A}_{D^+D^-}}{A_{D^+D^-}} = \frac{V_{cb}^* V_{cd}}{V_{cb} V_{cd}^*}. \tag{13.105}$$

Note that all the hadronic parts of the amplitudes cancel out in the ratio.

Using equations (13.90), and combining equation (13.104) and (13.105), we obtain

$$\lambda_{D^+D^-} = \frac{q}{p}\frac{\overline{A}_{D^+D^-}}{A_{D^+D^-}} = \frac{V_{tb}^* V_{td}}{V_{tb} V_{td}^*}\frac{V_{cb}^* V_{cd}}{V_{cb} V_{cb}^*} = -e^{-2i\beta}, \tag{13.106}$$

and consequently,

$$\mathcal{I}m(\lambda_{D^+D^-}) = \sin 2\beta. \tag{13.107}$$

This result demonstrates the power of *CP* asymmetries in measuring CKM parameters. The experimental measurement translates, up to a discrete ambiguity, directly into the value of a CKM parameter, β.

13.C.3 CP Violation from K Decays

CP violation was discovered in neutral kaon decays via measurements of ϵ_K, defined in equation (13.78), using the following two observables:

- *CP* violating in the $K \to \pi \ell \nu$ decays, defined in equation (13.86), provides a measurement of $\mathcal{R}e(\epsilon_K)$:

$$\delta_L = 2\mathcal{R}e(\epsilon_K). \tag{13.108}$$

- *CP* violating in the $K_L \to \pi\pi$ decay, defined in equation (13.92), provides a measurement of $|\epsilon_K|$:

$$\mathcal{A}^{\text{mass}}_{\pi^+\pi^-} = |\epsilon_K|^2. \tag{13.109}$$

The experimental central values, $\delta_L = 3.32 \times 10^{-3}$ and $\mathcal{A}^{\text{mass}}_{\pi^+\pi^-} = 4.97 \times 10^{-6}$, give

$$|\epsilon_K| = 2.23 \times 10^{-3}, \qquad \arg(\epsilon_K) = 43.5^{\text{o}}. \tag{13.110}$$

The Standard Model calculation for $2\mathcal{R}e(\epsilon_K)$ gives

$$2\mathcal{R}e(\epsilon_K) = \frac{G_F^2 m_W^2}{12\pi^2} \times f_K^2 B_K \times \frac{m_K}{\Delta m_K} \times \frac{J_{\text{CKM}}}{|\lambda_u|^2} \times \Big(\mathcal{R}e[\lambda_u\lambda_c^*][S_{cc} - S_{ct}] + \mathcal{R}e[\lambda_u\lambda_t^*][S_{tt} - S_{ct}]\Big), \tag{13.111}$$

where we use the shorthand notations

$$S_{ij} = S(x_i, x_j), \qquad \lambda_i = V_{is} V_{id}^*, \tag{13.112}$$

and equation (9.11), which implies $J_{\text{CKM}} = \mathcal{I}m(\lambda_u\lambda_c^*) = -\mathcal{I}m(\lambda_u\lambda_t^*)$.

For Further Reading

- There are many reviews and books about flavor physics, such as Branco, Lavoura, and Silva [29], Grossman and Tanedo [31], and Ramond [63].
- For a review on flavor and BSM, see, for example, Isidori, Nir, and Perez [44].

Problems

Question 13.1: Algebra

1. Starting from $\mathcal{H}$ as defined in equation (13.41), derive equations (13.44) and (13.48).
2. Any nondegenerate eigenstate of the Hamiltonian must be also an eigenstate of all operators that commute with the Hamiltonian. Thus, if *CP* were a good symmetry, all nondegenerate mass eigenstates must be also *CP* eigenstates. Show that if we impose *CP*, we obtain equations (13.49) and (13.50).
3. In the limit of $|\Gamma_{P\overline{P}}| \ll |M_{P\overline{P}}|$, derive equation (13.51).
4. Derive equations (13.62) under the assumptions stated in the text.
5. Starting from equations (13.60), derive equation (13.88) under the assumptions stated in the text.
6. Using equation (13.74), find the expression for $\mathcal{A}_f$ defined in equation (13.81) in terms of r_f, ϕ_f, and δ_f. Then apply the $r_f \ll 1$ approximation to get equation (13.82).
7. Derive equation (13.91).

Question 13.2: Determining m_c and m_t from neutral meson mixing

Historically, the first estimates of m_c and m_t came from Δm_K and Δm_B, respectively.

1. Determine m_c using the theoretical expression for and the experimental value of Δm_K, as well as the values of the CKM elements. Set $m_u = 0$.
2. Explain why it is practically impossible to get an estimate for m_t using Δm_K.
3. Determine m_t using the theoretical expression and the experimental value of Δm_B, and the values of the CKM elements. Set $m_c = 0$.
4. Compare your results to the direct measurements of m_c and m_t and comment on any difference.

Question 13.3: Mixing in the SM+χ model

Consider the SM+χ model defined in question 9.8 in chapter 9. This model exhibits tree-level, nondiagonal couplings of the Z-boson. The nondiagonal Zbd coupling can be written as

$$-\mathcal{L}_{Zbd} = -\hat{U}_{bd} \frac{g}{2\cos\theta_W} \overline{d_L}\gamma^\mu b_L Z_\mu, \qquad (13.113)$$

where $\hat{U}_{bd}$, defined in question 9.8, is dimensionless and complex. We assume $|\hat{U}_{bd}| \ll 1$, and thus we can approximate the 3×3 CKM matrix as unitary and use its values as extracted within the Standard Model.

1. In this model, there is a tree-level contribution to $B^0 - \overline{B}^0$ mixing. Similar to the Standard Model amplitude in equation (13.63), we write

$$\mathcal{A}^{\rm T}_{B^0 \to \overline{B}^0} = C_{\rm T} (\overline{d_L} \gamma_\mu b_L)(\overline{d_L} \gamma^\mu b_L). \tag{13.114}$$

   Draw the tree-level, Z-exchange diagrams that contributes to the mixing and evaluate $C_{\rm T}$. (You can neglect effects of order $m_{\rm B}^2/m_{\rm Z}^2$.) How does the amplitude depend on $\hat{U}_{db}$?
2. Use the definition Δ_d in equation (13.32) to write

$$M^{\rm T}_{B\overline{B}} = M^{\rm SM}_{B\overline{B}} (\Delta_d - 1), \tag{13.115}$$

   find $\Delta_d - 1$ in terms of $\hat{U}_{bd}$ and the Standard Model parameters.
3. The bounds on the real and imaginary parts of Δ_d are given in equation (13.35). Using $\left|\Delta_d - 1\right| < 0.2$ and $S(x_t, x_t) \approx 2.5$, obtain the bound on $|\hat{U}_{bd}|$.
4. We write the mass terms that involve the χ-field as

$$m_{\chi\chi} \overline{\chi_L} \chi_R + (v/\sqrt{2}) Y_{\chi i} \overline{D_{Li}} \chi_R, \qquad i = 1, 2, 3. \tag{13.116}$$

   We assume that $m_{\chi\chi} \gg v$ and $Y_{\chi i} \sim 1$. Argue that, under these assumptions, it is expected that

$$|\hat{U}_{bd}| \sim \frac{v^2}{m^2_{\chi\chi}}. \tag{13.117}$$

   Recall that a non-Hermitian matrix M can be diagonalized by the bi-unitary transformation $V_L M V_R^\dagger$. To find V_L (V_R), it is simpler to consider the Hermitian matrix $MM^\dagger$ ($M^\dagger M$), as it is diagonalized by V_L (V_R).
5. Given the bound on $|\hat{U}_{bd}|$, what can you say about $m_{\chi\chi}$? Comment about the numerical value.

Question 13.4: CP violation in $B \to \pi^+\pi^-$

The time-dependent CP asymmetry in $B \to \pi^+\pi^-$ probes the unitarity triangle angle α. We define $A \equiv \mathcal{A}(B^0 \to \pi^+\pi^-)$ and $\overline{A} \equiv \mathcal{A}(\overline{B}^0 \to \pi^+\pi^-)$.

1. Draw the tree-level decay diagrams for A and write the CKM dependence of both A and $\overline{A}$.
2. Argue that, in the tree-level approximation, $\overline{A}/A$ is a pure phase, and express it in terms of CKM elements.
3. In the approximation that $|\overline{A}/A| = 1$ and $|q/p| = 1$, $\lambda_{\pi^+\pi^-}$ is also a pure phase. Show that the phase of $\lambda_{\pi^+\pi^-}$ is 2α.

4. There are also contributions from one-loop diagrams to $B \to \pi^+\pi^-$. Draw a representative diagram, keeping all internal quarks, and show that this loop-amplitude provides a weak phase that is different from the phase of the tree-level diagram.
5. Estimate the relative error on the extracted value of α introduced by neglecting the loop amplitude.

Question 13.5: CP violation in $B \to \psi K_S$

1. Given that $|\Gamma_{B\overline{B}}/M_{B\overline{B}}| \ll 1$, we define

$$\frac{M^*_{B\overline{B}}}{|M_{B\overline{B}}|} \equiv e^{i\phi_B}. \tag{13.118}$$

Using equation (13.70), find ϕ_B in terms of the ratio of CKM matrix elements.

2. Write the underlying quark transition for $B^0 \to \psi K^0$. Draw the tree-level diagram. What is the CKM dependence of this diagram?

The $B^0 \to \psi K_S$ decay can be considered as $B^0 \to \psi K^0$, followed by $K^0 \to \pi\pi$, which (neglecting CP violation in $K^0 - \overline{K}^0$ mixing) projects it onto the K_S state:

$$A_{\psi K_S} = A_{\psi K} \times A(K \to \pi\pi). \tag{13.119}$$

3. Neglecting loop contribution to the decay, we define

$$\frac{\overline{A}_{\psi\overline{K}}}{A_{\psi K}} = \exp[-2i\phi_{\psi K}]. \tag{13.120}$$

Find $\phi_{\psi K}$ in terms of a ratio of CKM matrix elements.

4. Neglecting loop contribution to the decay, we define

$$\frac{A(\overline{K} \to \pi\pi)}{A(K \to \pi\pi)} = \exp[-2i\phi_{K\to\pi\pi}]. \tag{13.121}$$

Find $\phi_{K\to\pi\pi}$ in terms of a ratio of CKM matrix elements.

5. We define

$$\frac{\overline{A}_{\psi K_S}}{A_{\psi K_S}} = \exp[2i\phi_{\psi K_S}]. \tag{13.122}$$

Find $\phi_{\psi K_S}$ in terms of a ratio of CKM matrix elements.

6. Argue that, under the assumptions made in items 1, 3, and 4, $|\lambda_{\psi K_S}| = 1$.
7. Given that $|\lambda_{\psi K_S}| = 1$, we can use the expression in equation (13.88). Show that $\mathcal{I}m(\lambda_{\psi K_S}) = \sin 2\beta$.

We learn that $S_{\psi K_S}$ measures the angle β of the unitarity triangle.

Question 13.6: Comparing CP violation in $B \to \psi K_S$ and $B \to \phi K_S$

1. Write the underlying quark transition for $\overline{B}^0 \to \phi \overline{K}^0$. Explain why there is no tree-level diagram.
2. The leading one-loop diagrams for $\overline{B}^0 \to \phi \overline{K}^0$ are gluonic penguin diagrams. Draw a representative diagram and keep all internal quarks in the loop. Using CKM unitarity, eliminate the t-term and write the amplitude in a way similar to equation (13.3).
3. Argue that, to a good approximation, we can write the ratio of amplitude as a pure phase:

$$\frac{\overline{A}_{\phi K_S}}{A_{\phi K_S}} = \exp[2i\phi_{\phi K_S}]. \tag{13.123}$$

   Find $\phi_{\phi K_S}$ in terms of a ratio of CKM matrix elements.
4. Argue that, to a good approximation, $|\lambda_{\phi K_S}| = 1$.
5. Using equation (13.90), we learn that $S_{\psi K_S}$ and $S_{\phi K_S}$ each measures an angle of the unitarity triangle. Show that $\mathcal{I}m(\lambda_{\psi K_S}) = \mathcal{I}m(\lambda_{\phi K_S}) = \sin 2\beta$.
6. The standard model predicts that to a good approximation, $S_{\psi K_S} = S_{\phi K_S}$. Find in the Particle Data Group (PDG) their measured values and check whether the central values are within 2σ of each other.
7. Assume that in the future, experiments establish a statistically significant difference between $S_{\psi K_S}$ and $S_{\phi K_S}$. The following are three possible explanations for such a discrepancy. For each explanation, discuss whether it can resolve the discrepancy.
   (a) There are Standard Model contributions that we neglected.
   (b) There is a BSM contribution to $B^0 - \overline{B}^0$ mixing, with a weak phase that differs from the Standard Model one.
   (c) There is a BSM contribution to the gluonic penguin, with a weak phase that differs from the Standard Model one.

Question 13.7: The neutral K-meson system

We consider the following neutral K-meson decays:

$$K^0 \to \pi\pi, \qquad K^0 \to \pi\pi\pi, \tag{13.124}$$

where we keep the electric charges of the pions implicit. We assume that the isospin and CP symmetries are respected, which turn out to be very good approximations.

1. Explain why, in the CP limit, the two eigenstates, K_S and K_L, must also be CP eigenstates.
2. Explain why the CP-odd state cannot decay into two pions.
3. Examining the data, which of the two states, K_S or K_L, can be identified as the CP-even state?

4. Taking into account the experimental values of m_π and m_K, explain why we should expect

$$\Gamma(K^0 \to \pi\pi) \gg \Gamma(K^0 \to \pi\pi\pi). \qquad (13.125)$$

5. The $K^0 \to \pi\pi$ is the dominant decay mode of K^0. Check it explicitly by examining BR($K_S \to \pi\pi$). Explain why this fact, together with equation (13.125), imply that $\Gamma_S \gg \Gamma_L$.
6. Experimentally, $y_K \approx 1$. How would the value of y_K change if the kaon were slightly heavier (with the pion mass unchanged)?

Question 13.8: CP violation in $K \to \pi\pi$ and $K \to \pi\ell\nu$

For this question, we introduce the following notations:

$$y_{12} \equiv \frac{|\Gamma_{K\overline{K}}|}{\Gamma}, \qquad x_{12} \equiv \frac{2|M_{K\overline{K}}|}{\Gamma}, \qquad \phi_{12} \equiv \arg\left(\frac{M_{K\overline{K}}}{\Gamma_{K\overline{K}}}\right). \qquad (13.126)$$

Using these notations, equations (13.48) can be written as

$$xy = x_{12}y_{12}\cos\phi_{12}, \qquad x^2 - y^2 = x_{12}^2 - y_{12}^2, \qquad (13.127)$$

where x ad y are defined is equations (13.52). We further define

$$\lambda_{\pi\pi} = |\lambda_{\pi\pi}|e^{i\phi_{\pi\pi}}. \qquad (13.128)$$

As discussed in question 13.7, the two pion states make the dominant contributions to the width difference. We can then assume that

$$\mathcal{I}m\left(\Gamma_{K\overline{K}}\frac{\overline{A}_{\pi\pi}}{A_{\pi\pi}}\right) = 0. \qquad (13.129)$$

The $K \to \pi\pi$ decay is dominated by a single CKM phase, so we can also assume that

$$|\overline{A}_{\pi\pi}/A_{\pi\pi}| = 1. \qquad (13.130)$$

1. Using equations (13.44), (13.127), and (13.126), show that

$$(x^2 + y^2)\,|q/p|^2 = x_{12}^2 + y_{12}^2 + 2x_{12}y_{12}\sin\phi_{12}. \qquad (13.131)$$

2. Using equations (13.129) and (13.130) as well, show that

$$x^2\cos^2\phi_{\pi\pi} - y^2\sin^2\phi_{\pi\pi} = x_{12}^2\cos^2\phi_{12}. \qquad (13.132)$$

3. In the Standard Model, both $M_{K\overline{K}}$ and $\Gamma_{K\overline{K}}$ are dominated by the first two generations, and thus we expect $\phi_{12} \ll 1$. Under this assumption, show that

$$|q/p| - 1 = \frac{(y/x)\tan\phi_{12}}{1 + (y/x)^2}, \qquad \tan\phi_{\pi\pi} = \frac{-\tan\phi_{12}}{1 + (y/x)^2}. \qquad (13.133)$$

4. Use equation (13.133) to show that

$$\frac{y}{x} = \frac{1 - |q/p|}{\tan \phi_{\pi\pi}}. \tag{13.134}$$

5. Show that equation (13.134) translates into the following relation between measured observables:

$$\frac{[\mathcal{R}e(\epsilon_K)]^2}{|\epsilon_K|^2} = \frac{y^2}{x^2 + y^2}. \tag{13.135}$$

6. Use the experimental values of δ_L, $\mathcal{A}_{\pi\pi}^{\text{mass}}$, y, and x to test the theoretical prediction of equation (13.135), and comment about the comparison.

Question 13.9: Time scales in meson oscillations

In the following question, we consider the four neutral meson systems, $P = K, D, B$, and B_s. There are various timescales involved in neutral meson mixing: Δm, $\Delta\Gamma$, and Γ. Understanding the hierarchy (or lack of hierarchy) between them leads to insights and simplifications. We use the dimensionless quantities x and y, as defined in equation (13.52), to understand the possible hierarchies between the relevant timescales.

Evaluate $\mathcal{P}(P^0 \to P^0)$ and $\mathcal{P}(P^0 \to \overline{P}^0)$ in the following limits:

1. $y = 0$ and $x \ll 1$ ("slow oscillation"). Keep the leading-order terms in x in each case. Explain the result. Which of the four neutral meson systems fall into this category?
2. $y = 0$ and $x \gg 1$ ("fast oscillation"). Argue that in this case, you can average over the oscillations. Give the time-averaged oscillation to first-order in $1/x$. Explain the result. Which of the four neutral meson systems fall into this category?
3. $y = 0$ and $x \sim 1$. In this case, we have to use the full formula. Which of the four neutral meson systems fall into this category?
4. Derive a generalization of equation (13.62), still keeping $\lambda_f = 0$, but taking into account finite x and y.
5. $|y| \ll x \lesssim 1$. Argue that in this case, the effect of the width difference is quadratic in y. Which of the four neutral meson systems fall into this category?
6. $y \sim x \ll 1$. Argue that in this case, the width difference is as important as the mass difference. Which of the four neutral meson systems fall into this category?
7. $1 - |y| \ll 1$. Argue that in that case, independent of the value of x, you can think about the state as the long-lived one at late times. What can be considered "late times"? Which of the four neutral meson systems fall into this category?

There are additional limits that are not realized in the four neutral meson systems, and thus we do not elaborate on them.

14

Neutrinos

The Standard Model has an accidental $U(1)_e \times U(1)_\mu \times U(1)_\tau$ symmetry, also known as *lepton flavor symmetry*. Experiments established, however, that lepton flavor is not conserved, which provides unambiguous evidence for physics beyond the renormalizable Standard Model.

Predictions that depend on accidental symmetries of the Standard Model are violated in generic extensions of it. In chapter 12, we encountered $d = 6$ terms that violate the approximate custodial symmetry of the scalar sector of the Standard Model and are consequently constrained by electroweak precision measurements (EWPM). In chapter 13, we encountered $d = 6$ terms that violate the approximate flavor symmetries of the Standard Model and are consequently constrained by measurements of flavor-changing neutral current (FCNC) processes. In this chapter, we encounter $d = 5$ terms that violate the accidental lepton number and lepton flavor symmetries of the Standard Model and discuss how they can be experimentally probed.

14.1 Introduction

The accidental lepton flavor symmetry of the Standard Model predicts that lepton flavor is conserved. Thus, for example, the $\mu \to e\gamma$ process is forbidden since it violates both the muon number and electron number by 1 unit. Indeed, this and many other lepton flavor violating processes that have been searched for in experiments were not observed. Yet neutrino oscillation experiments have observed lepton flavor transitions, thus violating lepton flavor conservation. These experiments involve a source that produces neutrinos of a given flavor, as well as a detector, placed at a large distance from the source, that determines their flavor upon arrival. In this way, $\nu_\mu \to \nu_\tau$ transitions were established for neutrinos produced in the atmosphere of the Earth (atmospheric neutrinos) and $\nu_e \to \nu_a$ $(a = \mu, \tau)$ transitions were established for neutrinos produced in the sun (solar neutrinos). These results were independently confirmed using neutrinos produced in nuclear reactors and particle accelerators.

To test the conservation of total lepton number, a very different type of experiment is needed. These experiments search for $nn \to ppee$ processes, known as *neutrinoless double-beta decay*. So far no such process has been observed, and thus there is no experimental indication that total lepton number is violated.

The observed lepton flavor violation is well explained if neutrinos have masses and there is lepton mixing, similar to the mixing in the quark sector. The smallness of neutrino masses compared to the charged fermions ($m_\nu \lesssim 1$ eV), as well as the fact that they have neither strong nor electromagnetic interaction, have made the search for neutrino oscillations very challenging, and the discovery of neutrino masses a triumph of experimental particle physics.

When an accidental symmetry is experimentally shown to be violated, one first should consider the lowest dimension nonrenormalizable terms that break that symmetry. This is done in section 14.2, where we show that $d=5$ terms break the lepton flavor symmetry. The Standard Model extended to include the $d=5$ terms is an effective low-energy theory. In section 14.3, we present a renormalizable theory that can generate the effective $d=5$ terms. In Appendix 14.A, we discuss the way that the neutrino parameters are probed.

14.2 The νSM

14.2.1 Defining the νSM and the Lagrangian

In section 8.1 in chapter 8, we defined the Standard Model. The model that we call the νSM has the same symmetry and particle content as the Standard Model, but it extends the Standard Model Lagrangian, given in section 8.2, to include the most general $d=5$ terms. In fact, within the Standard Model effective field theory (SMEFT), there is a single class of dimension-five terms that depend on Standard Model fields and obey the Standard Model-imposed symmetries. These terms involve two $SU(2)$-doublet lepton fields and two $SU(2)$-doublet scalar fields:

$$\mathcal{L}_{\nu\text{SM}} = \mathcal{L}_{\text{SM}} + \frac{z_{ij}^{\nu}}{\Lambda}\phi^T \phi L_{Li}^T L_{Lj} + \text{h.c.}, \tag{14.1}$$

where $\mathcal{L}_{\text{SM}}$ is the Lagrangian of the renormalizable Standard Model (see equation (8.17)), z_{ij}^{ν} is a symmetric complex 3×3 matrix of dimensionless couplings, Λ is a high-mass scale, $\Lambda \gg v$, and, as explained in section 1.2.2 in chapter 1, for the sake of simplicity, we write L_L^T instead of $\overline{L_L^c}$.

14.2.2 The Neutrino Spectrum

With ϕ acquiring a vacuum expectation value (VEV), $\langle \phi^0 \rangle = v/\sqrt{2}$ (see equation (7.18) in chapter 7), $\mathcal{L}_{\nu\text{SM}}$ in equation (14.1) has a term that corresponds to Majorana masses for the neutrinos. For a single generation, we would have

$$\mathcal{L}_{\nu\text{SM,mass}} = \frac{1}{2} m_\nu \nu^T \nu, \qquad m_\nu = \frac{v^2}{\Lambda} z^{\nu}. \tag{14.2}$$

For three generations, we have a 3×3 mass matrix:

$$\mathcal{L}_{\nu\text{SM,mass}} = \frac{1}{2}(m_\nu)_{ij}\nu_i^T\nu_j, \qquad (m_\nu)_{ij} = \frac{v^2}{\Lambda}z_{ij}^\nu. \tag{14.3}$$

The matrix m_ν can be brought to a diagonal and a real form by a unitary transformation:

$$V_{\nu L}^* m_\nu V_{\nu L}^\dagger = \hat{m}_\nu = \text{diag}(m_1, m_2, m_3), \tag{14.4}$$

where $V_{\nu L}$ is a 3×3 unitary matrix. Majorana mass matrices are always symmetric. While the diagonalization of a general mass matrix M involves a general bi-unitary transformation, as in equation (7.98) in chapter 7, for a symmetric mass matrix the diagonalization is by a unitary matrix and its transpose, as in equation (14.4). (Denoting the unitary matrix by V^*, rather than by V, simplifies the discussion that follows.)

We denote the corresponding neutrino mass eigenstates by ν_1, ν_2, and ν_3. What oscillation experiments have determined are neutrino mass differences:

$$\Delta m_{ij}^2 \equiv m_i^2 - m_j^2, \tag{14.5}$$

with $i = 1, 2, 3$ and $i \neq j$. By convention,

$$|\Delta m_{3i}^2| > \Delta m_{21}^2 > 0. \tag{14.6}$$

In other words, the convention is such that states ν_1 and ν_2 are the ones separated by the smallest $|\Delta m_{ij}^2|$ with $m_2 > m_1$.

The values extracted from neutrino oscillation experiments are

$$\begin{aligned} \Delta m_{21}^2 &= (7.53 \pm 0.18) \times 10^{-5}\ eV^2, \\ |\Delta m_{32}^2| &= (2.453 \pm 0.033) \times 10^{-3}\ eV^2. \end{aligned} \tag{14.7}$$

The following two features of the neutrino spectrum are still not experimentally determined:

- The absolute mass scale of the neutrinos (any of the m_is).
  The possibilities range from quasi-degenerate neutrino masses to highly hierarchical (with the lightest possibly being massless).
- The mass ordering (sign of Δm_{3i}^2).
  It is not yet known experimentally whether m_3 is heavier ("normal ordering") or lighter ("inverted ordering") than m_2 and m_1.

While the measurements of the neutrino mass differences do not tell us the individual masses of the neutrinos, they do provide lower bounds on two mass eigenvalues: There is at least one neutrino mass heavier than $\sqrt{|\Delta m_{32}^2|}$:

$$m_{\text{heaviest}} \geq \sqrt{|\Delta m_{32}^2|} \sim 0.05 \text{ eV}, \tag{14.8}$$

and there is at least one additional mass heavier than $\sqrt{\Delta m_{21}^2} \sim 0.009$ eV.

As for an upper bound on neutrino masses, experiments and cosmological observations set it at a scale of order 1 eV. We present the experimental upper bounds in Appendix 14.B. We do not elaborate much on the cosmological bounds, but we will mention them in section 15.2.2 in chapter 15.

14.2.3 The Neutrino Interactions

The Standard Model interactions whose structure is most significantly modified due to the addition of the dimension-five terms are the charged current weak interactions in the lepton sector. Similar to the quark sector, the W-couplings to lepton mass eigenstate pairs are no longer universal but characterized by a mixing matrix.

To understand how the leptonic mixing matrix arises, we follow the steps of the analysis of the quark sector in section 8.4.4 of chapter 8. In the interaction basis, the charged current weak interaction is given by (see equation (8.48))

$$\mathcal{L}_{W,\ell} = -\frac{g}{\sqrt{2}} \sum_{\ell=e,\mu,\tau} \left(\overline{\ell_L} \not{W}^- \nu_{\ell L} + \text{h.c.} \right). \tag{14.9}$$

Transforming to the mass basis, we obtain

$$\mathcal{L}_{W,\ell} = -\frac{g}{\sqrt{2}} \left(\overline{\ell_\alpha} P_L \not{W}^- U_{\alpha i} \nu_i + \text{h.c.} \right), \tag{14.10}$$

where $\alpha = e, \mu, \tau$ denotes the charged lepton mass eigenstates and $i = 1, 2, 3$ denotes the neutrino mass eigenstates. The lepton-mixing matrix U is unitary. Starting from an arbitrary interaction basis, it is given by

$$U = V_{eL} V_{\nu L}^\dagger, \tag{14.11}$$

where $V_{\nu L}$ is defined in equation (14.4) and V_{eL} is defined in equation (7.98) in chapter 7. While V_{eL} and $V_{\nu L}$ are both basis dependent, the combination $V_{eL} V_{\nu L}^\dagger$ is basis independent. Explicitly, we denote the entries of U as follows:

$$U = \begin{pmatrix} U_{e1} & U_{e2} & U_{e3} \\ U_{\mu 1} & U_{\mu 2} & U_{\mu 3} \\ U_{\tau 1} & U_{\tau 2} & U_{\tau 3} \end{pmatrix}. \tag{14.12}$$

The matrix U plays a role in the lepton sector similar to the role of the Cabibbo-Kobayashi-Maskawa (CKM) matrix V in the quark sector. It is usually called the *leptonic mixing matrix* or the *Pontecorvo-Maki-Nakagawa-Sakata* (PMNS) *matrix*. We conclude that the leptonic charged current weak interactions in the mass basis are neither universal nor diagonal. Instead, they involve the unitary mixing matrix U.

As for the other Standard Model interactions, they are either not affected at all or modified in a negligible way:

- Since the neutrinos couple to neither the gluon nor the photon, neither the strong nor the electromagnetic (EM) interactions are modified.

Table 14.1. The νSM lepton interactions

Interaction	Force Carrier	Coupling	Leptons	Properties
Electromagnetic	γ	eQ	$\overline{\ell_i}\ell_i$	Universal
Neutral current weak	Z^0	$(g/\cos\theta_W)(T_3 - s_W^2 Q)$	$\overline{\ell_i}\ell_i, \overline{\nu_i}\nu_i$	Universal
Charged current weak	$W^{\pm}$	$(g/\sqrt{2})U_{ij}$	$\overline{\ell_i}\nu_j$	Unitary U
Yukawa	h	$\sqrt{2}m_\ell/v, 2\sqrt{2}m_\nu/v$	$\overline{\ell_i}\ell_i, \nu_i\nu_i$	Diagonal

- The Z-mediated neutral current weak interactions are modified in a trivial way. Transforming from the interaction basis (see equation (7.63)) to the mass basis, we find that the neutral current weak interactions remain diagonal and universal:

$$\mathcal{L}_{Z,\nu} = \frac{g}{2\cos\theta_W}\overline{\nu_i}\not{Z}\nu_i. \tag{14.13}$$

- The dimension-five terms generate Yukawa interactions of the Higgs boson with the neutrinos:

$$\mathcal{L}_{h,\nu} = \frac{2m_i}{v}h\nu_i^T\nu_i + \frac{m_i}{v^2}hh\nu_i^T\nu_i. \tag{14.14}$$

  - Similar to the charged fermions, in the mass basis, the neutrino Yukawa couplings are diagonal ($Y_{ij}^\nu = 0$ for $i \neq j$), proportional to the neutrino masses ($y_i/y_j = m_i/m_j$), and CP conserving ($\mathcal{I}m(y_i/m_i) = 0$).
  - Since the neutrino Yukawa couplings come from terms that involve two Higgs-doublet fields, the relation to the neutrino masses is modified by a combinatorial factor of 2, compared to the charged fermions: $y_i = 2(\sqrt{2}m_i/v)$.
  - The size of the Yukawa couplings is tiny, however, of the order of $m_i/v \lesssim 10^{-11}$, leading to unobservably small effects.

The νSM-lepton interactions are summarized in table 14.1, which is the νSM modification of the Standard Model interactions given in table 7.2 in chapter 7.

14.2.4 Global Symmetries and Parameters

The dimension-five terms in equation (14.1) break the $U(1)_e \times U(1)_\mu \times U(1)_\tau$ accidental symmetry of the Standard Model. With the addition of only $d = 5$ terms, all that remains of the $G_{\rm SM}^{\rm global}$ symmetry of the Standard Model (see equation (8.60)) is baryon number symmetry:

$$G_{\nu\rm SM}^{\rm global} = U(1)_B. \tag{14.15}$$

However, this symmetry is anomalous and broken by nonperturbative effects. In addition, it is broken by dimension-six terms.

The total lepton number, conventionally denoted by L, is the sum of the three lepton flavor numbers, $U(1)_L = U(1)_{e+\mu+\tau}$. The νSM breaks not only the lepton flavor symmetries separately, but also their sum (namely, the total lepton number). Experiments, however, thus far have established the violation of the electron, muon, and tau number separately, but not the violation of the total lepton number.

The number of physical parameters in the quark sector remains unchanged from the Standard Model: six quark masses and four CKM parameters, of which one is imaginary. This is not the case for the lepton sector. We proceed to count and identify the physical parameters in the νSM that add to those of the Leptonic Standard Model (LSM) (see section 7.5.3 in chapter 7). The Lagrangian of equation (14.1) involves the 3×3 matrix Y^e (see equation (8.17) in chapter 8) with 9 real and 9 imaginary parameters, and the symmetric 3×3 matrix z^ν with 6 real and 6 imaginary parameters. The kinetic and gauge terms have a $U(3)_L \times U(3)_E$ symmetry, which is completely broken by the Y^e and z^ν terms. Thus, the number of physical lepton flavor parameters is $30 - 2 \times 9 = 12$, out of which 9 are real and 3 are imaginary. Out of the 9 real parameters 3 are the charged lepton masses m_e, m_μ, m_τ and 3 are the neutrino masses m_1, m_2 and m_3. We conclude that the 3×3 lepton mixing matrix U depends on 3 real mixing angles and 3 phases.

14.2.5 The PMNS Matrix

A convenient parameterization of the PMNS matrix U is the following:

$$U = \begin{pmatrix} c_{12}c_{13} & s_{12}c_{13} & s_{13}e^{-i\delta} \\ -s_{12}c_{23} - c_{12}s_{23}s_{13}e^{i\delta} & c_{12}c_{23} - s_{12}s_{23}s_{13}e^{i\delta} & s_{23}c_{13} \\ s_{12}s_{23} - c_{12}c_{23}s_{13}e^{i\delta} & -c_{12}s_{23} - s_{12}c_{23}s_{13}e^{i\delta} & c_{23}c_{13} \end{pmatrix} \times \operatorname{diag}(1, e^{i\alpha_1}, e^{i\alpha_2}), \tag{14.16}$$

where $s_{ij} \equiv \sin\theta_{ij}$ and $c_{ij} \equiv \cos\theta_{ij}$.

The lepton mixing matrix U depends on three phases. The phase δ arises in a way similar to the Kobayashi–Maskawa (KM) phase of the quark-mixing matrix. The phases α_1 and α_2 are called *Majorana phases,* and have no analog in the CKM matrix V (see equation (9.3)). The reason that U depends on three phases while V depends on only a single phase lies in the fact that the Lagrangian of equation (14.1) leads to Majorana masses for neutrinos. Consequently, there is no freedom in changing the mass basis by redefining the neutrino phases, as such a redefinition will introduce phases to the neutrino mass terms. While redefinitions of the six quark fields allow one to remove five nonphysical phases from V, redefinitions of the three charged lepton fields allow one to remove only three nonphysical phases from U. The two additional physical phases in U are called *Majorana phases* because they appear as a result of the assumed Majorana nature of neutrinos. They affect total lepton number–violating processes.

The theoretical discussions of mixing in the quark and the lepton sector are very similar. The way that experiments probe these mixings, however, are very different. For particles that have electromagnetic (and strong) interactions—as is the case for quarks and charged leptons—experiments identify the mass eigenstates. For particles that have only weak interactions—as is the case for the neutrinos—experiments identify the interaction eigenstates. Explicitly, the neutrinos are identified as ν_e, ν_μ, or ν_τ according to whether their production or detection are associated with an e, μ, or τ lepton, respectively.

The difference in the experimental probes suggests an alternative way to think about lepton mixing. The point that is common to quarks and leptons is that there are three relevant matrices in each sector—two mass matrices and one matrix for the W-couplings—and only two can be simultaneously diagonal. For quarks, it is convenient to work in the mass basis where M_u and M_d are diagonal but V is not. For leptons, it is often more convenient to work in an interaction basis, where M_e and the W-couplings are diagonal, but M_ν is not.

Also, note a subtle difference between V and U in the convention for the order of indexes. In V, the first index corresponds to the $T_3 = +1/2$ component of the doublet and the second index to the $T_3 = -1/2$ component. In U, it is the other way around.

Neutrino flavor transitions have been observed for solar, atmospheric, reactor, and accelerator neutrinos. Five flavor parameters have been measured: two mass differences (see equation (14.7)) and three mixing angles (we describe the experimental determination of the lepton-mixing parameters in Appendix 14.A):

$$\begin{aligned}\sin^2\theta_{12} &= 0.307 \pm 0.013,\\ \sin^2\theta_{23} &= 0.546 \pm 0.021,\\ \sin^2\theta_{13} &= 0.0220 \pm 0.0007.\end{aligned} \tag{14.17}$$

The present status of our knowledge of the absolute values of the various entries in the lepton-mixing matrix can be summarized as follows:

$$|U| = \begin{pmatrix} 0.80-0.84 & 0.52-0.59 & 0.14-0.16 \\ 0.23-0.48 & 0.46-0.67 & 0.65-0.78 \\ 0.30-0.53 & 0.50-0.70 & 0.61-0.75 \end{pmatrix}. \tag{14.18}$$

The CP-violating Dirac phase δ is constrained by measurements of the ν_e appearance from atmospheric and accelerator experiments, $\delta = 1.36^{+0.20}_{-0.16}$. The two Majorana phases, α_1 and α_2, are still not experimentally determined.

14.2.6 Testing the νSM

While all neutrino-related experimental results to date can be accommodated in the νSM, there are other models that are consistent with the data. Here, we briefly mention a few tests that can probe the νSM:

- In the νSM, the total lepton number is broken and the neutrinos have Majorana masses. Experimentally, however, we do not know the nature of the neutrino mass. If experiments establish lepton number violation via the observation of neutrinoless double beta decays, it will confirm that aspect of the νSM.
- So far, lepton flavor violation was observed only in neutrino oscillation experiments. Yet it could also be searched for in decays involving charged leptons, such as $\mu \to e\gamma$, $Z \to \tau\mu$, and $h \to \tau e$. The νSM predicts that these decay rates are tiny, well below any possible experimental sensitivity (see question 14.3 at the end of

this chapter). If experiments observe lepton flavor violation in charged lepton, Z-boson, or h-boson decays, that will indicate that there are sources of lepton flavor violation beyond the νSM.

- In the νSM, there are three charged leptons and three neutrinos, and consequently, the 3×3 PMNS matrix is unitary. If experiments establish that U is not unitary, additional leptons must exist in nature.

14.2.7 The Scale Λ

In the νSM, the scale of neutrino masses depends on the scale Λ in the coefficients of the dimension-five terms. So long as experiments involve energies $E \ll \Lambda$, what is probed is the combination z^ν/Λ. Thus, there is an ambiguity in the definitions of Λ and z^ν. The separation of the coefficient of a $d=5$ term to a dimensionless coupling and a scale is meaningful when we discuss a full high-energy theory that generates the effective term. What we refer to as the scale of a nonrenormalizable term is Λ/z^ν (or, if z^ν is a matrix, as in equation (14.1), $\Lambda/z^\nu_{\rm max}$, where $z^\nu_{\rm max}$ is the largest eigenvalue of z^ν).

The effective low-energy Lagrangian of equation (14.1), where $\Lambda \gg v$ by definition, predicts that the neutrino masses are much lighter than the weak scale:

$$m_{1,2,3} \lesssim v^2/\Lambda \ll v. \tag{14.19}$$

The fact that experiments find that $m_\nu \ll m_Z$ makes the notion that neutrino masses are generated by $d=5$ terms very plausible. In fact, all fermions of the Standard Model except for the top quark are light relative to m_Z. The lightness of charged fermions is related to the smallness of the corresponding Yukawa couplings. The question of why Yukawa couplings are small may find an answer in a more fundamental theory—namely, beyond the Standard Model (BSM). The neutrinos, however, are not only much lighter than m_Z, but also lighter by at least six orders of magnitude than all charged fermions. This extreme lightness of the neutrinos is explained if their masses are generated by $d=5$ terms.

The Standard Model cannot be a valid theory above the Planck scale, $\Lambda \lesssim M_{\rm Pl}$. We thus expect that $m_i \gtrsim v^2/M_{\rm Pl} \sim 10^{-5}$ eV. A more relevant scale might be the Grand Unified Theory (GUT) scale. As discussed in section 11.4 in chapter 11, in GUTs, a unifying simple group is spontaneously broken to the Standard Model gauge group at a scale of $\Lambda_{\rm GUT} = \mathcal{O}(10^{16}$ GeV$)$. If the $d=5$ terms are generated at $\Lambda_{\rm GUT}$, then with $z^\nu \sim 1$, we expect $m_\nu \sim 10^{-2}$ eV.

Conversely, an experimental lower bound on neutrino masses provides an upper bound on Λ/z^ν. Note, however, that the combination of a measurement of Λ/z^ν and the assumption that z^ν is generated by perturbative physics (and therefore $z^\nu \lesssim 1$), translates into an upper bound on Λ itself. Using the lower bound of equation (14.8) and the relation of equation (14.19), we conclude that the Standard Model cannot be a valid theory above the scale:

$$\Lambda \lesssim \frac{v^2}{m_\nu} \sim 10^{15}\ \text{GeV}, \tag{14.20}$$

and in particular up to the Planck scale. Furthermore, this upper bound is intriguingly close to the GUT scale.

14.3 The NSM: The Standard Model with Singlet Fermions

In section 14.2, we introduced the νSM and showed that the addition of nonrenormalizable, dimension-five terms to the Standard Model Lagrangian generates neutrino masses and, moreover, explains why they are much lighter than the charged fermions. Nonrenormalizable terms must arise from a more fundamental theory. For the full high energy theory, we again write only renormalizable terms. One uses the term *UV-completion* for a full high-energy theory that leads to the effective theory. In general, there is a number of possible ultraviolet (UV)–completions for a given effective theory.

In this section, we present the NSM, a model that serves as an example of a full high-energy theory that generates the νSM at low energy. This extension amounts to adding heavy gauge-singlet fermions to the Standard Model. The way that these singlet fields generate masses to the light neutrinos is called the *seesaw mechanism*. The reason for this name, as explained next, is that the heavier the singlet fermions, the lighter the neutrinos are.

The Lagrangian $\mathcal{L}_{d=5}$ of equation (14.1) can come not only from the NSM, but also from other high-energy theories. In particular, there are three types of seesaw models that differ by the representation of the heavy fields that one adds to the Standard Model:

- Type I seesaw: $(1,1)_0$ fermion fields. (This is the NSM discussed in this chapter.)
- Type II seesaw: $(1,3)_{-1}$ scalar fields. (See question 5.)
- Type III seesaw: $(1,3)_0$ fermion fields. (See question 4.)

All three types of seesaw models predict that the light neutrinos have Majorana masses and their mass scale is inversely proportional to the mass scale of the new heavy particles and, in particular, much lower than the electroweak-breaking scale.

14.3.1 Defining the NSM

The NSM is defined as follows:

1. The symmetry is a local

$$SU(3)_C \times SU(2)_L \times U(1)_Y. \tag{14.21}$$

2. There are three fermion generations ($i = 1, 2, 3$), each consisting of six representations:

$$Q_{Li}(3,2)_{+1/6},\quad U_{Ri}(3,1)_{+2/3},\quad D_{Ri}(3,1)_{-1/3},\quad L_{Li}(1,2)_{-1/2},$$
$$E_{Ri}(1,1)_{-1},\quad N_{Ri}(1,1)_0. \tag{14.22}$$

3. There is a single scalar multiplet:

$$\phi(1,2)_{+1/2}. \tag{14.23}$$

The presence of a scalar allows spontaneous symmetry breaking (SSB). We define the NSM as the model that exhibits the following breaking pattern:

$$SU(3)_C \times SU(2)_L \times U(1)_Y \rightarrow SU(3)_C \times U(1)_{EM}. \qquad (14.24)$$

14.3.2 The NSM Lagrangian

The NSM has the same gauge group, the same scalar content, and the same pattern of SSB as the Standard Model. In the fermion sector, all the Standard Model representations are included. The only difference is the addition of the fermionic N_{Ri} fields. These fields are usually called *right-handed neutrinos*. Since the imposed symmetry is the same, all the terms that appear in the Standard Model Lagrangian also appear in the NSM Lagrangian. However, the NSM Lagrangian has several additional terms. These are all the terms that involve the N_{Ri} fields. We can write

$$\mathcal{L}_{\text{NSM}} = \mathcal{L}_{\text{SM}} + \mathcal{L}_N, \qquad (14.25)$$

where $\mathcal{L}_{\text{SM}}$ is the Lagrangian of the renormalizable Standard Model (see equation (8.17)).

Our task is to find the specific form of $\mathcal{L}_N$. We note the following points in this regard:

- Given that the N_R fields are singlets of the gauge group, we have $D^\mu N_R = \partial^\mu N_R$.
- Since the N_R fields are singlets of the gauge group they can have Majorana mass terms.
- The combination $\overline{L_L} N_R$ transforms as $(1,2)_{+1/2}$ under the gauge group and thus can have a Yukawa coupling to the scalar doublet.

We thus obtain the most general form for the renormalizable terms in $\mathcal{L}_N$:

$$\mathcal{L}_N = i\overline{N_{Ri}}\partial\!\!\!/ N_{Ri} - \left(\frac{1}{2}M^N_{ij} N^T_{Ri} N_{Rj} + Y^\nu_{ij}\overline{L_{Li}}\tilde{\phi} N_{Rj} + \text{h.c.}\right), \qquad (14.26)$$

where $\tilde{\phi}$ is defined in equation (8.14) in chapter 8, M^N is a symmetric 3×3 complex matrix of dimension-one mass terms, and Y^ν is a general 3×3 complex matrix of dimensionless Yukawa couplings.

14.3.3 The NSM Spectrum

As for the bosonic spectrum, it remains unchanged from the Standard Model. As for the fermionic spectrum, we note that, since the N_R fields are singlets of the full gauge group, they are also singlets of the unbroken subgroup—namely, they transform as $(1)_0$ under $SU(3)_C \times U(1)_{EM}$. This means that the spectrum of the charged fermions (quarks and charged leptons) also remains unchanged from the Standard Model.

As for the neutrinos (the ν_L components of the $SU(2)$-doublet leptons and the N_R fields), taking into account SSB, we find the following mass terms in $\mathcal{L}_N$:

$$-\mathcal{L}_{N,\,\text{mass}} = \frac{1}{2}M^N_{ij} N^T_{Ri} N_{Rj} + \frac{Y^\nu_{ij} v}{\sqrt{2}}\overline{\nu_{Li}} N_{Rj} + \text{h.c.} \qquad (14.27)$$

This gives a 6×6 neutrino mass matrix, which can be decomposed into four 3×3 blocks as follows (see equation (2.20)):

$$M_\nu = \begin{pmatrix} 0 & m_D \\ m_D^T & M^N \end{pmatrix}, \qquad m_D = \frac{vY^\nu}{\sqrt{2}}. \tag{14.28}$$

To obtain the six neutrino mass eigenstates, we need to diagonalize M_ν.

Note that the neutrino sector of the NSM has six mass eigenstates, in contrast to the charged lepton, up-quark, and down-quark sectors, where there are three mass eigenstates in each. The reason is that charged fermions have Dirac masses and each mass eigenstate has 4 degrees of freedom (DoF). The neutrinos have Majorana masses, and each mass eigenstate has only 2 DoF. The total number of DoF is thus the same in each of the four sectors.

Before we analyze the neutrino spectrum of the three-generation NSM, we gain some intuition by analyzing the simpler, one-generation NSM, where we have only one copy of the L and N fields, and thus two neutrino mass eigenstates, which we write as

$$\nu_1 = \cos\theta\ \nu_L + \sin\theta\ N_R^c, \qquad N_1 = -\sin\theta\ \nu_L + \cos\theta\ N_R^c. \tag{14.29}$$

Without loss of generality, we choose a basis where the symmetric 2×2 matrix M_ν of equation (14.28) is real: $M_D = M_D^T$ and M^N are real numbers. Inspired by both phenomenology and theoretical model building, we assume that $M^N \gg v$. Since $M_D \lesssim v$, it follows that $M^N \gg m_D$. To leading order in m_D/M^N, we obtain the following mass eigenvalues and mixing angle:

$$m_1 = \frac{m_D^2}{M^N}, \qquad M_1 = M^N, \qquad \sin\theta = \frac{m_D}{M^N}. \tag{14.30}$$

The neutrino mass terms in the Lagrangian are written as

$$-\mathcal{L}_{N,\,\text{mass}} = \frac{1}{2} M_1 N_1^T N_1 + \frac{1}{2} m_1 \nu_1^T \nu_1 + \text{h.c.} \tag{14.31}$$

These results demonstrate two points about the seesaw mechanism:

- The light mass, m_1, is inversely proportional to the heavy one, M_1. This is the reason why the mechanism that generates masses for the light neutrinos via their Yukawa couplings to heavy neutrinos is called the "seesaw" mechanism: the higher one side of the seesaw is, the lower the other side is.
- The mixing angle between the two states is very small. The light state is almost a pure upper component of an $SU(2)$-doublet ν_L, while the heavy one is almost a pure $SU(2)$-singlet N_R.

We now return to the three-generation case. Without loss of generality, we can choose a basis where M^N of equation (14.28) is diagonal and real:

$$M^N = \hat{M}_N = \text{diag}(M_1, M_2, M_3). \tag{14.32}$$

Unlike the Standard Model, which has a single dimensionful parameter, v, the NSM has four dimensionful parameters: v, M_1, M_2, M_3. We assume that the Majorana masses of the heavy states have the same scale, denoted as m_N, which is much higher than the weak scale:

$$m_N \sim M_1 \sim M_2 \sim M_3 \gg v. \tag{14.33}$$

To diagonalize M_ν, we first use the unitary matrix K to block-diagonalize it:

$$KM_\nu K^T = \begin{pmatrix} m_\nu & 0 \\ 0 & \hat{M}_N \end{pmatrix}. \tag{14.34}$$

To leading order in v/m_N, we have

$$K = \begin{pmatrix} 1 & -m_D \hat{M}_N^{-1} \\ \hat{M}_N^{-1} m_D^T & 1 \end{pmatrix}, \qquad m_\nu = -m_D \hat{M}_N^{-1} m_D^T, \tag{14.35}$$

where we omitted terms of order v^2/m_N^2. The lower-right block in equation (14.34) is already diagonal. The upper-left block, m_ν, can be diagonalized by a further unitary transformation:

$$\hat{m}_\nu = V_{\nu L}^* m_\nu V_{\nu L}^\dagger = \mathrm{diag}(m_1, m_2, m_3). \tag{14.36}$$

We thus learn the following points:

- There are three heavy Majorana neutrinos of masses M_1, M_2, and M_3. We call these states N_1, N_2, and N_3. These mass eigenstates are approximately $SU(2)$-singlet states but have a small, $\mathcal{O}(v/m_N)$, $SU(2)$-doublet component. The masses are, by assumption, much larger than the electroweak scale.
- There are three light neutrinos of masses m_1, m_2, and m_3, of the order v^2/m_N. We call these states ν_1, ν_2, and ν_3. These mass eigenstates are approximately $SU(2)$-doublet states but have a small, $\mathcal{O}(v/m_N)$, $SU(2)$-singlet component. The masses are much smaller than the electroweak scale.

We conclude that the spectrum of the NSM has three scales:

- The masses of all the massive bosons and the charged fermions are of the order v.
- The masses of the approximately singlet neutrinos are heavy, of the order m_N.
- The masses of the approximately doublet neutrinos are light, of the order v^2/m_N.

14.3.4 The NSM Interactions

The neutrino sector of the NSM has neither strong nor electromagnetic interactions. The ν_i and N_i states, however, do have weak and Yukawa interactions.

The charged current weak interactions are given by

$$\mathcal{L}_{W,\ell} = -\frac{g}{\sqrt{2}} \overline{\ell} P_L \not{W}^- (U_{\ell\nu}^W \; U_{\ell N}^W) \begin{pmatrix} \nu \\ N \end{pmatrix} + \mathrm{h.c.} \tag{14.37}$$

Here, $U^W = (U^W_{\ell\nu}\ U^W_{\ell N})$ is the 3×6 lepton mixing matrix connecting the three charged lepton mass eigenstates ($\ell = e, \mu, \tau$) to the six neutrino mass eigenstates ($\nu = \nu_1, \nu_2, \nu_3$ and $N = N_1, N_2, N_3$). The entries of the $U^W_{\ell N}$ block are small, of $\mathcal{O}(v/m_N)$. Thus, to zeroth order in v/m_N, only the light neutrinos couple to the W-boson, with the mixing matrix being

$$U^W_{\ell\nu} = V_{eL} V^\dagger_{\nu L}, \tag{14.38}$$

which is the same as the mixing matrix (equation (14.11)) of the νSM.

The neutral current weak interactions are given by

$$\mathcal{L}_{Z,\nu} = \frac{g}{2c_W}(\overline{\nu}\,\overline{N})\not{Z}\begin{pmatrix} U^Z_{\nu\nu} & U^Z_{\nu N} \\ U^Z_{N\nu} & U^Z_{NN} \end{pmatrix}\begin{pmatrix} \nu \\ N \end{pmatrix}. \tag{14.39}$$

Here, U^Z is a 6×6 mixing matrix among the six neutrino mass eigenstates ($\nu_1, \nu_2, \nu_3, N_1, N_2$, and N_3). The entries of the $U^Z_{\nu N}$ and $U^Z_{N\nu}$ blocks are of $\mathcal{O}(v/m_N)$, while the entries of the U^Z_{NN} block are of $\mathcal{O}(v^2/m_N^2)$. As for $U^Z_{\nu\nu}$, it is very close to a unit matrix up to effects of $\mathcal{O}(v^2/m_N^2)$. Thus, to zeroth order in $v/m_N \sim m_\nu/v$, only the light neutrinos couple to the Z-boson, with a mixing matrix of

$$U^Z_{\nu\nu} = \mathbf{1} + \mathcal{O}\left(\frac{m_\nu^2}{v^2}\right). \tag{14.40}$$

We conclude that nondiagonal and nonuniversal effects in the Z-couplings to the light neutrinos are unobservably small effects, and thus, to a very good approximation, the light neutrinos couplings are universal, as in the νSM.

While the deviations from diagonality and universality are quantitatively very small, it is worth emphasizing a qualitative point with regard to these deviations. As explained in section 9.4 in chapter 9, the universality of the Z-couplings holds only if all fermions of a given chirality, color, and charge come from the same $SU(2) \times U(1)$ representation. In the NSM, neutrinos come from two types of $SU(2) \times U(1)$ representations, $(2)_{-1/2}$ and $(1)_0$, and therefore, there is no universality in the couplings of the Z-boson to neutrinos. The NSM demonstrates, then, that the absence of a tree-level FCNC is a rather special feature of the Standard Model, which is violated in general by BSM physics.

The Yukawa interactions, as can be learned from the Lagrangian $\mathcal{L}_N$ of equation (14.26), connects the heavy N_i and light ν_j states. Thus, unlike the Yukawa interactions of the Standard Model, in the NSM the Higgs boson has off-diagonal couplings. As explained in section 9.4 in chapter 9, the special Standard Model features that lead to diagonality are, first, that for a given fermion sector, all fermions are chiral and charged and thus there are no bare mass terms; and, second, that the scalar sector has a single Higgs doublet. In the neutrino sector of the NSM, the first condition is violated and the Higgs boson has neutrino flavor-changing couplings.

The Yukawa and weak interactions of the N_i states imply that they are unstable and decay to a light lepton (neutral or charged) and a boson (h, Z, or W). By assumption, however, the N_i particles are heavy, and thus they cannot be produced in experiments and their interactions cannot be directly probed. We thus do not study their interactions any further here. We note, however, that the Yukawa interactions of the N_i particles might be the source of the baryon asymmetry of the universe, a scenario that is known by the name of *leptogenesis*. See section 15.3.3 in chapter 15 for a discussion.

14.3.5 The Low-Energy Limit of the NSM

In this section, we show that the low-energy limit of the NSM reproduces the νSM. Thus, the NSM constitutes a UV-completion of the νSM. We do not present a formal derivation (based on integrating out the heavy modes) of this statement; instead, we provide an intuitive explanation.

The first point concerns the field content of the NSM and the νSM. Since the N_i neutrinos are very heavy, they are not directly produced in current experiments. (Given that they are almost pure gauge-singlets, it would be difficult to produce them, even if it were kinematically possible to do so.) Thus, at low energy, we can only observe their virtual effects. All other fields are at or below the electroweak scale. Their list is the same as the νSM list of fields.

The second point concerns the mass of the light neutrinos. Comparing equation (14.2) to equation (14.30), we observe that they would assume the same form if we were to identify

$$\frac{z^\nu}{\Lambda}=\frac{(Y^\nu)^2}{M^N}. \tag{14.41}$$

We can then further identify Λ as the mass of the heavy neutrino, $\Lambda=M_N$, and z^ν as the square of the Yukawa coupling, $z^\nu=(Y^\nu)^2$. In the three-generation case, we have

$$\frac{z^\nu_{ij}}{\Lambda}=\left[Y^\nu(M^N)^{-1}Y^{\nu T}\right]_{ij}. \tag{14.42}$$

We can use Feynman diagrams to come to the same conclusion. While we usually use these diagrams to calculate amplitudes, we can also use them to describe masses. A charged lepton mass is described by the diagram in figure 14.1(a), a neutrino mass in the νSM is described by the diagram in figure 14.1(b), and a neutrino mass in the NSM is described by the diagram in figure 14.1(c). For the latter, since m_N is very large, we can replace the propagator by a point-like vertex, which would result in figure 14.1(b). We

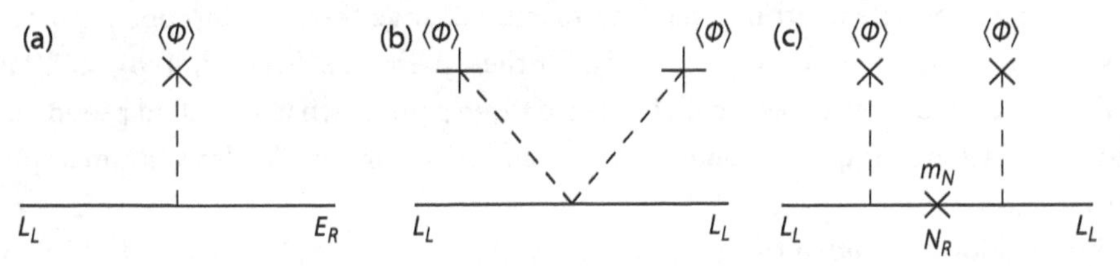

FIGURE 14.1. Lepton masses: (*a*) The electron mass. (*b*) The neutrino mass in the νSM. (*c*) the neutrino mass in the NSM.

learn that, treating the heavy fields as nondynamical, we get the relevant diagram in the effective theory.

The third point concerns the lepton interactions. In the νSM, the neutrinos are purely $SU(2)$-doublets. Their charged current weak interactions involve the mixing matrix U defined in equation (14.11), and their neutral current weak interactions are universal (see equation (14.13)). In the NSM, they have a small $SU(2)$-singlet component. However, to the zeroth order in v/m_N, the singlet component in the light neutrinos vanishes. Their charged current weak interaction mixing matrix is $U^W_{\ell\nu}$ as given by equation (14.38), which is the same as equation (14.11). Their neutral current weak interaction mixing matrix is $U^Z_{\nu\nu}$ as given by equation (14.40), which is universal to leading order in m_ν/v.

We learn that the NSM is indeed a possible UV completion of the νSM.

14.3.6 The Case of $m_N \ll v$: Sterile Neutrinos

So far, we have worked under the assumption given in equation (14.33)—that is, $m_N \gg v$. Here, we comment on the consequences of an opposite hierarchy ($m_N \ll v$). In this case the N_i DoF cannot be integrated out, and the low-energy limit of the theory does not generate the νSM.

Let us first discuss the special case of $M^N = 0$. Setting M^N to zero requires that we postulate a global symmetry, in addition to the gauge symmetry of the Standard Model. A good candidate model, where M^N is forbidden but Y^ν is allowed, is one where we impose lepton number symmetry and assign the N_i fields with $L = 1$. (Note that, while the lepton number is an anomalous symmetry, $U(1)_{B-L}$ is nonanomalous and achieves the same renormalizable Lagrangian—namely, $\mathcal{L}_{\rm NSM}$ with $M^N = 0$.)

With $M^N = 0$, the neutrinos are massive Dirac particles, similar to the charged fermions. This should be the case, since in this model, the neutrinos carry a conserved charge (L, or $B - L$). We emphasize, however, that to extend the Standard Model in a way that would make the neutrinos Dirac fermions, one needs not only to add matter fields, but also impose a global symmetry. Moreover, in the NSM, the lightness of the doublet neutrinos is explained by a new, high scale of physics. In contrast, the lightness of Dirac neutrinos requires that their dimensionless Yukawa couplings are set to be tiny.

For $m_N \neq 0$ but $m_N \ll v$, we have to consider two cases. First, if $m_N \gtrsim Y^\nu v$, there are six very light Majorana mass eigenstates. Light states that are dominantly electroweak singlets are called *sterile neutrinos*, while those that are dominantly electroweak doublets are called *active neutrinos*. Sterile neutrinos can significantly change the neutrino phenomenology, but so far, there is no conclusive evidence for their existence. Second, if $m_N \ll Y^\nu v$, the six Majorana mass eigenstates are divided into three pairs of what are called pseudo-Dirac neutrinos. Each pair is quasi-degenerate, with an average mass of order $y^\nu v$ and a splitting of order m_N.

The question of whether the neutrinos are Majorana or Dirac fermions is not yet experimentally decided. If a neutrinoless double beta decay is observed (see the discussion in Appendix 14.B.2), it will prove their Majorana nature.

14.4 Open Questions

Experimental data prove that neutrinos are massive and there is lepton mixing. Specifically, experiments measured two mass differences and three mixing angles. Yet there is more to learn experimentally about neutrino masses and mixing, including the following:

1. Is the neutrino spectrum hierarchical or quasi-degenerate?
2. Is the neutrino mass ordering normal or inverted?
3. Is *CP* violated in the PMNS matrix?

While the νSM is an attractive possibility to encompass the low-energy description of the neutrino sector, there are several open questions where experiments might further support or falsify the νSM:

4. Are the neutrinos Dirac or Majorana fermions?
5. Are there sterile neutrinos (i.e., other light states that mix with the active neutrinos)?
6. Are there dimension-six operators that significantly affect the neutrino interactions?

If the νSM continues to be supported by experiments, the following question will still be open:

7. What is the full high-energy theory that generates the νSM as its low-energy limit?

This list of open questions explains why experimental neutrino physics is a very active field of research.

Appendix

14.A Neutrino Oscillations

In this appendix, we explain how neutrino masses and mixing are probed. Our analysis here is mostly model independent, but for the sake of concreteness, we will use the νSM. The results can be easily modified for other models.

Fluxes of neutrinos are produced in both natural sources, such as the sun or cosmic rays hitting the atmosphere, and artificial sources, such as accelerators or nuclear reactors. These sources are used to study neutrino masses and mixing, providing sensitivities to large regions of the parameter space.

14.A.1 Neutrino Oscillations in a Vacuum

In experiments, neutrinos are often produced and detected by charge current weak interactions. Thus, the states that are relevant to production and detection are the $SU(2)_L$-doublet partners of the charged lepton mass eigenstates (namely, ν_e, ν_μ, and ν_τ). On the other hand, the eigenstates of free propagation in spacetime are the mass eigenstates,

ν_1, ν_2, and ν_3. In general, the interaction eigenstates (also known as *flavor eigenstates*) are different from the mass eigenstates:

$$|\nu_\alpha\rangle = U^*_{\alpha i}|\nu_i\rangle \qquad (\alpha = e, \mu, \tau;\ i = 1, 2, 3), \tag{14.43}$$

where U is the PMNS matrix defined in equation (14.11). Consequently, flavor is not conserved during propagation in spacetime and, in general, we may produce ν_α but detect $\nu_\beta \neq \nu_\alpha$.

The probability $P_{\alpha\beta}$ of producing neutrinos of flavor α and detecting neutrinos of flavor β is calculable in terms of the following inputs:

- The neutrino energy E
- The distance between source and detector L
- The mass differences $\Delta m^2_{ij} \equiv m^2_i - m^2_j$
- The parameters, mixing angles and the Dirac phase, of the mixing matrix U

Starting from equation (14.43), we can write the expression for the evolution of a state that started as a ν_α. We denote the time-evolved state as $|\nu_\alpha(t)\rangle$ and obtain

$$|\nu_\alpha(t)\rangle = U^*_{\alpha i}|\nu_i(t)\rangle, \tag{14.44}$$

where $|\nu_\alpha(0)\rangle \equiv |\nu_\alpha\rangle$ and

$$|\nu_i(t)\rangle = e^{-iE_i t}|\nu_i(0)\rangle. \tag{14.45}$$

The probability of a state that is produced as ν_α to be detected as ν_β is given by

$$P_{\alpha\beta} = \left|\langle\nu_\beta|\nu_\alpha(t)\rangle\right|^2. \tag{14.46}$$

In all cases of interest, the neutrinos are ultra-relativistic, and we then approximate

$$E_i = p + \frac{m_i^2}{2p}, \qquad E_i - E_j = \frac{\Delta m^2_{ij}}{2p}, \tag{14.47}$$

where p is the momentum of the neutrinos. Explicit calculation (see question 14.1) gives

$$\begin{aligned} P_{\alpha\beta} = \delta_{\alpha\beta} &- 4\sum_{j>i} \mathcal{R}e\left(U_{\alpha i}U^*_{\beta i}U^*_{\alpha j}U_{\beta j}\right)\sin^2\left(\frac{\Delta m^2_{ij}L}{4E}\right) \\ &+ 2\sum_{j>i} \mathcal{I}m\left(U_{\alpha i}U^*_{\beta i}U^*_{\alpha j}U_{\beta j}\right)\sin\left(\frac{\Delta m^2_{ij}L}{2E}\right), \end{aligned} \tag{14.48}$$

where we use the approximations $p = E$ and $L = t$ in the final result.

Applying this calculation to a two-generation case, where there is a single mixing angle, θ (and no relevant phase) and a single mass difference,

$$U = \begin{pmatrix} \cos\theta & \sin\theta \\ -\sin\theta & \cos\theta \end{pmatrix}, \qquad \Delta m^2 = m_2^2 - m_1^2, \tag{14.49}$$

Table 14.2. Neutrino oscillation experiments.

Source	E[MeV]	L[km]		Δm^2[eV^2]
Solar (vacuum oscillations)	1	10^8	$\Longrightarrow$	$10^{-11}-10^{-9}$
Short baseline reactor	1	$0.1-1$	$\Longrightarrow$	$10^{-3}-10^{-2}$
Long baseline reactor	1	10^2	$\Longrightarrow$	$10^{-5}-10^{-3}$
Atmospheric	10^3	10^1-10^4	$\Longrightarrow$	$10^{-5}-1$
Short baseline accelerator	10^3-10^4	0.1	$\Longrightarrow$	$\gtrsim 0.1$
Long baseline accelerator	10^4	10^2-10^3	$\Longrightarrow$	$10^{-3}-10^{-2}$
Source	n_0[cm^{-3}]	r_0[cm]		Δm^2[eV^2]
Solar (MSW)	6×10^{25}	7×10^9	$\Longrightarrow$	$10^{-9}-10^{-5}$

Note: The column Δm^2 gives the range to which the corresponding class of experiments is sensitive. For solar neutrinos, flavor transitions can occur as a result of vacuum oscillations (VO), presented in section 14.A.1, or of the MSW effect, presented in section 14.A.2.

we obtain, for $\alpha\neq\beta$,

$$P_{\alpha\beta}=\sin^2 2\theta\,\sin^2 x,\qquad x\equiv\frac{\Delta m^2 L}{4E}. \tag{14.50}$$

We learn that the time evolution of neutrinos that are produced in a flavor eigenstate exhibits oscillations (as a function of time or, equivalently, distance) between the various flavor eigenstates. This phenomenon is known as *neutrino oscillations.*

The expression for $P_{\alpha\beta}$ in equation (14.50) depends on two parameters that are related to the experimental design (E and L), and two that are parameters of the Lagrangian (Δm^2 and $\sin 2\theta$). To allow observation of neutrino oscillations and measurements of the neutrino parameters, nature better provides a value of $\sin^2 2\theta$ that is not tiny. Furthermore, one has to design the experiment appropriately:

- If $E/L\gg\Delta m^2$, then $P_{\alpha\beta}\to 0$, and there is no sensitivity to the neutrino parameters.
- If $E/L\sim\Delta m^2$, then $P_{\alpha\beta}$ is sensitive to both Δm^2 and $\sin\theta$.
- If $E/L\ll\Delta m^2$, then the oscillations average out, and $P_{\alpha\beta}=\frac{1}{2}\sin^2 2\theta$ is sensitive to θ, but not to Δm^2.

To obtain a more quantitative intuition, it is useful to write

$$x\approx 1.27\left(\frac{\Delta m^2}{\text{eV}^2}\right)\left(\frac{L}{\text{km}}\right)\left(\frac{\text{GeV}}{E}\right). \tag{14.51}$$

Typical values of E and L for the various classes of experiments and the corresponding range of Δm^2 that can be probed are given in table 14.2.

It is interesting to understand the differences between neutrino oscillations and neutral meson oscillations (as discussed in appendix 13.A in chapter 13):

- Mesons decay while the neutrinos are stable (at least on the timescale of experiments). This brings the meson decay width into the analysis of the time evolution of mesons; see equation (13.62) in chapter 13.
- Neutrino oscillations depend on the mixing angle. In neutral meson mixing, we consider particle-antiparticle oscillation. In this case, *CPT* requires that the two diagonal elements in the mass matrix are equal, and thus the mixing angle is maximal at $\pi/4$.
- The argument of the time-dependent oscillations is $(\Delta m)t$ for mesons (see equation (13.62)), and $(\Delta m^2)t/(2E)$ for neutrinos (see equation (14.50)). This apparent difference, however, is nothing but the effect of relativistic time dilation. Using, for the quasi-degenerate mesons, $\Delta m = \Delta m^2/(2m)$, and for the ultrarelativistic neutrinos, $t = \tau/\gamma$ (with τ the proper time) and $\gamma = E/m$, the dependencies on time and mass become the same.

14.A.2 The MSW Effect

To describe the propagation of neutrinos in matter, modifications to the oscillation formalism are necessary. Matter generates a potential to the neutrinos that affects their propagation. This effect is similar to the effect of a medium on photon propagation, which results in a refraction index. For photons, the interaction leads to a speed of light in matter that is smaller than in vacuum. Furthermore, it generates dispersion where the velocity depends on the energy. For neutrinos propagating in matter, the velocity almost does not change, but flavor oscillations are significantly affected.

The smallness of the relevant cross sections makes most effects of neutrino scattering off the medium negligible. This is, however, not the case for forward scattering, where there is no energy or momentum exchange between the neutrinos and the medium. The effect of forward scattering is to induce effective masses for the neutrinos. The resulting modification to $P_{\alpha\beta}$ of equation (14.50) is known as the *Mikheyev-Smirnov-Wolfenstein* (MSW) *effect*.

For the current measurements of neutrino flavor transitions, the relevant effects come from neutrino propagation in the sun or inside Earth. In both cases, matter consists of electrons, protons, and neutrons. In particular, there are neither muons nor tau nor antileptons in the medium.

All neutrinos have the same (universal) neutral current weak interactions. In contrast, in a medium that contains electrons, but neither muons nor tau-leptons, only ν_e's have charged current weak interactions with matter. The effective potential induced by the charged current weak interactions of ν_e is a vector potential and it is given by (see question 14.7)

$$V_\mu = (V_C, 0, 0, 0), \qquad V_C = \sqrt{2} G_F n_e. \tag{14.52}$$

$$V_C \approx 7.6 \frac{n_e}{n_p + n_n} \left(\frac{\rho}{10^{14} \text{g/cm}^3} \right) \text{eV}, \tag{14.53}$$

where n_i $(i=e, p, n)$ stands for the number density and ρ is the mass density. At the solar core, $\rho \sim 100\,\text{g/cm}^3$, which gives rise to $V_C \sim 10^{-12}$ eV, while at the Earth's core, $\rho \sim 10\,\text{g/cm}^3$, which gives rise to $V_C \sim 10^{-13}$ eV.

Current data suggest that $m_\nu \gtrsim 10^{-3}$ eV, and thus $m_\nu \gg V_C$. One may then naively think that matter effects are irrelevant. Matter effects, however, arise from vector interactions, while masses are scalar operators. Consequently, the right comparison to make is between m_ν^2 and EV_C, where E is the neutrino energy. Since $E \gg m_\nu$, matter effects can be important. To see how this enhancement of matter effects arises, consider the dispersion relation of the neutrino in the vacuum, $p_\mu p^\mu = m^2$. In matter, it is modified as follows:

$$(p_\mu - V_\mu)(p^\mu - V^\mu) = m^2. \tag{14.54}$$

Writing the dispersion relation as $E^2 = p^2 + m_m^2$ and using equation (14.52), we learn that the effective mass in matter, m_m, is given by

$$m_m^2 = m^2 + A_e, \qquad A_e \equiv 2EV_C, \tag{14.55}$$

where we neglected terms of the order of V_C^2. It is the vector nature of the weak interaction that makes matter effects relevant.

For the sake of simplicity, we proceed in our analysis of the matter effects in a two-neutrino framework. We consider ultrarelativistic neutrinos, where we approximate the Hamiltonian as $\mathcal{H} = \sqrt{p^2 + m^2} \approx p + m^2/(2p)$. In the vacuum, in the mass basis (ν_1, ν_2), the Hamiltonian can be written as

$$\mathcal{H} = p + \frac{1}{2p}\begin{pmatrix} m_1^2 & 0 \\ 0 & m_2^2 \end{pmatrix}. \tag{14.56}$$

In the interaction basis (ν_e, ν_μ), we have

$$\mathcal{H} = p + \frac{m_1^2 + m_2^2}{4p} + \frac{1}{4p}\begin{pmatrix} -\Delta m^2 \cos 2\theta & \Delta m^2 \sin 2\theta \\ \Delta m^2 \sin 2\theta & \Delta m^2 \cos 2\theta \end{pmatrix}. \tag{14.57}$$

In matter that contains only electrons, protons, and neutrons, the Hamiltonian in the interaction basis is modified from its vacuum form of equation (14.57):

$$\mathcal{H} = p + \frac{m_1^2 + m_2^2}{4p} + \frac{1}{4p}\begin{pmatrix} 2A_e - \Delta m^2 \cos 2\theta & \Delta m^2 \sin 2\theta \\ \Delta m^2 \sin 2\theta & \Delta m^2 \cos 2\theta \end{pmatrix}, \tag{14.58}$$

where A_e is the effective potential defined in equation (14.55), we used the leading-order relation $p = E$, and we omitted the part in the Hamiltonian that is proportional to the unit matrix in flavor space (which plays no role in the oscillations). We can define the effective mass difference and mixing angle in matter:

$$\Delta m_m^2 = \sqrt{(\Delta m^2 \cos 2\theta - A_e)^2 + (\Delta m^2 \sin 2\theta)^2}, \tag{14.59}$$

$$\tan 2\theta_m = \frac{\Delta m^2 \sin 2\theta}{\Delta m^2 \cos 2\theta - A_e},$$

where the subindex m stands for matter.

The oscillation probability in matter with a constant n_e is simply obtained from equation (14.50) by replacing Δm^2 and θ with Δm_m^2 and θ_m:

$$P_{\alpha\beta} = \sin^2 2\theta_m \sin^2 x_m, \qquad x_m = \frac{\Delta m_m^2 L}{4E}. \tag{14.60}$$

The following points are worth mentioning regarding equation (14.60):

- If mixing in vacuum vanishes, $(\theta = 0)$, so does mixing in matter, $(\theta_m = 0)$.
- The vacuum result is reproduced for $A_e = 0$, as it should be.
- For $|A_e| \ll \Delta m^2 \cos 2\theta$, the matter effect is a small perturbation to the vacuum result.
- For $|A_e| \gg \Delta m^2 \cos 2\theta$, the neutrino mass is a small perturbation to the matter effect. In that case, the oscillations are highly suppressed since the effective mixing angle is very small.
- For $A_e = \Delta m^2 \cos 2\theta$, the mixing is maximal, $\theta_m = \pi/4$ (namely, it is on resonance).

When the matter density is not constant, there are further modifications to the oscillation formalism. For a neutrino propagating in varying matter densities, the effective neutrino mass and mixing change while propagating. Then, the flavor composition of the neutrinos along their path is a function of the density profile of the medium. We do not elaborate on this point here, but only mention that it is a very important effect for solar neutrinos.

14.B Direct Probes of Neutrino Masses

The measurements of neutrino oscillations in vacuum and in matter probe mass differences (Δm_{ij}^2), but not the absolute mass scale (m_i or m_j). Here, we discuss experimental ways to measure the neutrino mass. Specifically, we discuss kinematic tests for neutrino masses and neutrinoless double beta decay. To date, these measurements have provided only upper bounds.

14.B.1 Kinematic Tests

In decays that produce neutrinos, the decay spectra are sensitive to neutrino masses. For example, in the $\pi \to \mu\nu$ decay, the muon momentum is fixed (up to tiny width effects) by the masses of the pion, muon, and neutrino. To first-order in m_ν^2/m_π^2, the muon momentum in the pion rest frame is given by

$$|\vec{p}\,| = \frac{1}{2m_\pi}\left(m_\pi^2 - m_\mu^2 - \frac{m_\pi^2 + m_\mu^2}{m_\pi^2 - m_\mu^2}\, m_\nu^2\right). \tag{14.61}$$

Since the correction to the massless neutrino limit is proportional to m_ν^2, the kinematic tests are not very sensitive to small neutrino masses. The current best bounds obtained using kinematic tests are the following:

$$\begin{aligned} m_\nu &< 18.2 \text{ MeV} \quad \text{from } \tau \to 5\pi + \nu\,, \\ m_\nu &< 190 \text{ KeV} \quad \text{from } \pi \to \mu\nu\,, \\ m_\nu &< 1.1 \text{ eV} \quad \text{from } {}^3\text{H} \to {}^3\text{He} + e + \overline{\nu}\,. \end{aligned} \tag{14.62}$$

The combination of oscillation experiments, which are sensitive to the neutrino mass differences and mixing angles, and kinematic tests, which are sensitive to the neutrino masses themselves, implies that all three neutrino masses are bounded by

$$m_i < 1.1 \text{ eV}, \tag{14.63}$$

for $i = 1, 2, 3$.

14.B.2 Neutrinoless Double-Beta ($0\nu2\beta$) Decay

Neutrino Majorana masses violate the lepton number by two units. Therefore, if neutrinos have Majorana masses, we expect that there are also $\Delta L = 2$ processes. The smallness of the neutrino masses indicates that such processes have very small rates. Therefore, the only practical way to look for a $\Delta L = 2$ process is when the corresponding $\Delta L = 0$ process is forbidden or highly suppressed. Neutrinoless double-beta ($0\nu2\beta$) decay, where the single beta decay is forbidden, is such a process. An example of this is

$${}^{32}_{76}\text{Ge} \to {}^{34}_{76}\,\text{Se} + 2e^-\,. \tag{14.64}$$

The only physical background to $0\nu2\beta$ decay is from double-beta decay with two neutrinos ($2\nu2\beta$).

The $0\nu2\beta$ decays are sensitive to the following combination of neutrino parameters:

$$m_{\beta\beta} = \sum_{i=1}^{3} m_i U_{ei}^2. \tag{14.65}$$

The best bound derived from $0\nu2\beta$ decay is $m_{\beta\beta} < 0.26$ eV. We emphasize the following points:

- If the neutrinos are Dirac particles, the lepton number is conserved, and their masses do not contribute to $0\nu2\beta$ decays.
- The $0\nu2\beta$ decay rate depends not only on $m_{\beta\beta}^2$ but also on some nuclear matrix elements. Those matrix elements introduce theoretical uncertainties in extracting $m_{\beta\beta}$ from the signal, or in deriving an upper bound on $m_{\beta\beta}$ if no signal is observed.
- The $0\nu2\beta$ decay is sensitive to other $\Delta L = 2$ operators, and not only to the neutrino Majorana masses. Thus, the relation between the $0\nu2\beta$ decay rate and the neutrino mass is model dependent.

For Further Reading

There are many reviews and books about neutrino physics, including the following:

- For a general review, see Gonzalez-Garcia and Nir [45], Grossman [46], and de Gouvea [47].
- For a review of neutrinoless double beta decays, see Vogel [48].
- For a review of experimental aspects, see Conrad [49].
- For a review of astrophysical aspects of neutrinos, see Beacom [50].

Problems

Question 14.1: Algebra

1. Restore all the indexes that are implicit in equation (14.1).
2. Show that equation (14.3) arises from equation (14.1) when the Higgs field is replaced by its VEV, equation (7.18) in chapter 7.
3. Derive equation (14.30).
4. Derive equation (14.48). It is useful to recall that

$$P_{\alpha\beta} = \left|\langle \nu_\beta | \nu_\alpha(t)\rangle\right|^2 = \left|\sum_i \langle \nu_\beta | \nu_i\rangle \langle \nu_i | \nu_\alpha(t)\rangle\right|^2. \tag{14.66}$$

5. Derive expressions for the difference between the T-conjugate processes,

$$\Delta_T \equiv P(\nu_e \to \nu_\mu) - P(\nu_\mu \to \nu_e), \tag{14.67}$$

   and the difference between the CP-conjugate processes,

$$\Delta_{CP} \equiv P(\nu_e \to \nu_\mu) - P(\bar{\nu}_e \to \bar{\nu}_\mu). \tag{14.68}$$

   Show that in the limit of CP conservation, which you can take as the case where U is real, $\Delta_T = \Delta_{CP} = 0$.
6. Explain why CPT invariance implies that $\Delta_T - \Delta_{CP} = 0$, and check that this is indeed the case.
7. For propagation in matter where

$$x_m \ll 1, \qquad r_c \equiv \frac{|A_e|}{\Delta m^2 \cos 2\theta} \ll 1, \tag{14.69}$$

   we can approximate $\sin x_m \sim x_m$ in equation (14.60). Show that in this case, to first-order in r_c, the transition probability in matter, $P_{\alpha\beta}$ of equation (14.60), is equal to that of equation (14.50).

Question 14.2: Neutrino lifetime

In the Standard Model, the neutrinos are massless, and thus stable. In the νSM, however, they are massive and nondegenerate, and thus the two heavier states decay. For simplicity, consider a two-generation model with $m_1 = 0$, $m_2 = 1$ eV and a mixing angle $\theta = \pi/4$. The two leading decay modes for ν_2 are

$$\nu_2 \to \nu_1 \gamma, \qquad \nu_2 \to \nu_1 \bar{\nu}_1 \nu_1. \tag{14.70}$$

1. Explain why there are no tree-level contributions to ν_2 decay.
2. Draw the diagrams for the decay modes in equation (14.70).
3. Estimate the partial decay width for each of the two modes, as well as the lifetime of ν_2.
4. In the NSM, there is a tree-level contribution to one of the two modes in equation (14.70). Draw the relevant tree-level diagram and explain why the other mode does not have such a contribution.
5. Estimate the tree-level and the one-loop amplitudes for $\nu_2 \to \nu_1 \bar{\nu}_1 \nu_1$ and compare them.
6. The best bound on the neutrino lifetime comes from the observation of a burst of neutrinos from SN1987A, a supernova that occurred about 170,000 light-years away. The neutrino signal lasted $\mathcal{O}(10\text{ s})$. Had the neutrinos decayed on their way to Earth, the neutrino signal would have spanned a significantly longer time window. Explain why.
7. Estimate a bound on the lifetime of the neutrino by requiring that it is longer than the time that it took the neutrinos to travel to Earth from SN1987A. Make sure to convert the lab-frame time to the proper time using the fact that the neutrino energies in the lab were of the order of 10 MeV.
8. Compare the theoretical calculation of the neutrino lifetime in the νSM and the observational bound from SN1987A. Does the SN1987A bound provide a meaningful constraint on the parameters of the νSM?

Question 14.3: FCNC charged lepton decays

In the Standard Model, $\mu \to e\gamma$ is forbidden due to the accidental lepton flavor symmetry. In the νSM, however, the symmetry is broken, so we expect $\mu \to e\gamma$ to occur.

1. Explain why $\mu \to e\gamma$ is not allowed at tree level.
2. Draw the leading one-loop diagrams for $\mu \to e\gamma$ in the νSM.
3. Find the ratio

$$\frac{\Gamma(\mu \to e\gamma)}{\Gamma(\mu \to e\nu\bar{\nu})} \tag{14.71}$$

in terms of couplings and the neutrino masses and mixing angles.

4. Based on the available data on the masses and mixing angles, give a numerical prediction of the ratio in equation (14.71). Check your estimate against the experimental bound and comment on it.

Question 14.4: The type III seesaw model

Consider an extension of the one-generation Standard Model where we add a right-handed fermion field $N_R(1,3)_0$. This mechanism to generate neutrino masses is called the *type III seesaw*.

1. What are the electric charges of the various components of N?
2. We write

$$\mathcal{L} = \mathcal{L}_{\rm SM} + \mathcal{L}_N, \tag{14.72}$$

   where $\mathcal{L}_N$ includes all the terms that involve N_R. Write $\mathcal{L}_N$ explicitly (the analog of equation (14.26) for the NSM). Write the covariant derivative $D^\mu N_R$ explicitly. In the Yukawa term, make sure that you use ϕ or $\tilde{\phi}$ correctly. Denote the Majorana mass term for N_R by M_N. From this point on, we assume $M_N \gg v$.
3. Show that the light neutrino acquires a mass of the form $m_\nu = cv^2/M_N$, and then find c.
4. Both the neutral and the charged components of N mix with the Standard Model fields. The mixing generates a mass splitting between the charged and neutral components of N. Estimate the size of the splitting.
5. The mass splitting that you found in item 4 is rather small and, in fact, is negligible compared to those generated by one-loop effects. Discuss where these effects come from and then estimate their size.
6. Argue that the low-energy limit of the model is the one-generation νSM.

Question 14.5: Neutrino masses in the LRS model (type II seesaw)

Consider the left right symmetric (LRS) model of question 7.10 in chapter 7 with $N_g = 1$.

1. Are there bare mass terms for any fermions in the model?
2. Write the most general $\mathcal{L}_{\rm Yuk}$ that involves the lepton fields, L_L and L_R, and the three scalar fields, Φ, Δ_L and Δ_R.
3. We denote the VEVs of Φ by k_1 and k_2, as we did in question 7.10. We further write $v_L = \langle \Delta_L \rangle$ and $v_R = \langle \Delta_R \rangle$. Write the mass terms for the leptons in terms of k_1, k_2, v_L, and v_R. Use the notations of equation (7.150).
4. Write the mass eigenstates as

$$\nu = \cos\theta\, \nu_L + \sin\theta\, \nu_R^c, \qquad N = -\sin\theta\, \nu_L + \cos\theta\, \nu_R^c. \tag{14.73}$$

   Assume the hierarchy of VEVs as given in equation (7.151). Working to leading order in the small ratios of VEVs, find m_ν, m_N, and θ. Verify that

$$m_N \sim v_R, \qquad m_\nu \sim \max\left(v_L, \frac{k^2}{v_R}\right), \qquad \theta \sim \frac{k}{v_R}. \tag{14.74}$$

5. An analysis of the scalar potential shows that generically, $k^2 \equiv k_1^2 + k_2^2 \sim v_L v_R$. Using this relation, show that

$$m_\nu \sim \frac{k^2}{m_N}, \tag{14.75}$$

which is the seesaw result.
6. Take $N_g = 2$. In the LRS model, there are two tree-level, vector-boson-mediated Feynman diagrams that contribute to muon decay. Draw the two diagrams.
7. Argue that the interference between the two amplitudes is very small and can be neglected.
8. Find the ratio of the two amplitudes in terms of the coupling constants and the masses of the vector bosons, and then express it in terms of the VEVs. Denote this ratio by $r_{R/L}$.
9. What bound on the VEVs can you deduce from the experimental bound $r_{R/L} < 0.033$?

Question 14.6: Solar neutrinos

Matter effects in the sun play a major role in suppressing the ν_e flux arriving to Earth from the sun. In the early days of solar neutrino experiments, however, the possibility that vacuum oscillations between the sun and the Earth play a role was viable. In this question, we explore this scenario. For simplicity, consider a two-generation model. We denote the mass difference as Δm^2. For the neutrino energy, take $E_\nu \sim 1$ MeV.

1. Estimate Δm^2, which corresponds to an oscillation length of the order of the distance between the sun and the Earth.
2. A consequence of this scenario is a seasonal variation of the oscillation probability. Estimate the relative effect of the seasonal variation as a function of Δm^2, E_ν, as well as the parameters of the Earth-sun system.
3. The flux of electron neutrinos arriving to Earth is about a third of the flux produced in the sun. Assuming vacuum oscillations, and that the data is collected over several years, estimate the mixing angle and put a bound on Δm^2.

Question 14.7: Matter effects

In the analysis of matter effects on neutrino propagation, we average over the background, effectively reducing an interaction term into an effective mass term. The starting point is the low-energy effective Lagrangian for the leptonic W interaction:

$$\mathcal{L} = \mathcal{L}_{\text{free}} + \mathcal{L}_{\text{int}}, \tag{14.76}$$

where

$$\mathcal{L}_{\text{free}} = \overline{\nu_L}(i\partial\!\!\!/ - m)\nu_L, \qquad \mathcal{L}_{\text{int}} = \sqrt{2}G_F(\overline{\nu_L}\,\gamma_\mu\,\nu_L)\,(\bar{e}\,\gamma^\mu(1-\gamma_5)\,e)\,. \qquad (14.77)$$

Averaging over the background matter amounts to

$$\bar{e}\,\gamma_\mu(1-\gamma_5)\,e \to n_e\,\langle\mathcal{M}_\mu\rangle_{\text{bg}} \qquad \mathcal{M}_\mu \equiv \langle e(\boldsymbol{p},\boldsymbol{\lambda})|\bar{e}\,\gamma_\mu\,(1-\gamma_5)\,e|e(\boldsymbol{p},\boldsymbol{\lambda})\rangle, \quad (14.78)$$

where $\boldsymbol{p}$ and $\boldsymbol{\lambda}$ denote the momentum and polarization vectors of the background electrons, respectively. For the electron number density n_e, and for the averaged value in matter of any operator x, we have

$$n_e = \sum_\lambda \int \frac{d^3p}{(2\pi)^3} f_e(\boldsymbol{p},\boldsymbol{\lambda}), \qquad \langle x\rangle_{\text{bg}} = \frac{1}{n_e}\sum_\lambda \int \frac{d^3p}{(2\pi)^3} f_e(\boldsymbol{p},\boldsymbol{\lambda})\,x(\boldsymbol{p},\boldsymbol{\lambda}), \quad (14.79)$$

where $f_e(\boldsymbol{p},\boldsymbol{\lambda})$ is the electron distribution function. The result of this *averaging* leads to an effective free Lagrangian for the neutrinos:

$$\mathcal{L} = \overline{\nu_L}\left[i\partial\!\!\!/ - m - V_\mu\gamma^\mu\right]\nu_L. \qquad (14.80)$$

1. Show that

$$V_\mu = \sqrt{2}G_F\,n_f\left[\left\langle\frac{p_\mu}{E_e}\right\rangle_{\text{bg}} - m_e\left\langle\frac{s_\mu}{E_e}\right\rangle_{\text{bg}}\right], \qquad (14.81)$$

where $\langle\ldots\rangle_{\text{bg}}$ is defined in equation (14.79). The spin vector s_μ is defined as follows:

$$s_\mu \equiv \left(\frac{\boldsymbol{p}\cdot\boldsymbol{\lambda}}{m_e}, \boldsymbol{\lambda} + \frac{\boldsymbol{p}\,(\boldsymbol{p}\cdot\boldsymbol{\lambda})}{m_e\,(m_e+E_e)}\right) \qquad (14.82)$$

and satisfies $s^2 = -1$ and $s_\mu\,p^\mu = 0$. You can assume that the background electrons are free, such that the following plane wave expansion for the field operators can be used:

$$\langle e(\boldsymbol{p},\boldsymbol{\lambda})|\bar{e}\,\gamma_\mu\,(1-\gamma^5)\,e|e(\boldsymbol{p},\boldsymbol{\lambda})\rangle = \frac{1}{2E_e}\bar{u}_e(\boldsymbol{p},\boldsymbol{\lambda})\,\gamma_\mu\,(1-\gamma^5)\,u_e(\boldsymbol{p},\boldsymbol{\lambda}). \quad (14.83)$$

Moreover, you can use the following identity:

$$e(\boldsymbol{p},\boldsymbol{\lambda})\,\bar{e}(\boldsymbol{p},\boldsymbol{\lambda}) = \frac{1}{2}(p\!\!\!/ + m_e)\,(1+\gamma^5 s\!\!\!/)\,, \qquad (14.84)$$

and thus

$$\begin{aligned}\mathcal{M}_\mu &= \langle e(\boldsymbol{p},\boldsymbol{\lambda})|\bar{e}\,\gamma_\mu\,(1-\gamma^5)\,e|e(\boldsymbol{p},\boldsymbol{\lambda})\rangle \\ &= \frac{1}{2E_e}\bar{e}(\boldsymbol{p},\boldsymbol{\lambda})\,\gamma_\mu\,(1-\gamma^5)\,e(\boldsymbol{p},\boldsymbol{\lambda}) \\ &= \frac{1}{4E_e}\mathrm{Tr}\left[\gamma_\mu\,(1-\gamma^5)\,(p\!\!\!/ + m_e)\,(1+\gamma^5 s\!\!\!/)\right]. \end{aligned} \qquad (14.85)$$

2. For a uniform unpolarized background, show that

$$V_\mu = (V_C, 0, 0, 0), \qquad V_C = \sqrt{2} G_F n_e, \tag{14.86}$$

thus confirming equation (14.52).

3. Using equation (14.80) and assuming $E \gg m_\nu$ and $E \gg V_C$, show that

$$E^2 = p^2 + m_m^2, \qquad m_m^2 = m^2 + 2EV_C, \tag{14.87}$$

thus confirming equation (14.55). Here, E is the neutrino energy and p is its momentum.

15

Cosmological Tests

In chapter 14, we described the experimental evidence showing that the Standard Model is not the full theory of nature: neutrino flavor transitions imply that the Standard Model needs to be extended. To date, this is the only effect beyond the Standard Model (BSM) that has been established by particle physics experiments. Yet there is additional data that cannot be explained within the Standard Model, which comes from observations related to astrophysics and cosmology. In this chapter, we describe these topics.

15.1 The Interplay of Particle Physics and Cosmology

Cosmology and astrophysics provide tests of the Standard Model. It should not come as a surprise that one can gain insights into particle physics from astrophysics, as it often involves long distances and high densities that cannot be reached in terrestrial experiments. Think, for example, of the sun: its burning goes through weak interactions, and the propagation of neutrinos through its high-density plasma can (and did) reveal new features of neutrinos. It should be even less of a surprise that cosmology can test the Standard Model: various features of the present universe were determined by events that involved a huge range of energy scales and, in particular, energy scales well above the weak scale and beyond the direct reach of any terrestrial experiments. Two examples of cosmology-related observations that are consistent with our present understanding of atomic, nuclear, and particle physics are the Cosmic Microwave Background (CMB) radiation and Big Bang Nucleosynthesis (BBN).

In this chapter, we focus on two features of the universe that are inconsistent with the Standard Model: dark matter and the baryon asymmetry of the universe (BAU). Unlike the case of neutrino masses, it is not enough to consider higher-dimension terms to simply explain these inconsistencies. New degrees of freedom (DoF) have to be introduced to explain the relevant observations. In both cases, the prospects of directly probing the new physics in future experiments are unclear. Some of the candidate theories introduce DoF with masses that are smaller, or not much larger, than the weak scale, and that can, in principle, be directly discovered in experiments. Other candidate theories involve very heavy new particles and thus demonstrate that directly probing the solutions to these puzzles may be beyond the reach of direct experimental probes.

To follow the particle physics aspects of this section, some knowledge of cosmology is needed. We provide a very basic introduction to cosmology in Appendix 15.A.

15.2 Dark Matter

The term *dark matter* refers to massive particles with the following properties:

- Stable or, at least, have lifetimes that are much longer than the age of the universe.
- Interacting weakly, or not interacting at all, with the Standard Model particles. Specifically, they carry neither color nor electromagnetic charge.

15.2.1 The Observational Evidence

The evidence for dark matter comes from astrophysics and cosmology. There is a variety of relevant observations, and they span a variety of scales, ranging from galaxies to the universe as a whole:

- Rotation curves in galaxies. Stars at distance r from the center of a galaxy have a rotational velocity v_r given by

$$v_r^2 = G_N \frac{M(r)}{r}, \tag{15.1}$$

  where G_N is the Newton constant and $M(r)$ is the mass of all objects within radius r, assuming that the mass does not depend on the angular coordinates. Let us define radius R, within which the bulk of the mass of the galaxy is contained. The rotational velocity of stars farther away, at $r > R$, falls off as $v_r^2 \propto 1/r$. Observations, however, are inconsistent with this expectation. When we take R to be the radius within which the bulk of the luminous matter in the galaxy is contained, we observe that $v_r(r)$ is roughly a constant for stars much farther from the core of the galaxy. This can be explained if $M(r) \propto r$ at these distances, which requires that, in addition to the luminous matter in the core of the galaxy, there is a halo of dark matter, with a mass density of $\rho(r) \propto 1/r^2$. (Of course, at some even larger radii, the mass density must fall much faster in order to be consistent with a finite mass of the galaxy.)
- Gravitational lensing. Massive objects warp spacetime around them, causing light to deflect. As a result, light from distant sources bends when passing near massive objects. Strong lensing allows light from a distant source to arrive at observers on Earth after traveling in multiple paths. Consequently, a source can appear as, for example, an Einstein ring. The radius-squared of the ring is proportional to the mass inside it. Even the much more common, but less dramatic, weak lensing distorts light sources in a way that allows us statistically to map the mass distribution between these sources and Earth. Thus, gravitational lensing measures the total mass of various objects. In some cases, the result is that the luminous mass accounts for only a small part of the total mass. A prime example of evidence for

dark matter is the Bullet Cluster, which actually consists of two clusters that have collided. We observe components of the Bullet Cluster in three ways: the galaxies via optical images, the hot gas via X-ray, and the mass distribution via gravitationally lensed images of background galaxies. The combination of these observations implies that there is a significant component of dark matter in the Bullet Cluster.

- The CMB and the large-scale structure. Measurements of anisotropies in the CMB and of the distribution of galaxies, when fitted to cosmological parameters, provide quantitative information about the energy density of dark matter:

$$\Omega_{\rm DM} = 0.265 \pm 0.007, \tag{15.2}$$

which is about five times larger than the energy density of baryonic matter:

$$\Omega_b = 0.0493 \pm 0.0006. \tag{15.3}$$

Another feature of dark matter, deduced from the analysis of structure formation in the universe, is that most dark matter should be "cold" (i.e., nonrelativistic at the start of galaxy formation).

We conclude that there must be dark matter, but since we observe it only though its gravitational effects, the question remains of what particles constitute the dark matter.

15.2.2 Neutrinos Cannot Be the Dark Matter

The νSM does have dark matter particles among its list of elementary particles. These are the neutrinos, which carry neither color charge nor electromagnetic charge, and are stable on the cosmological scale. Can they constitute the dark matter of the universe? The answer is negative. We present three reasons for this answer, each related to a different scale: the universe, the large-scale structure, and galaxies:

- Cosmology. Due to the weak interactions, neutrinos have been in thermal equilibrium at sufficiently high temperatures. Thus, we can calculate their number density today and find (see question 15.2 at the end of the chapter):

$$\frac{[n(\nu_i)]_0}{[n_\gamma]_0} = \frac{3}{11} \to [n(\nu_i)]_0 \approx 110/\text{cm}^3, \tag{15.4}$$

where we use $[n_\gamma]_0 \approx 410/\text{cm}^3$. The neutrino contribution to the mass density of the universe is given by

$$\Omega_\nu = \frac{\sum_i m_i [n(\nu_i)]_0}{\rho_c}. \tag{15.5}$$

Within the currently allowed range for neutrino masses, the maximum contribution arises in case the neutrinos are quasi-degenerate, with masses at the experimental upper bound of $m_\nu \lesssim 1$ eV for each mass eigenstate (see equation (14.63) in chapter 14). In that situation, we have

$$\Omega_\nu = 0.07 \times \frac{\sum_i m_i}{3\ \text{eV}} \Longrightarrow \Omega_\nu \lesssim 0.25\, \Omega_{\text{DM}}. \tag{15.6}$$

If the neutrino masses are hierarchical with normal ordering, $\Sigma m_\nu \sim 0.06$ eV and $\Omega_\nu \approx 0.005 \Omega_{\text{DM}}$. We conclude that the neutrinos cannot be the dominant component of the dark matter.

- Galaxy clusters. The large-scale structure implies that most of the dark matter particles should be cold (i.e., nonrelativistic at the time of structure formation). Neutrinos, however, constitute hot dark matter (i.e., they were relativistic at the time of structure formation). Thus, neutrinos cannot constitute a dominant component of the dark matter.
- Galaxies. The Tremaine-Gunn bound is a generic lower bound on the mass of fermionic dark matter. The idea is that the lighter the dark matter is, the larger the number of particles needed to get the required total mass. For fermions, the higher the density is, the more high-energy states are occupied. For a sufficiently small mass, the dark matter particles occupy states with a velocity that is greater than the escape velocity of the galaxy, and these states are not confined to the galaxy. The lower bound is given by

$$m(\text{fermionic DM}) \gtrsim 70\ \text{eV}. \tag{15.7}$$

Given that $m_\nu < 1.1$ eV (see equation (14.63) in chapter 14), neutrinos are excluded from being the dark matter at the galactic scale.

15.2.3 The χSM

Given the arguments listed in section 15.2.2, the dark matter particles must come from beyond the list of Standard Model particles. There are many theoretical ideas as to what the dark matter particles can be and what mechanism determines their relic abundance. The most intensively studied class of models has been that of weakly interacting massive particles (WIMPs). ("Weak" refers here to an interaction that is of a strength similar to the Standard Model weak interactions, not necessarily to W- or Z-mediated interaction.) In this section, we explain the motivation for WIMPs as dark matter particles. We do so by presenting a specific effective WIMP model that we call the χSM.

The χSM is defined as follows:

1. The symmetry, as well as the pattern of spontaneous symmetry breaking (SSB), are as follows:

$$SU(3)_C \times SU(2)_L \times U(1)_Y \times Z_2 \to SU(3)_C \times U(1)_{\text{EM}} \times Z_2. \tag{15.8}$$

2. All the Standard Model fields are even under the Z_2 symmetry.
3. There is a single real scalar field that is added to the Standard Model which is odd under the Z_2 symmetry:

$$\chi(1,1)_{0,-}. \tag{15.9}$$

The χSM has the same gauge group, the same pattern of SSB, and the same fermion content as the Standard Model. Since all Standard Model fields are even under the Z_2 symmetry, all the terms that appear in the Standard Model Lagrangian also appear in the χSM Lagrangian. The χSM Lagrangian, however, has several additional terms, which are all the terms that involve the χ field. We can write

$$\mathcal{L}_{\chi SM} = \mathcal{L}_{SM} + \mathcal{L}_{\chi}. \tag{15.10}$$

The most general form of the renormalizable terms in $\mathcal{L}_{\chi}$ is the following:

$$\mathcal{L}_{\chi} = \frac{1}{2}(\partial_{\mu}\chi)(\partial^{\mu}\chi) - \frac{1}{2}m_{\chi}^2\chi^2 - \frac{1}{4}\lambda_{\chi}\chi^4 - \frac{Z_{\chi\phi}}{2}\chi^2\phi^{\dagger}\phi, \tag{15.11}$$

where ϕ is the Standard Model Higgs field and m_{χ} is the mass of χ. We assume that $m_{\chi}^2 > 0$, to ensure that χ does not acquire a vacuum expectation value (VEV). We are interested in the interactions between χ and the Standard Model fields, which is given by $Z_{\chi\phi}$.

The χ particle is a dark matter candidate:

- It is stable because it is the only (and, in particular, the lightest) Z_2-odd particle.
- It carries neither color nor an electromagnetic charge.

What remains to be understood is whether the parameters of $\mathcal{L}_{\chi}$ can lead to the observed relic abundance, as quoted in equation (15.2).

Let us follow the time evolution of the number density of χ. We assume that it has been in thermal equilibrium in the early universe. There are three relevant parameters: the number density, n_{χ}; the annihilation cross section of χ-pairs into Standard Model particles, σ_A; and the temperature-dependent Hubble constant, H. The condition for freeze-out, stated in equation (15.49), is $\Gamma_i = H$, where in this case, $\Gamma_i = n_{\chi}\langle\sigma_A v_{\chi}\rangle$ (v_{χ} is the relative velocity of the incoming χ-particle) and we consider only the case where $T_F < m_{\chi}$. We distinguish three stages:

1. $T > m_{\chi}$. The number density of χ is equal to the equilibrium number density (see equation (15.45)):
$$n_{\chi} \propto T^3. \tag{15.12}$$
2. $T_F < T < m_{\chi}$. When T falls below m_{χ}, but before freeze-out occurs, the number density is equal to the equilibrium number density and is exponentially suppressed (see equation (15.46)):
$$n_{\chi} \propto (m_{\chi}T)^{3/2}\exp(-m_{\chi}/T). \tag{15.13}$$
3. $T \leq T_F$. At freeze-out, we roughly have
$$n_{\chi}^F\langle\sigma_A v_{\chi}\rangle \sim H \sim \frac{T_F^2}{m_{\rm Pl}}. \tag{15.14}$$

Later, the number density n_{χ} is only diluted by the expansion of the universe:

$$n_{\chi}(T < T_F) \propto T^3. \tag{15.15}$$

To obtain the current energy density of χ, it is convenient to consider the ratio $\rho_\chi^0/\rho_\gamma^0$. At T_F, the photon number density is $n_\gamma^F \propto T_F^3$. For $T < T_F$, the number densities of both the dark matter particles and photons scale like T^3, but in addition, the photon energy scales like $E_\gamma \sim T$. We further use the fact that T_F/m_χ is approximately independent of m_χ. We obtain

$$\frac{\rho_\chi^0}{\rho_\gamma^0} \sim \frac{\rho_\chi^F}{\rho_\gamma^F}\frac{T_F}{T_0} \propto \frac{1}{T_0\, m_{\rm Pl}\langle\sigma_A v_\chi\rangle}. \tag{15.16}$$

We learn that the observed energy density of the dark matter provides a measurement of the χ annihilation cross section, such that $\langle\sigma_A v_\chi\rangle \propto 1/(T_0 m_{\rm Pl})$. When inserting the numerical factors that are implicit in this evaluation and relaxing some of the simplifying assumptions, the result is

$$\frac{\Omega_\chi}{0.26} \approx \frac{1\text{pb}}{\langle\sigma_A v_\chi\rangle} \approx \frac{(20\text{ TeV})^{-2}}{\langle\sigma_A v_\chi\rangle}, \tag{15.17}$$

where pb stands for a picobarn.

We can translate equation (15.17) into a requirement for the χSM parameters. Using $\sigma_A = \sigma(\chi\chi \to \phi\phi^\dagger)$ and assuming $m_\chi \gg m_h$, we obtain (see question 15.3, and in particular equation (15.58))

$$\langle\sigma_A v_\chi\rangle = \frac{1}{16\pi}\frac{Z_{\chi\phi}^2}{m_\chi^2}. \tag{15.18}$$

Equation (15.17) then gives

$$\frac{m_\chi}{Z_{\chi\phi}} \approx 3\text{ TeV}. \tag{15.19}$$

A few comments are in order:

- Our working assumption that $m_\chi \gg m_h$ is satisfied so long as $Z_{\chi\phi}$ is not tiny.
- The required cross section indicates that $m_\chi/Z_{\chi\phi}$ is not far above the weak scale.
- The required annihilation cross section for the dark matter is of the order of a weak interaction cross section. This intriguing coincidence is known as the *WIMP miracle*. It might indicate that the physics of the dark matter is related to the weak scale and, in particular, to models that aim to solve the Higgs fine-tuning problem. Yet, given the absence of experimental evidence for new degrees of freedom (DoF) at the TeV scale, it is still unknown whether this is the case.
- The χSM, as well as many other dark matter models, predict new states that are at the weak scale. These models motivate a large variety of experimental searches. Thus, the search for dark matter is a field of intensive experimental activity.

15.3 Baryogenesis

There is more matter than antimatter in the universe. All structures in the observable universe—stars, galaxies, clusters—are made of matter particles: protons, neutrons, and electrons. Antimatter particles (in short, "antiparticles") come only from late time processes and do not make any structure. This inequality between the number densities of matter particles and antiparticles is called, in physics jargon, *baryon asymmetry*. *Baryogenesis* refers to dynamical generation of the baryon asymmetry. As we discuss in this section, baryogenesis via Standard Model processes cannot explain the observed baryon asymmetry, which must then come from BSM processes.

15.3.1 The Observational Evidence

The BAU is usually expressed as a ratio of number densities:

$$\eta_B \equiv \left.\frac{n_B - n_{\bar{B}}}{n_\gamma}\right|_0 = (6.21 \pm 0.16) \times 10^{-10}, \tag{15.20}$$

where n_B, $n_{\bar{B}}$, and n_γ are the number densities of baryons, antibaryons, and photons, respectively. An equivalent way of presenting the asymmetry is in terms of the baryonic fraction of the critical energy density:

$$\Omega_B = 0.0493 \pm 0.0006. \tag{15.21}$$

The value of the BAU is inferred in two independent ways. The first is via BBN. This topic in cosmology predicts the abundances of the light elements, H, D, ^{3}He, ^{4}He, and ^{7}Li. These abundances depend on a single parameter, which is η_B. The second way is from measurements of CMB radiation. The fact that the two determinations agree gives great confidence in the value of the BAU.

As for leptons, the electron asymmetry should cancel the proton asymmetry in order to guarantee the observed electromagnetically neutral universe. The total lepton asymmetry is unknown, however, since the primordial neutrinos (the cosmic neutrino background, CνB) have not been observed to date. Moreover, if the neutrinos are Majorana fermions, it is not clear how to define the lepton asymmetry, since lepton number is not a conserved quantity.

Standard cosmology implies that the universe has started from a state with equal numbers of baryons and antibaryons. In particular, both have vanishing number densities at the end of inflation. It follows that the BAU must have been generated dynamically, a scenario that is known as *baryogenesis*.

15.3.2 Sakharov Conditions

Andrei Sakharov formulated three necessary conditions for baryogenesis (i.e., for a dynamical generation of a baryon asymmetry):

1. Baryon number violation: If the baryon number were conserved—$\Delta B = 0$ in all processes—then a universe that starts with $\eta_B = 0$ (namely, total baryon charge zero) will remain with $\eta_B = 0$ at all times and, in particular, cannot evolve into a state with $\eta_B \neq 0$.

2. C and CP violation: If C and/or CP were conserved, then processes that change the baryon number by ΔB would proceed at the same rate as the C- or CP-conjugate processes, which change the baryon number by $-\Delta B$, resulting in no net asymmetry.
3. Out of equilibrium dynamics: In thermal equilibrium, processes that generate a baryon asymmetry, and the inverse processes that erase this asymmetry, proceed at the same rate, and no net asymmetry is left.

These necessary ingredients all exist in the Standard Model:

1. The baryon number is an anomalous accidental symmetry of the Standard Model. Consequently, baryon number–violating processes are forbidden at any order in perturbation theory, but they do occur nonperturbatively via the so-called sphaleron processes. These processes involve nine left-handed quarks (three of each generation) and three left-handed leptons (one from each generation). They violate lepton number and baryon number by 3 units each, but they conserve $B-L$. The physics of sphalerons is an advanced quantum field theory (QFT) topic beyond the scope of this book, and we do not elaborate on it.
2. The Standard Model violates C, as it is a chiral theory. The Standard Model violates CP via the phase in the Cabibbo-Kobayashi-Maskawa (CKM) matrix.
3. Departure from thermal equilibrium occurs at the electroweak phase transition.

Quantitatively, however, the Standard Model fails to explain the observed asymmetry:

1. Baryon number violation does not pose a problem. At zero temperature, the amplitude of the baryon number-violating processes is proportional to $e^{-8\pi^2/g^2}$, which is too small to have any observable effect. At high temperatures ($T \gg T_{\rm EWPT}$, where $T_{\rm EWPT}$ is the temperature where the electroweak phase transition starts to take place, of the order of the weak scale), however, these transitions take place faster than the expansion rate of the universe. Consequently, there is no suppression of the baryon asymmetry due to B violation.
2. CP violation provides a strong suppression. The BAU is a CP-violating observable. As explained in section 9.2.4 in chapter 9, CP violation in the Standard Model is proportional to X_{CP}, as defined in equation (9.15). In fact, η_B is proportional to the dimensionless quantity $X_{CP}/T_{\rm EWPT}^{12}$. Given the values of the quark masses and CKM parameters, one obtains a suppression factor of the order of 10^{-20}. Thus, the Kobayashi–Maskawa (KM) mechanism cannot account for the observed baryon asymmetry of $\eta_B \sim 10^{-9}$. We learn that baryogenesis implies that there must be new sources of CP violation, beyond the phase of the CKM matrix of the Standard Model.
3. The nonequilibrium condition provides another source of suppression. For successful baryogenesis, the electroweak phase transition should be first-order, which

could have been the case if m_h were less than about 60 GeV. Given the experimental value, $m_h \sim 125$ GeV, within the Standard Model, the electroweak phase transition is a smooth crossover, and most of the generated asymmetry is washed out. We learn that baryogenesis implies that there must be BSM physics that either provides an instance of departure from thermal equilibrium additional to the electroweak phase transition or modifies the electroweak phase transition itself.

We learn that baryogenesis requires extending the Standard Model in two ways. BSM physics must introduce new sources of *CP* violation and either provide a departure from thermal equilibrium in addition to the electroweak phase transition or modify the electroweak phase transition to make it first-order. There are many proposals of how to extend the Standard Model such that successful baryogenesis is achieved. Next, we briefly describe one of them: baryogenesis via leptogenesis or, in short, leptogenesis.

15.3.3 Leptogenesis

The general idea of leptogenesis is that the BAU originates from processes in the lepton sector. A lepton asymmetry is produced at temperatures above the weak scale, and the $B+L$–violating sphaleron interactions convert a large fraction of it into a baryon asymmetry. Thus, any model of leptogenesis above the weak scale is also a model of baryogenesis.

Here, we consider a specific model that can lead to successful leptogenesis: the NSM presented in section 14.3 in chapter 14. For simplicity, we consider a two-generation NSM and assume that $M_N \gg v$. While the NSM is motivated by the seesaw mechanism for light neutrino masses, the presence of the Yukawa (Y^ν) and mass (M^N) terms of equation (14.26) has an additional intriguing consequence: the physics of the singlet fermions N_R leads to leptogenesis. The three Sakharov conditions for leptogenesis are fulfilled in the NSM as follows:

1. Lepton number violation: Since a lepton number cannot be consistently assigned to the N_R fields in the presence of both Y^ν and M^N, the Lagrangian terms in equation (14.26) violate L.
2. *CP* violation: Since there are irremovable phases in the Lagrangian of equation (14.26), it provides new sources of *CP* violation.
3. Departure from thermal equilibrium: The interactions of the N_i particles are only of the Yukawa type. These interactions can be slower than the expansion rate of the universe, in which case the singlet fermions decay out of equilibrium.

Thus, in the NSM, all the conditions necessary for leptogenesis are qualitatively fulfilled. The question of whether the NSM can successfully explain the observed baryon asymmetry is thus a quantitative one.

We now consider leptogenesis via the decays of N_1, the lightest of the singlet fermions N_i. For simplicity, we assume that the decay is into a single-flavor α:

$$N_1 \to L_\alpha \phi, \qquad N_1 \to \overline{L}_\alpha \phi^\dagger. \tag{15.22}$$

The baryon asymmetry can be written as follows:

$$\eta_B = \eta_{\rm in} \times C_{\rm sphal} \times \eta_{\rm eff} \times \epsilon_{CP}. \tag{15.23}$$

The four factors on the right hand side of equation (15.23) represent the following physics aspects:

- $\eta_{\rm in} = n_{N_1}/n_\gamma$ represents the number density of N_1 at temperature $T \gg M_1$, where they are assumed to be in thermal equilibrium before they decay.
- ϵ_{CP} is the CP asymmetry in N_1 decays. That is, on average, $1/\epsilon_{CP}$ decays of N_1 generate one more $\bar{L}$ than Ls.
- $\eta_{\rm eff}$ is the efficiency factor that encodes the deviation from thermal equilibrium. It reduces the asymmetry by $0 \leq \eta_{\rm eff} \leq 1$. In particular, $\eta_{\rm eff} = 0$ is the limit of N_1 in perfect equilibrium, so no asymmetry is generated.
- $C_{\rm sphal}$ describes the dilution of the asymmetry due to the processes that redistribute the asymmetry that was produced in the lepton sector into the baryon sector.

These factors can be calculated, with ϵ_{CP} and $\eta_{\rm eff}$ depending on the Lagrangian parameters. The final result can be written (with some simplifying assumptions) as

$$\eta_B \sim 10^{-2}\, \eta_{\rm eff}\, \epsilon_{CP}. \tag{15.24}$$

The CP asymmetry is generated by the diagrams of figure 15.1. We define $x \equiv M_1/M_2$. In the small x limit, we get

$$\epsilon_{CP} = \frac{\Gamma(N_1 \to L_\alpha \phi) - \Gamma(N_1 \to \overline{L}_\alpha \phi^\dagger)}{\Gamma(N_1 \to L_\alpha \phi) + \Gamma(N_1 \to \overline{L}_\alpha \phi^\dagger)} = \frac{3x}{16\pi} \frac{1}{(Y^{\nu\dagger} Y^\nu)_{11}} \mathcal{I}m \left\{ \left[(Y^{\nu\dagger} Y^\nu)_{12} \right]^2 \right\}. \tag{15.25}$$

The washout effect is given by

$$\eta_{\rm eff} \sim \frac{10^{-3}\ {\rm eV}}{\tilde{m}}, \tag{15.26}$$

where $\tilde{m}$ is given by

$$\tilde{m} \equiv (Y^{\nu\dagger} Y^\nu)_{11} \frac{v^2}{M_1}. \tag{15.27}$$

The plausible range for $\tilde{m}$ is the one suggested by the light neutrino masses, $10^{-3} - 1$ eV (see chapter 14).

Putting together equations (15.24), (15.25), (15.26), and (15.27), we conclude that successful leptogenesis requires

$$\frac{M_1}{M_2} \frac{\mathcal{I}m[(Y^{\nu\dagger} Y^\nu)_{12}^2]}{(Y^{\nu\dagger} Y^\nu)_{11}} \frac{10^{-3}\ {\rm eV}}{\tilde{m}} \sim 10^{-8}. \tag{15.28}$$

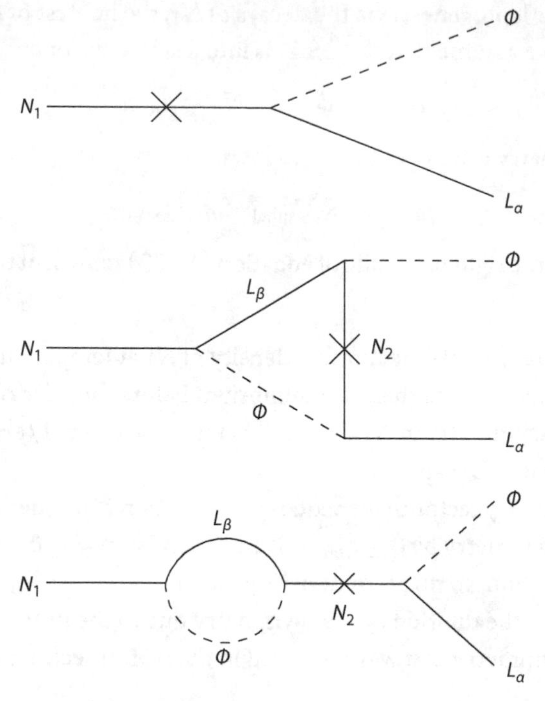

FIGURE 15.1. The diagrams contributing to the *CP* asymmetry ϵ_{CP}.

The condition in equation (15.28) can be satisfied in a large part of the parameter space of the NSM, simultaneous with fitting the neutrino mass and mixing parameters.

Leptogenesis, while theoretically a very attractive scenario, is very difficult to test experimentally. The new DoF are heavy and have neither strong, nor electromagnetic, nor weak interactions (except for the small mixing with the $SU(2)$-doublet neutrinos), and thus they are unlikely to be produced and detected in experiments. The idea of leptogenesis might remain a plausible but unproved explanation of the baryon asymmetry.

15.4 Open Questions

The Standard Model, as well as its νSM extension, fail to account for dark matter and the BAU. Thus, particle cosmology provides clues for the theory that extends the νSM. A large variety of experiments are seeking to shed further light on the relevant questions:

- What is the nature—mass, spin, and couplings—of the dark matter particles?
- What are the *CP*-violating source and the mechanism of departure from thermal equilibrium that drive baryogenesis?

On the first question, there are three classes of experiments. Denoting a generic dark matter particle by χ and a Standard Model particle by ψ, direct searches look for signals coming from processes of the type $\chi + \psi \to \chi + \psi$, indirect searches look for signals coming from $\chi + \chi \to \psi + \psi$, and collider experiments look for signals coming from $\psi + \psi \to \chi + \chi$.

On the second question, experiments that seek to find further clues for leptogenesis aim to establish *CP* violation in the lepton sector via measurement of neutrino oscillations, and to establish the Majorana nature of neutrino masses via neutrinoless double-beta decay. Experiments that seek to find clues for electroweak baryogenesis aim to find evidence of modified electroweak phase transition by measuring Higgs couplings or searching for new DoF that couple to the Higgs boson, and to find *CP* violation in processes involving the Higgs boson.

Appendix

15.A Introduction to Cosmology

In this appendix, we briefly review some basic ingredients of cosmology that are relevant to particle physics. In particular, we focus on issues that are needed to follow our discussion of baryogenesis and of dark matter in the main text.

The Big Bang model of cosmology is well established by now. It is based on the observation that the universe is expanding, from which it follows that the early universe was dense and hot. This implies that for cosmology, we should use the tools of classical general relativity, thermodynamics, and statistical mechanics.

The relevant chronological history of the universe starts in an epoch of inflation, during which almost all the energy in the universe was stored in a field called *inflaton*. Inflation ends by reheating, where the energy stored in the inflaton field converts to the known (and maybe some yet unknown) particles that were in thermal equilibrium at a very high temperature. The highest temperature is called the *reheat temperature*, $T_{\rm RH}$. The value of this temperature must be above a few MeV, and it can be much larger. We assume $m_W \ll T_{\rm RH} \ll M_{\rm Pl}$.

Particle physics starts to play a significant role after reheating. All the known particles were in thermal equilibrium at $T \gg m_W$, and therefore the time evolution of their number and energy densities is independent of the initial conditions. This fact makes cosmology very predictive. So far as particle physics is concerned, the details of inflation are unimportant.

After reheating, the universe expanded and cooled. At various times (or, equivalently, temperatures), particles dropped out of equilibrium. The unstable particles decayed and the universe evolved into its present state, with protons, neutrons, electrons, neutrinos, photons, and probably at least one more type of particle, dark matter, whose exact properties are yet unknown.

The more formal discussion that follows aims to obtain the number densities of particles in the universe. These number densities depend on the particle interactions (hence the connection to particle physics). In section 15.A.1, we discuss the metric of the universe, as that is our tool to incorporate the expansion. In section 15.A.2, we use thermodynamics to calculate the densities.

15.A.1 The Dynamical Metric

In principle, given the matter and energy distribution in the universe, one can derive the metric of the universe as a solution of Einstein equations. In practice, this is an impossible task for general mass and energy distributions. What makes this derivation feasible are two assumptions about these distributions: homogeneity and isotropy in space. While these may seem unjustified based on local observations that show a lot of structure, they do provide a very good approximation at scales above 100 Mpc. With these approximations, we can determine the metric of the universe that is valid at large scales, which is what is relevant for cosmology.

In general, the geometry of the universe can be open, flat, or closed. Inflation predicts, and observations confirm, that the universe is very close to flat. The metric for a homogeneous, isotropic, and flat universe is given by

$$ds^2 = -dt^2 + a^2(t)\left[dx^2 + dy^2 + dz^2\right], \tag{15.29}$$

where $a(t)$ is called the *scale factor*. By definition, $a(t_0) = 1$. (In cosmology, subindex zero indicates "at the present time.") The effect of the expansion is entirely encoded in the scale factor $a(t)$. From this point on, we usually keep the time dependence of various parameters implicit.

We define the Hubble constant:

$$H \equiv \frac{\dot{a}}{a}. \tag{15.30}$$

Since H is time-dependent, the term *Hubble constant* is somewhat misleading. It is, however, space-independent. One often replaces H with the dimensionless parameter h:

$$h \equiv \frac{H_0}{100\,\mathrm{km/(s\,Mpc)}}, \tag{15.31}$$

where H_0 is the present-day Hubble constant. Observations find $h \approx 0.7$.

The energy density of the universe has contributions from radiation, matter, and vacuum energy. The dependence of the energy density ρ_X ($x = r, m, v$ for radiation, matter, and vacuum, respectively) on the scale factor is different for each of these three components:

$$\rho_X \propto a^{-n} \text{ with } n = \begin{cases} 3 & \text{matter,} \\ 4 & \text{radiation,} \\ 0 & \text{vacuum energy.} \end{cases} \tag{15.32}$$

"Matter" refers to nonrelativistic species, in which case $n = 3$ is a simple geometrical factor. "Radiation" refers to ultrarelativistic particles (either massless or not), in which case there is an extra factor of $1/a$ due to the red-shift of the energy. "Vacuum energy" refers to the case of no dilution. This case does not correspond to known particles but, for example, to the cosmological constant.

The total energy density of a flat universe is called the *critical density*, and is given by

$$\rho_c = \frac{3H^2}{8\pi G_N}, \tag{15.33}$$

where G_N is the Newton constant. Numerically, the present critical density is given by

$$\rho_c = 10.53672(24)\, h^2\, \text{keV cm}^{-3}. \tag{15.34}$$

We define

$$\Omega_i = \frac{\rho_i}{\rho_{\text{crit}}}, \qquad \Omega = \sum_i \Omega_i. \tag{15.35}$$

For a flat universe, $\Omega = 1$. In this case, we can invert the relation of equation (15.33) and obtain H as a function of the critical density:

$$H^2 = \frac{8\pi G_N \rho_c}{3}. \tag{15.36}$$

Thus, the energy density determines the expansion rate H.

Due to the different time dependencies for the various components, Ω_i changes in time, even though $\Omega = 1$ does not. Specifically, as time goes on and a increases, Ω_r and Ω_m decrease (the former faster than the latter), while Ω_v increases correspondingly to keep $\Omega = 1$. Observations imply the following present values:

$$\Omega_r \approx 5.4 \times 10^{-5}, \qquad \Omega_m \approx 0.31, \qquad \Omega_v \approx 0.69. \tag{15.37}$$

Given a situation where the universe has only one dominant component, the solution for $a(t)$ is simple: $a(t) \propto t^{2/3}$ for matter domination, $a(t) \propto t^{1/2}$ for radiation domination, and $a(t) \propto e^{Ht}$ for vacuum energy domination. In all cases, $a(t)$ is a monotonic rising function of t (i.e., the universe is expanding).

For completeness, we also mention the equation of state, which gives the relation between the energy density ρ and the pressure p:

$$p = \omega \rho, \tag{15.38}$$

where $\omega = n/3 - 1$, with n defined in equation (15.32), and thus

$$\omega_r = 1/3, \qquad \omega_m = 0, \qquad \omega_v = -1. \tag{15.39}$$

15.A.2 Thermodynamics in the Universe

It is possible to calculate the relic abundance of all particles that have been in thermal equilibrium at some stage of the history of the universe. (By *relic abundance*, we are referring to the present number density.) This applies to all particles with $m \ll T_{\text{RH}}$ and sufficiently large couplings and, in particular, assuming that $T_{\text{RH}} \gg m_W$, to all the Standard Model particles.

The momentum distribution of a specific type of particle in thermal equilibrium is given by

$$f(p) = \frac{1}{e^{[E(p)-\mu]/kT} \pm 1}, \tag{15.40}$$

where the plus (minus) sign corresponds to fermions (bosons), k is the Boltzmann factor, and μ is the chemical potential, which corresponds to an asymmetry between particles and antiparticles. The number and energy densities are given by

$$n_i = \frac{g_i}{(2\pi)^3} \int f_i(p) d^3p, \qquad \rho_i = \frac{g_i}{(2\pi)^3} \int E(p) f_i(p) d^3p, \tag{15.41}$$

where g_i is the number of DoF. Given equation (15.32), it is clear that the early universe was radiation dominated. In that era, the total energy density is given by the energy density in relativistic particles. We obtain (see question 15.1)

$$\rho = \sum_i \rho_i = \frac{\pi^2}{30} g_* T^4, \tag{15.42}$$

where g_* counts the effective number of relativistic DoF that are in thermal equilibrium:

$$g_* = \sum_{\text{bosons}} g_i + \frac{7}{8} \sum_{\text{fermions}} g_i. \tag{15.43}$$

Using equation (15.36), we obtain a relation between the expansion rate and the temperature in the radiation-dominated era,

$$H^2 = \frac{8\pi^3 G_N g_*}{90} T^4. \tag{15.44}$$

We are interested in the relic abundance of stable particles, and we are distinguishing the cases that $\mu_i = 0$ and $\mu_i \neq 0$. The $\mu_i = 0$ case applies when there is no asymmetry between particles and antiparticles (or when the asymmetry can be neglected). In particular, it applies to particles that are their own antiparticles (such as the photon or Majorana neutrinos). For high temperatures, $T \gg m_i$, the number density at thermal equilibrium is given by

$$n_i = K g_i \frac{\zeta(3)}{\pi^2} T^3, \tag{15.45}$$

where $K = 3/4(1)$ for fermions (bosons), and the Riemann zeta function has the value of $\zeta(3) \approx 1.2$. For low temperatures, $T \ll m_i$, the number density at thermal equilibrium is given by

$$n_i = g_i \left(\frac{m_i T}{2\pi} \right)^{3/2} e^{-m_i/T}. \tag{15.46}$$

The exponential suppression of the number density when $T \ll m_i$ is called *Boltzmann suppression*.

The $\mu_i \neq 0$ case applies when there is an asymmetry between particles and antiparticles. We limit our discussion to fermions that carry a conserved charge. In this case, the excess (summed over all particles carrying the relevant charge) is diluted only by the expansion of the universe. Assuming that there is just one type of particle that carries the charge, we obtain

$$\Delta n_i \equiv n_i - n_{\bar{i}} = \frac{g_i \mu_i^3}{6}, \tag{15.47}$$

If the would-be number density for the $\mu_i = 0$ case is much smaller than the asymmetry ($n_i(\mu_i = 0) \ll \Delta n_i$), then all the antiparticles (which, by definition, have smaller number density than the particles) annihilate and $\Delta n_i = n_i$.

The next step is to obtain the number density of particles after they drop out of thermal equilibrium. While dropping out is a gradual process, which can be followed by solving the Boltzmann equations, it is often a good approximation to think of the transition from thermal equilibrium to complete decoupling as instantaneous. We define a freeze-out temperature, T_F, such that for $T > T_F$, we take the number density to be that of a thermal equilibrium, while for $T < T_F$, the number density is diluted only by the expansion of the universe.

The transition from equilibrium to decoupling depends on the interplay between the strength of the interaction and the expansion of the universe. We define the interaction rate of particle i as Γ_i. There are two kinds of interaction: scattering and decays. For decays, Γ_i corresponds to the decay width of the particle. For scattering,

$$\Gamma_i = \sum_j n_j \langle \sigma_{ij} v \rangle, \tag{15.48}$$

where n_j is the number density of the particle j on which i scatters, σ_{ij} is the cross section of the scattering process, and v is the typical relative velocity between i and j. The cross section, in general, depends on the specific energies of the particles that collide, and thus we average over it. We use $\langle \sigma_{ij} v \rangle$ to indicate thermal averaging.

The freeze-out temperature is then taken as the temperature at which

$$\Gamma_i(T_F) = H(T_F). \tag{15.49}$$

The intuition is that for $\Gamma_i \gg H$, the expansion can be neglected and thus a particle is at equilibrium. In the opposite limit, $\Gamma_i \ll H$, the typical time between interactions is longer than the age of the universe. Another way to understand the choice of equation (15.49) as the definition of freeze-out is to require that after freeze-out and until today, the number of interactions has been less than 1. The number of interactions is given by

$$n_{\text{int}}^i = \int_{t_F}^{t_0} \Gamma_i(t) dt = \int_{a_F}^{a_0} \frac{\Gamma_i(a)}{H} d(\log a) \approx \left. \frac{\Gamma_i}{H} \right|_{t=t_F}, \tag{15.50}$$

where, in the last stage, we assume that Γ_i and H do not have a very strong dependence on a. The definition of the freeze-out time is t_F, such that $n_{\text{int}}^i = 1$.

We learn that the stronger the interaction, the lower T_F is and the smaller the relic abundance is. While the number density depends strongly on the scattering cross section, the temperature is not very sensitive to it. For interaction strength of the order of the weak interaction, we have

$$T_F \sim m/20, \tag{15.51}$$

with only logarithmic sensitivity of T_F on Γ_i.

After freeze-out, the momentum distribution for a stable particle is given, to a good approximation, by equation (15.40), but with T as a parameter that is not the temperature in the thermodynamic meaning of the word. At present, the temperature of the photons, usually called the temperature of the universe, is given as $T_0 = 2.7255(6)$ K. This temperature is just the parameter that characterizes the distributions. Note also that after decoupling, and with this understanding of temperature, various stable particles can have different temperatures.

The freeze-out condition shows explicitly where particle physics plays a role: It is the decay rate and the scattering cross section that eventually determine the number densities of particles. Measuring the relic abundance is a probe of the interaction rate, that depends on the Lagrangian parameters.

15.A.3 Observables

There are several observables that are used to determine the cosmological model and its parameters. Here, we briefly list the ones that are used to determine the relevant number and energy densities.

- CMB. When the temperature of the universe dropped below about 1 eV, neutral atoms formed. At that point, the radiation, which had been in thermal equilibrium with the plasma at earlier times, froze out, and it follows its blackbody spectrum until today. These photons have been observed, providing pivotal evidence for the Big Bang theory. The CMB encodes a lot of information. In particular, there are very small, $\mathcal{O}(10^{-5})$, deviations from a perfect isotropic blackbody spectrum. These deviations are sensitive to the details of the universe when the photon freeze-out occurred. Examples that are relevant for this discussion include sensitivity to the energy density of dark matter, $\Omega_{\rm DM}$, to the ratio of baryon to photon number densities, η_B, and to neutrino masses, $\sum_i m_i$.
- BBN. At very high temperatures, neutrons and protons were in a plasma with almost identical number densities. When the temperature dropped below about 1 MeV, two processes began to take place. The neutrons started to decay, and protons and neutrons started to form bound states. Detailed calculations show that almost all the neutrons that did not decay ended up in ^{4}He, with much smaller amounts of D, ^{3}He, and ^{7}Li. The relic number densities of these light elements depend on a single parameter, which is η_B.

- Large-scale structure. At large distance scales, we expect the universe to be isotropic and homogeneous. At smaller scales, we do have structure: namely, galaxies and clusters. To probe the structure of the universe, surveys of galaxies are performed and their distributions extracted. This provides information about the deviation from isotropy and homogeneity, which is sensitive to parameters like the velocity of dark matter.
- Supernova survey. Faraway supernovae provide another class of cosmological observations. Since the intrinsic luminosity of supernovae is known, they can be used, to a good approximation, as standard candles. Using their red-shift, one can also measure their velocity and thus check the distance-to-velocity relations. This dependence can be used to extract the fraction of dark energy in the universe, Ω_ν.

For Further Reading

There are many textbooks and reviews about cosmology. Here are a few examples:

- More details and rigorous derivations of general results in particle cosmology are presented in Kolb and Turner [51], Carroll [52], Trodden and Carroll [53], Rubakov and Gorbunov [54], Gorbunov and Rubakov [55], and Piattella [56].
- For a review of dark matter, see Lisanti [57].
- For a review of neutrinos in cosmology, see Dolgov [58].
- For a review of leptogenesis, see Davidson, Nardi, and Nir [59].

Problems

Question 15.1: Algebra

1. Using equation (15.14), derive equation(15.16).
2. Derive equation (15.50):

$$\int_{t_F}^{t_0} \Gamma_i(t)dt = \int_{a_F}^{a_0} \frac{\Gamma_i(a)}{H} d(\log a). \qquad (15.52)$$

3. Explain the factor of 7/8 in equation (15.43). To do so, show that the energy density per fermionic DoF is 7/8 that of a bosonic one. It is useful to recall the following identity:

$$\int_0^\infty dx \frac{x^n}{e^x - 1} = \frac{1}{(1-2^{-n})} \int_0^\infty dx \frac{x^n}{e^x + 1}. \qquad (15.53)$$

4. Using equations (15.40), (15.41), and (15.53), and

$$\int_0^\infty dx \frac{x^{(n)}}{e^x - 1} = n! \times \zeta(n+1), \qquad (15.54)$$

with $\zeta(3) \approx 1.2$, $\zeta(4) = \pi^4/90$, derive equations (15.42) and (15.45).

Question 15.2: The neutrino temperature

In this question, you are asked to derive equation (15.4). At very early times, the neutrinos and the photon were in thermal equilibrium with each other and with other DoF. Then the neutrinos decoupled, and later the remaining DoF. This chain of events led to $T_\nu \neq T_\gamma$.

1. Neutrinos decouple at a temperature $T_{\nu,F}$, when $\Gamma_{\rm int} \sim H \sim T^2/m_{\rm Pl}$. The neutrino cross section is given by $\langle \sigma v\rangle \sim G_F^2 T^2$, and the neutrino number density is given by $n \sim T^3$ (see equation (15.45)). Thus, $\Gamma_{\rm int} = n\,\langle \sigma v\rangle \simeq G_F^2 T^5$. Find $T_{\nu,F}$ in terms of G_F and $M_{\rm Pl}$. Show that numerically, $T_{\nu,F} \approx 1$ MeV.

At $T_{\nu,F}$, the photon and the neutrino temperatures are the same ($T_\nu = T_\gamma$). Below $T_{\nu,F}$, the neutrinos are diluted by the expansion of the universe ($T_\nu \sim a^{-1}$). The photon temperature, however, goes through a more complicated evolution that has to do with e^+e^- annihilation.

2. Show that, after e^+e^- annihilation (at temperature $T_{e,\rm ann} \sim m_e$), the effective number of DoF changes from $g = 11/2$ to $g = 2$.
3. For particles in thermal equilibrium at that era, $g(aT)^3$ remains constant (where g is the effective number of relativistic DoF). Show that this implies that aT_γ increases by $(11/4)^{1/3}$ as a result of e^+e^- annihilation.
4. Derive equation (15.4); that is, show that today,

$$\frac{T_{\gamma,0}}{T_{\nu,0}} = \left(\frac{11}{4}\right)^{1/3} \approx 1.40 \Longrightarrow T_{\nu,0} \approx 1.95\ \text{K}. \tag{15.55}$$

Use $T_{\gamma,0} \approx 2.73$ K.

Question 15.3: The WIMP cross section

Consider the χSM and the Lagrangian in equation (15.11). We study the $\chi\chi \to hh$ scattering process. We assume that $m_\chi \gg m_h$, and approximate $m_h = 0$ in the calculation. We further consider the case of nonrelativistic collision (i.e., in the center of mass, $|\vec{p}_\chi| \ll m_\chi$).

1. There are three tree-level diagrams for the $\chi\chi \to hh$ process: one with a Higgs propagator, one with a χ propogator, and one with no propagator. Draw each one, and argue that the diagrams with propagators can be neglected.
2. Show that

$$\sigma(\chi\chi \to hh) = \frac{Z_{\chi\phi}^2}{16\pi m_\chi^2}. \tag{15.56}$$

Remember to include symmetry factors due to Wick contractions and outgoing identical particles.

3. Consider all the processes of the form $\chi + \chi \to f_{SM}+f_{SM}$, where f_{SM} is an Standard Model particle, and argue that the most important ones are $\chi\chi \to hh$, $\chi\chi \to W^+W^-$, and $\chi\chi \to ZZ$.
4. Show that, in our approximation,

$$\sigma v \equiv \sum \sigma(\chi\chi \to f_{SM} f_{SM}) v \approx 4\sigma(\chi\chi \to hh) v, \tag{15.57}$$

and thus that

$$\langle \sigma v \rangle = \frac{1}{4\pi} \frac{Z_{\chi\phi}^2}{m_\chi^2}. \tag{15.58}$$

Note that, when σv is independent of the kinematics, $\sigma v = \langle \sigma v \rangle$.
5. Using the required value of $\langle \sigma v \rangle$ of equation (15.17), find $m_\chi / Z_{\chi\phi}$; that is, verify equation (15.19).

Question 15.4: Leptogenesis

In this question, we discuss some of the properties of the *CP* asymmetry that is generated by the decay of N_1, equation (15.25).

1. Work out the dependence on the Yukawa couplings separately for the numerator and the denominator in equation (15.25). What assumptions did you have to make to recover the result of equation (15.25)?
2. The *CP* asymmetry vanishes as $M_2 \to \infty$. This seems to contradict two examples that we have given in which loop effects grow with the mass of the internal fermion, electroweak precision measurements (EWPM) (see equation (12.11) in chapter 12) and the Glashow-Iliopoulos-Maiani (GIM) mechanism for FCNC (see section 13.2 in chapter 13). Explain the difference between the three cases; that is, explain why, for leptogenesis, the effect decreases with the mass of the internal fermion, while in the EWPM and the GIM mechanism cases, it increases.

WHAT'S NEXT?

By the time you read this chapter, you should have gained a basic understanding of the Standard Model and, in particular, the underlying principles that are used to construct it and the many ways in which it has been tested. All the Standard Model particles have been directly observed, and all the Standard Model parameters have been measured (albeit with varying degrees of precision). Given the Standard Model's numerous successes, one must ask: what is next in our experimental and theoretical endeavors to understand the basic laws of nature?

The search for a more fundamental theory is motivated by the few problems that experiments and observations pose, as well as by the curiosity-driven quest to explore new territory (i.e., short distances and high energies). Inspired by these motives, the research in high-energy physics has to proceed in a coordinated way on both the theoretical and the experimental front.

On the experimental side, we can roughly divide the efforts into three complementary frontiers.

- The energy frontier: By having colliders with higher energy, we are hoping to find new particles with masses that are beyond the reach of current colliders but within the reach of future ones. To date, the highest center of mass energy achieved at colliders, 13 TeV, occurred at the Large Hadron Collider (LHC). There are plans to build a more powerful accelerator, with up to 100 TeV in energy. Clearly, discovering new elementary particles will be a clean indication of Beyond the Standard Model (BSM) physics and will provide direct information about the new particles and their couplings. This information then can be used to find the BSM theory.
- The luminosity frontier: By having higher luminosity and collecting more data, we can improve the precision of the rates of various processes, as well as measure for the first time processes that are more rare. If theoretical precision matches the experimental one, we can probe in these ways small deviations from the Standard Model predictions. Discovering such derivations will signal BSM physics. While such information will not be directly related to particles and their couplings, the reach—in terms of high energy and short distance—is much higher than that of direct searches.
- The astroparticle frontier: By looking at cosmology and astronomy, we hope to find answers to questions like the nature of dark matter and the mechanism of baryogenesis. Experiments are looking for detection of the dark matter, as well

as other cosmological parameters, with the goal of improving our understanding of astrophysical events and cosmology, and at the same time searching for BSM physics.

On the theory side, the effort can also be divided into three main directions of research:

- Model building: There is an effort to devise new models that address the open questions of the Standard Model. By doing so, one can make testable predictions. In particular, this can lead to new experiments, as well as new analyses within the working experiments.
- More precise calculations: In many cases, to benefit from more precise measurements, we need more precise calculations. In other words, the precision of the theoretical calculations should match the experimental precision. Some of these calculations are done perturbatively—that is, by going to higher loops or, nonperturbatively, by using Effective Field Theories (EFTs) or on the lattice.
- New experimental probes: One can suggest new ways to test the Standard Model in regions of parameter space that have not yet been explored without the framework of a specific BSM theory. Thus, one can hope to find anomalies in the data that will provide hints for BSM physics even without the guidance of a specific theory.

We do not know what will lead us to the next stage. Based on the history of physics, however, we hope that sooner or later, a deeper theory that extends the Standard Model will emerge. While the Standard Model itself is a very beautiful theory, based on symmetry principles, we await with great curiosity the wonders of the next level of understanding the basic laws of nature.

APPENDIX

Lie Groups

Symmetries play a crucial role in model building. You are already familiar with symmetries and some of their consequences. For example, nature is symmetric under the Poincaré group, which implies conservation of energy, momentum and angular momentum. To understand the interplay between symmetries and interactions, we need a mathematical tool called *Lie groups*. These are the groups that describe all continuous symmetries. In the following, we give definitions and quote various statements without proving them. There are many texts about Lie groups where these statements are proved and more details given.

A.1 Groups

Definition: A *group* G is a set $\{x_i\}$ (finite or infinite), with a multiplication law $\cdot$, subject to the following four requirements:

- Closure:

$$x_i \cdot x_j \in G \qquad \forall\, x_i, x_j. \tag{A.1}$$

- Associativity:

$$x_i \cdot (x_j \cdot x_k) = (x_i \cdot x_j) \cdot x_k \qquad \forall\, x_i, x_j, x_k. \tag{A.2}$$

- There is an identity element I (or e) such that

$$I \cdot x_i = x_i \cdot I = x_i \qquad \forall\, x_i. \tag{A.3}$$

- Each element has an inverse element x_i^{-1}:

$$x_i \cdot x_i^{-1} = x_i^{-1} \cdot x_i = I \qquad \forall\, x_i. \tag{A.4}$$

A group is specified by its multiplication table.

Definition: A group is *Abelian* if all its elements commute:

$$x_i \cdot x_j = x_j \cdot x_i \qquad \forall\, x_i, x_j. \tag{A.5}$$

A *non-Abelian* group is a group that is not Abelian (i.e., at least one pair of elements does not commute).

Let us give a few examples here:

- Z_2, also known as *parity*, is a group with two elements, I and P, such that I is the identity and $P^{-1} = P$. This completely specifies the multiplication table. This group is finite and Abelian. The multiplication table is

Z_2	I	P
I	I	P
P	P	I

(A.6)

- Z_N, with N a positive integer, is a generalization of Z_2. It contains N elements labeled from zero through $N-1$. The multiplication law is the same as addition modulo N: $x_i \cdot x_j = x_{(i+j)\text{mod } N}$. The identity element is x_0, and the inverse element is given by $x_i^{-1} = x_{(N-i)\text{mod } N}$. This group is also finite and Abelian.
- Multiplication of positive numbers. It is an infinite Abelian group. The identity is the number 1, the multiplication law is just standard multiplication, and the inverse of x is $1/x$.
- S_3, the group that describes the permutations of three elements. It contains six elements. This group is non-Abelian. In question A.1 at the end of this appendix, you are asked to find the six elements and their multiplication table.

A.2 Representations

Definitions:

- A *representation* is a realization of the multiplication law among matrices.
- Two representations are *equivalent* if they are related by a similarity transformation (i.e., by a unitary rotation).
- A representation is *reducible* if it is equivalent to a representation that is block diagonal.
- An *irreducible* representation (irrep) is a representation that is not reducible.
- An irrep that contains matrices of size $n \times n$ is said to be of *dimension n*.

Statements:

- Any reducible representation can be written as a direct sum of irreps.
- The dimension of all irreps of an Abelian group is one. For non-Abelian groups, there is at least one irrep that has a dimension larger than one.
- Any finite group has a finite number of irreps R_i. If N is the number of elements in the group, the irreps satisfy

$$\sum_i [\dim(R_i)]^2 = N. \tag{A.7}$$

- Infinite groups have an infinite number of irreps.

- For any group, there exists a *trivial* representation, such that all the matrices are just the number 1. This representation is also called the *singlet* representation. It is of particular importance for us.

Let us give a few examples of the statements that we have made here:

- Z_2: Its trivial irrep is $I = 1, P = 1$. The other irrep is $I = 1, P = -1$. Clearly, these two irreps satisfy equation (A.7).
- Z_N: An example of a nontrivial irrep is $x_k = \exp(i2\pi k/N)$ with $k = 0, 1, \ldots,$ $N - 1$.
- S_3: In question A.1, you are asked to work out its properties.

The groups that we are interested in are *transformation groups of physical systems*. Such transformations are associated with *unitary operators* in the Hilbert space. We often describe the elements of the group by the way that they transform physical states. When we refer to representations of the group, we mean either the appropriate set of unitary operators or, equivalently, the matrices that operate on the vector states of the Hilbert space.

A.3 Lie Groups and Lie Algebras

While finite groups are very important, the ones that are most relevant to particle physics and, in particular, to the Standard Model, are infinite groups. More specifically, the most relevant ones are Lie groups, which provide a formal way to discuss rotations in any real or abstract space. The various groups correspond to rotations in different spaces.

Definition: A *Lie group* is an infinite group whose elements are labeled by a finite set of N continuous real parameters α_ℓ, and whose multiplication law depends smoothly on the α_ℓs. The number N is called the *dimension* of the group.

Different groups have different N values. Yet the dimension of the group does not uniquely define it. Next, we discuss the classifications of groups.

Statement: An Abelian Lie group has $N = 1$. A non-Abelian Lie group has $N > 1$.

The first example of an Abelian Lie group is $U(1)$, which represents the addition of real numbers modulo 2π (i.e., rotation on a circle). Such a group has an infinite number of elements that are labeled by a single continuous parameter α. We can write the group elements as $M = \exp(i\alpha)$ and the multiplication law as

$$\exp(i\alpha) \cdot \exp(i\beta) = \exp[i(\alpha + \beta)]. \tag{A.8}$$

We can also represent the same group by $M = \exp(2i\alpha)$ or, more generally, as

$$M = \exp(i\alpha X), \tag{A.9}$$

with X being real. Each X generates an irrep of the group. For $U(1)$, we usually call X the "charge" and often denote it by q.

We are mainly interested in *compact* Lie groups. We do not define this term formally here, but we can use the $U(1)$ example to give an intuitive explanation of what it means. A group of adding with a modulo is compact, while just adding, without the modulo, would be noncompact. In the former, if you repeat the same addition a number of times, you may return to your starting point, while in the latter, this would never happen. In other words, in a compact Lie group, the parameters have a finite range, while in a noncompact group, their range is infinite. (Do not confuse that with the number of elements, which is infinite in either case.) Another example is rotations and boosts: rotations represent a compact group, while boosts do not.

Statement: The elements of any compact Lie group can be written as

$$M = \exp(i\alpha_\ell X_\ell), \tag{A.10}$$

such that α_ℓ are real numbers and X_ℓ are specific Hermitian matrices. This is the generalization of the Abelian case, equation (A.9), where X is a real number. We use the standard summation convention, $\alpha_\ell X_\ell \equiv \sum_\ell \alpha_\ell X_\ell$, and ℓ runs from 1 to N, where N is the dimension of the group.

Statement: The inverse element of M of equation (A.10) is

$$M^{-1} = \exp(-i\alpha_\ell X_\ell). \tag{A.11}$$

Definition: The X_ℓs are called the *generators* of the group.

Let us perform some algebra before we turn to our next definition. Consider two elements of a group, A and B, such that in A, only $\alpha_a \neq 0$, and in B, only $\alpha_b \neq 0$, and furthermore, $\alpha_a = \alpha_b = \lambda$:

$$A \equiv \exp(i\lambda X_a), \qquad B \equiv \exp(i\lambda X_b). \tag{A.12}$$

Since A and B are in the group, each of them has an inverse. Thus,

$$C = BAB^{-1}A^{-1} \equiv \exp(i\beta_c X_c) \tag{A.13}$$

is also in the group. Let us take λ to be a small parameter and expand around the identity. Clearly, if λ is small, all the β_c are also small. Keeping the leading-order terms, we get

$$C = \exp(i\beta_c X_c) \approx I + i\beta_c X_c, \qquad C = BAB^{-1}A^{-1} \approx I + \lambda^2 [X_a, X_b]. \tag{A.14}$$

In the $\lambda \to 0$ limit, we have

$$[X_a, X_b] = if_{abc} X_c, \qquad f_{abc} \equiv \frac{\beta_c}{\lambda^2}. \tag{A.15}$$

Note that f_{abc} is independent of λ (see question A.2), and furthermore, while λ and β_c are infinitesimal, the f_{abc}-constants do not diverge. This brings us to a new set of definitions:

Definition: f_{abc} are called the *structure constants* of the group.

Definition: The commutation relations of equation (A.15), $[X_a, X_b] = if_{abc} X_c$, constitute the *algebra* of the Lie group.

Definition: A *trivial algebra* is one where $f_{abc} = 0$ for all the elements. A *nontrivial algebra* is an algebra that is not trivial.

Statements:

- The algebra defines the local properties of the group, but not its global properties. Usually, the local properties are all we care about.
- The algebra is closed under commutation.
- Similar to our discussion of groups, one can define representations of the algebra (i.e., matrix representations of X_ℓ). In particular, each representation has its own dimension. (Do not confuse the dimension of the representation with the dimension of the group.)
- The generators satisfy the Jacobi identity:

$$[X_a, [X_b, X_c]] + [X_b, [X_c, X_a]] + [X_c, [X_a, X_b]] = 0. \tag{A.16}$$

- Each algebra has a trivial (singlet) representation: $X_\ell = 0$ for all ℓ. The trivial representation of the algebra generates the trivial representation of the group.
- Since an Abelian Lie group has only one generator, its algebra is always trivial. Thus, the algebra of $U(1)$ is the only Abelian Lie algebra.
- Non-Abelian Lie groups have nontrivial algebras.
- The generators of non-Abelian Lie groups are traceless.

The example of $SU(2)$ algebra is well known from quantum mechanics (QM):

$$[X_a, X_b] = i\varepsilon_{abc} X_c, \tag{A.17}$$

where ε_{abc} are the structure constants of the $SU(2)$ group. Usually in QM, X is called L, S, or J. The three matrices S_x, S_y, and S_z for a given spin S correspond to a given irrep of $SU(2)$. The $SU(2)$ group represents nontrivial rotations in a two-dimensional, complex space. Its algebra is the same as the algebra of the $SO(3)$ group, which represents rotations in the three-dimensional, real space. (We discuss these names next.)

We should explain what we mean when we say that "the group represents rotations in a space." The QM example makes this clear. Consider a finite Hilbert space of a particle with spin S. The matrices that rotate the direction of the spin are written in terms of exponentials of the S_i operators. For a spin-half particle, the S_i operators are written in terms of the Pauli matrices. For particles with spin different from $1/2$, the S_i operators will be written in terms of different matrices. We learn that the group represents rotations in some space, while the various representations correspond to different objects that can exist in that space.

There are three important irreps that have special names. The first one is the trivial—or *singlet*—representation that we already mentioned. Its importance stems from the fact that it corresponds to an object that is invariant under rotations. Rotation of a vector in a real space, which is not a singlet, does change its direction. The magnitude of the vector, which is a singlet, does not change under rotation.

The second important irrep is the *fundamental* representation. This is the smallest nontrivial irrep. For $SU(2)$, this is the spinor, or spin-half, representation. An important property of the fundamental representation is that it can be used to get all other representations. We return to this point later in this discussion. Here, we just remind you that this statement is well familiar from QM. One can get spin-1 by combining two spin-1/2s, and spin-3/2 by combining three spin-1/2s. Any non-Abelian Lie group has a fundamental irrep.

The third important irrep is the *adjoint* representation. It is made of the structure constants themselves. Think of a matrix representation of the generators. Each entry, T^c_{ij} is labeled by three indexes. One is the c index of the generator itself, that runs from 1 to N, such that N depends on the group. The other two indexes, i and j, are the matrix indexes that run from 1 to the dimension of the representation. One can show that each Lie group has one representation where the dimension of the representation is the same as the dimension of the group. This representation is obtained by defining

$$(X_c)_{ab} \equiv -if_{abc}. \tag{A.18}$$

In other words, the structure constants themselves satisfy the algebra of their own group. (See question A.3 for more details.) In $SU(2)$, the adjoint representation is that of spin-1. In question A.3, you are asked to check that ε_{ijk} is a set of three 3×3 matrices that is a representation of spin-1.

Before closing this section, we make several remarks about subalgebras and simple groups:

Definition: A *subalgebra* M is a set of generators that are closed under commutation.

Definition: Consider algebra L, with two subalgebras L_1 and L_2 such that for any $X \in L_1$ and $Y \in L_2$, $[X, Y] = 0$. The algebra L is *not simple*, and it can be written as a direct product: $L = L_1 \times L_2$.

Definition: A *simple* Lie algebra is an algebra that cannot be written as a direct product.

Since any algebra can be written as a direct product of simple Lie algebras, we can think about each of the simple algebras separately. A useful example is that of the $U(2)$ group. A $U(2)$ transformation corresponds to a rotation in two-dimensional, complex space. This group is not simple:

$$U(2) = SU(2) \times U(1). \tag{A.19}$$

Think, for example, about the rotation of a spinor. It can be separated into two parts: The $U(1)$ transformation is a phase multiplication of the spinor. The $SU(2)$ transformation is an internal rotation that affects the two spin components differently.

A.4 Roots and Weights

In this section, we discuss properties of the algebra and the representations. From this point on, we consider only irreps, and thus we do not distinguish between a representation and an irrep any longer.

Definition: The *Cartan subalgebra* is the maximal commuting subalgebra.

In practice, we obtain it by finding a basis where the number of the diagonal generators is maximal. In this basis, the Cartan subalgebra is defined to be the subset of all the diagonal generators. Obviously, these generators all commute with each other, and thus they constitute a subalgebra.

Definition: The number of generators in the Cartan subalgebra is called the *rank* of the algebra. At times, we also refer to it as the *rank* of the group.

Let us consider a few examples. Since the $U(1)$ algebra has only a single generator, it is of rank 1. $SU(2)$ is also rank 1. You can make one of its three generators (say S_3), diagonal, but not two of them simultaneously. $SU(3)$ is rank 2. We later elaborate on $SU(3)$ in much more detail.

Our next step is to introduce the terms *roots* and *weights*. We do that via an example. Consider the $SU(2)$ algebra. It has three generators. We usually choose S_3 to be in the Cartan subalgebra, and we can combine the two other generators, S_1 and S_2, to a raising and a lowering operator, $S^{\pm} = S_1 \pm iS_2$. Any representation can be defined by the eigenvalues under the operation of the generators in the Cartan subalgebra (in this case, S_3). For example, for the spin-1/2 representation, the eigenvalues are $-1/2$ and $+1/2$; For the spin-1 representation, the eigenvalues are -1, 0, and $+1$. Under the operation of the raising (S^+) and lowering (S^-) generators, an eigenstate of S_3 is moved to another. For example, for a spin-1 representation, we have $S^-|1\rangle \propto |0\rangle$.

Let us now consider a general Lie group of rank n.

Statement: Any representation is characterized by the possible eigenvalues of its eigenstates under the operation of the Cartan subalgebra: $|e_1, e_2, \ldots, e_n\rangle$.

Statement: We can assemble all the operators that are not in the Cartan subalgebra into lowering and raising operators. That is, when they act on an eigenstate, they either move it to another eigenstate or annihilate it.

Definition: The *weight vectors* (or simply *weights*) of a representation are the possible eigenvalues of the generators in the Cartan subalgebra.

Definition: The *roots* of the algebra are the various ways in which the generators move a state between the possible weights.

Statement: The weights completely describe the representation.

Statement: The roots completely describe the algebra.

Statement: The weights of the adjoint representation are the roots of the Lie algebra.

Note that both roots and weights live in an n-dimensional vector space, where n is the rank of the group. The number of roots is the dimension of the group. The number of weights is the dimension of the representation.

A.5 $SU(2)$

In this section, we review $SU(2)$. The fundamental irrep can be explicitly written in terms of the Pauli matrices, $X_a = \sigma_a/2$, with

$$\sigma_1 = \begin{pmatrix} 0 & 1 \\ 1 & 0 \end{pmatrix}, \qquad \sigma_2 = \begin{pmatrix} 0 & -i \\ i & 0 \end{pmatrix}, \qquad \sigma_3 = \begin{pmatrix} 1 & 0 \\ 0 & -1 \end{pmatrix}. \tag{A.20}$$

The Pauli matrices are traceless, and only one of them is diagonal, as is expected given that $SU(2)$ has rank 1. Given that, we can get the weight diagram of the fundamental representation. We take the two eigenvectors,

$$\begin{pmatrix}1\\0\end{pmatrix}, \qquad \begin{pmatrix}0\\1\end{pmatrix}, \tag{A.21}$$

and apply to them the generator in the Cartan subalgebra, X_3. We find the two weights, $\pm 1/2$, which can be drawn as

$$\begin{array}{cc} \bullet & \bullet \\ -1/2 & +1/2 \end{array} \quad . \tag{A.22}$$

The nonzero roots are the combination of generators that move us between the weights. Here, they are the known raising and lowering generators:

$$X_\pm = \sigma_\pm/2, \qquad \sigma_\pm = \frac{1}{2}(\sigma_1 \pm i\sigma_2). \tag{A.23}$$

They change the weight by ± 1, so the root diagram has one zero root (correspond to σ_3) and ± 1, which we can draw as

$$\begin{array}{ccc} \bullet & \bullet & \bullet \\ -1 & 0 & +1 \end{array} \quad . \tag{A.24}$$

This root diagram is also the weight diagram of the adjoint representation, also known as spin-1.

A.6 $SU(3)$

In this section, we review $SU(3)$. It is more complicated than $SU(2)$, and it allows us to demonstrate a few aspects of Lie groups that cannot be demonstrated with $SU(2)$. Moreover, it is relevant to particle physics.

We can gain some intuition about $SU(3)$ by thinking about it as rotations in a three-dimensional, complex space. Similar to $SU(2)$, the full symmetry of these rotations is called $U(3)$, and it can be written as a direct product of simple groups, $U(3) = SU(3) \times U(1)$. The $SU(3)$ algebra has eight generators. You can see it by recalling that rotation in a complex space is done by unitary matrices, and any unitary matrix can be written with a Hermitian matrix in the exponent. There are nine independent Hermitian 3×3 matrices. They can be separated to a unit matrix, which corresponds to the $U(1)$ part, and eight traceless matrices, which correspond to the $SU(3)$ part.

Similar to the use of the Pauli matrices for the fundamental representation of $SU(2)$, the fundamental representation of $SU(3)$ is usually written in terms of the Gell-Mann matrices:

$$X_a = \lambda_a/2, \tag{A.25}$$

with

$$\lambda_1 = \begin{pmatrix}0&1&0\\1&0&0\\0&0&0\end{pmatrix}, \qquad \lambda_2 = \begin{pmatrix}0&-i&0\\i&0&0\\0&0&0\end{pmatrix},$$

$$\lambda_3 = \begin{pmatrix} 1 & 0 & 0 \\ 0 & -1 & 0 \\ 0 & 0 & 0 \end{pmatrix}, \qquad \lambda_4 = \begin{pmatrix} 0 & 0 & 1 \\ 0 & 0 & 0 \\ 1 & 0 & 0 \end{pmatrix},$$

$$\lambda_5 = \begin{pmatrix} 0 & 0 & -i \\ 0 & 0 & 0 \\ i & 0 & 0 \end{pmatrix}, \qquad \lambda_6 = \begin{pmatrix} 0 & 0 & 0 \\ 0 & 0 & 1 \\ 0 & 1 & 0 \end{pmatrix},$$

$$\lambda_7 = \begin{pmatrix} 0 & 0 & 0 \\ 0 & 0 & -i \\ 0 & i & 0 \end{pmatrix}, \qquad \lambda_8 = \frac{1}{\sqrt{3}} \begin{pmatrix} 1 & 0 & 0 \\ 0 & 1 & 0 \\ 0 & 0 & -2 \end{pmatrix}. \tag{A.26}$$

We emphasize the following points here:

- The Gell-Mann matrices are traceless, as they should be.
- There are several $SU(2)$ subalgebras. One of them is manifest: λ_1, λ_2, and λ_3.
- $SU(3)$ is of rank 2: λ_3 and λ_8 constitute the Cartan subalgebra.

Having an explicit expression for the fundamental representation in our disposal, we can draw the weight diagram. To do so, we follow the steps that we used for $SU(2)$. We take the three eigenvectors:

$$\begin{pmatrix} 1 \\ 0 \\ 0 \end{pmatrix}, \quad \begin{pmatrix} 0 \\ 1 \\ 0 \end{pmatrix}, \quad \begin{pmatrix} 0 \\ 0 \\ 1 \end{pmatrix}, \tag{A.27}$$

and apply on them the two generators in the Cartan subalgebra, X_3 and X_8. We find the three weights:

$$\left(+\frac{1}{2}, +\frac{1}{2\sqrt{3}}\right), \quad \left(-\frac{1}{2}, +\frac{1}{2\sqrt{3}}\right), \quad \left(0, -\frac{1}{\sqrt{3}}\right). \tag{A.28}$$

We can plot this in a weight diagram in the $X_3 - X_8$ plane; see question A.5.

Once we have the weights, we can obtain the roots. They are the combinations of generators that move us between the weights. The two roots that are in the Cartan subalgebra are at the origin. The other six are those that move us among the three weights:

$$\left(\pm\frac{1}{2}, \pm\frac{\sqrt{3}}{2}\right), \qquad (\pm 1, 0). \tag{A.29}$$

You are asked to plot them in question A.5. This root diagram is also the weight diagram of the adjoint representation. In terms of the Gell-Mann matrices, we can see that the raising and lowering generators are proportional to

$$I_{\pm} = \frac{1}{2}(\lambda_1 \pm i\lambda_2), \qquad V_{\pm} = \frac{1}{2}(\lambda_4 \pm i\lambda_5), \qquad U_{\pm} = \frac{1}{2}(\lambda_6 \pm i\lambda_7). \tag{A.30}$$

The names I, U, and V are, at this point, just names, but they are useful in some physics cases that you may encounter later in your study.

A.7 Classification and Dynkin Diagrams

The $SU(3)$ example allows us to obtain additional formal results. In the case of $SU(2)$, it is clear what the raising and lowering operators are. The generalization to groups with a higher rank is as follows.

Definition: A *positive (negative) root* is a root whose first nonzero component is positive (negative). A raising (lowering) operator corresponds to a positive (negative) root.

Definition: A *simple root* is a positive root that is not the sum of other positive roots.

Statement: Every rank-k algebra has k simple roots. Which ones they are is a matter of convention, but their relative lengths and angles are fixed.

In fact, it can be shown that the simple roots fully describe the algebra. It can be further shown that there are only four possible angles and corresponding relative lengths between simple roots:

Angle	90°	120°	135°	150°
Relative length	N.D.	$1:1$	$1:\sqrt{2}$	$1:\sqrt{3}$.

(A.31)

(N.D. stands for "not determined.") These rules can be visualized using Dynkin diagrams. Each simple root is described by a circle. The angle between two roots is described by the number of lines connecting the circles:

90° ○ ○ 120° ○—○ 135° ○═● 150° ○≡●, (A.32)

where the solid circle in a link represents the larger root between the two.

There are seven classes of Lie groups. Four classes include an infinite number of Lie groups. The three remaining classes, called the *exceptional groups*, have only a finite number of Lie groups apiece. In equations (A.33) and (A.34) you can find all the sets. The number of circles is the rank of the group. Note that different names for the infinite groups are used in the physics and mathematics communities. Here, we give both names, but we use only the physics names from now on. In these diagrams, k is the rank of the group:

SU$(k+1)$ $[A_k]$ ○—○—··· —○—○

Sp$(2k)$ $[B_k]$ ○—○—··· —○═●

SO$(2k+1)$ $[C_k]$ ●—●—··· —●═○

SO$(2k)$ $[D_k]$ ○—○—··· —○—○ (with an additional ○ attached to the second-to-last circle) (A.33)

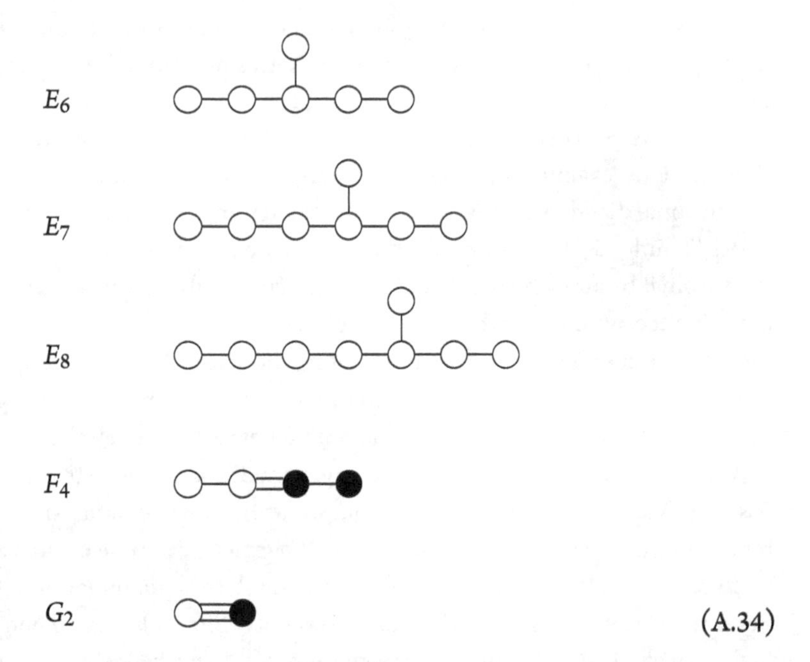

(A.34)

Consider, for example, $SU(3)$. The two simple roots are equal in length and have an angle of 120° between them. Thus, the Dynkin diagram is just ○—○.

Dynkin diagrams also provide a very good tool to tell us about the subalgebras of a given algebra. We do not describe the procedure in detail here, but only make the following simple point: removing a simple root always corresponds to a subalgebra. For example, removing simple roots, one obtains the following breaking pattern, which is relevant to Grand Unified Theory (GUT):

$$E_6 \to SO(10) \to SU(5) \to SU(3) \times SU(2). \tag{A.35}$$

Finally, we mention that the algebras of some small groups are identical:

$$SU(2) \simeq SO(3) \simeq Sp(2), \quad SU(4) \simeq SO(6), \quad SO(4) \simeq SU(2) \times SU(2),$$
$$SO(5) \simeq Sp(4). \tag{A.36}$$

A.8 Naming Representations

We are now back to discussing representations. How do we name an irrep? In the context of $SU(2)$, which is rank 1, there are three ways to do so.

1. We denote an irrep by its highest weight. For example, spin-0 denotes the singlet representation, spin-1/2 refers to the fundamental representation, where the highest weight is 1/2, and spin-1 refers to the adjoint representation, where the highest weight is 1.

2. We can define the irrep according to the dimension of the representation matrices, which is also the number of weights. Then the singlet representation is denoted by **1**, the fundamental by **2**, and the adjoint by **3**.

3. We can name the representation based on the number of times we can apply S_- to the highest weight without annihilating it. In this notation, the singlet is denoted as (0), the fundamental as (1), and the adjoint as (2).

Before we proceed, let us explain in more detail what we mean by "annihilating the state." Let us examine the weight diagram. In $SU(2)$, which is rank-1, this is a one-dimensional diagram. For example, for the fundamental representation, it has two entries, at $+1/2$ and $-1/2$. We now take the highest weight (in our example, $+1/2$), and move away from it by applying the root that corresponds to the lowering operator, -1. When we apply it once, we move to the lowest weight, $-1/2$. When we apply it once more, we move out of the weight diagram, thus annihilating the state. Thus, for the spin-1/2 representation, we can apply the root corresponding to S_- once to the highest weight before moving out of the weight diagram, and—in naming scheme 3—we call the representation (1).

We are now ready to generalize this to general Lie algebras. Either method 2 or method 3 is used. Method 2 is straightforward but somewhat problematic, as there could be several representations with the same dimension. We give an example of such a situation later.

In scheme 3, the notation used for a specific state is unambiguous. To use it, we must order the simple roots in a well-defined (even if arbitrary) order. Then we have a unique highest weight. We denote a representation of a rank-k algebra as a k-tuple, such that the first entry is the maximal number of times that we can apply the first simple root to the highest weight before the state is annihilated; the second entry refers to the maximal number of times that we can apply the second simple root on the highest weight before annihilation; and so on.

Take $SU(3)$ as an example. We order the Cartan subalgebra as X_3, X_8, and the two simple roots as

$$S_1 = \left(+\frac{1}{2}, +\frac{\sqrt{3}}{2}\right), \qquad S_2 = \left(+\frac{1}{2}, -\frac{\sqrt{3}}{2}\right). \tag{A.37}$$

Consider the fundamental representation, where we choose the highest weight to be $\left(1/2, 1/(2\sqrt{3})\right)$. Subtracting S_1 twice or subtracting S_2 once from the highest weight would annihilate it. Thus the fundamental representation is denoted by $(1, 0)$. You can work out the case of the adjoint representation and find that it should be denoted as $(1, 1)$. In fact, it can be shown that any pair of nonnegative integers forms a different irrep. (For $SU(2)$ with naming scheme 3, any nonnegative integer defines a different irrep.)

From now on, we limit our discussion mostly to $SU(N)$.

Statement: For any group, the singlet irrep is $(0, 0, \ldots, 0)$.

Statement: For any $SU(N)$ algebra, the fundamental representation is $(1, 0, 0, \ldots, 0)$.

Statement: For any $SU(N)$ algebra with $N \geq 3$, the adjoint representation is $(1, 0, 0, \ldots, 1)$.

Definition: For any irrep of an $SU(N)$, the *conjugate representation* is one where the order of the k-tuple is reversed.

For example, $(0, 1)$ is the conjugate of the fundamental representation, which is usually called the *antifundamental representation*. An irrep and its conjugate have the same dimension. In the naming scheme 2, they are called m and $\overline{m}$.

Definition: If $m = \overline{m}$, then m is called a *real irrep.*

Definition: If $m \neq \overline{m}$, then m and $\overline{m}$ are called a *complex irrep.*

For example, the singlet and the adjoint representation are real. Also, note that all the irreps of $SU(2)$ are real.

We now return to the notion that the groups that we are dealing with are transformation groups of physical states. These physical states are often just fields or particles. For example, when we talk about the $SU(2)$ group that is related to the spin transformations, the physical system that is being transformed is often that of a single particle with well-defined spin. We often say, for example, that a particle is in the spin-1/2 representation of $SU(2)$. What we mean is that, as a state in the Hilbert space, it transforms via the spin operator in the 1/2 representation of $SU(2)$. Similarly, when we say that the proton and the neutron form a doublet of isospin-$SU(2)$ (see section 10.4.1 in chapter 10), we mean that we represent p by the vector state $(1,0)^T$ and n by the vector state $(0,1)^T$, so the appropriate representation of the isospin generators is that of the 2×2 Pauli matrices. In other words, we loosely speak of "particles in a representation," when we mean "the representation of the group generators acting on the vector states that describe these particles."

Asking how many particles are in a given irrep is the equivalent of asking how many degrees of freedom (DoF) it has, or what the dimension of the irrep is. Here, we consider only $SU(N)$, and use naming convention 3. We only state the results.

- Consider an (α) representation of $SU(2)$. Its dimension is

$$N = \alpha + 1. \tag{A.38}$$

  The singlet (0), fundamental (1), and adjoint (2) representations have 1, 2, and 3 DoF, respectively.
- Consider an (α, β) representation of $SU(3)$. Its dimension is

$$N = (\alpha + 1)(\beta + 1)\,\frac{\alpha + \beta + 2}{2}. \tag{A.39}$$

  The singlet $(0,0)$, fundamental $(1,0)$, and adjoint $(1,1)$ representations have 1, 3, and 8 DoF, respectively.
- Consider an (α, β, γ) representation of $SU(4)$. Its dimension is

$$N = (\alpha + 1)(\beta + 1)(\gamma + 1)\,\frac{\alpha + \beta + 2}{2}\,\frac{\beta + \gamma + 2}{2}\,\frac{\alpha + \beta + \gamma + 3}{3}. \tag{A.40}$$

  The singlet $(0,0,0)$, fundamental $(1,0,0)$, and adjoint $(1,0,1)$ representations have 1, 4, and 15 DoF, respectively. Note that there is no $\alpha + \gamma + 2$ factor. Only a consecutive sequence of the label integers appears in any factor.
- The generalization to any $SU(N)$ is straightforward. It is easy to see that the fundamental of $SU(N)$ has N DoF and the adjoint has $N^2 - 1$ DoF.

In $SU(2)$, the number of DoF in a representation is unique. In a general Lie group, however, the case may be different. Yet it is often used to identify irreps. For example, in

$SU(3)$ we usually call the fundamental 3 and the adjoint 8. For the antifundamental, we use $\bar{3}$. In cases where there are several irreps with the same dimension, we often use a prime to distinguish them. For example, the dimension of both (4, 0) and (2, 1) is 15, which we denote by 15 and 15′, respectively.

We next look at how the count goes in terms of subgroup irreps. For example, any $SU(3)$ irreps can be written in terms of their $SU(2)$ subgroup irreps. Two useful decompositions are $3 = 2 + 1$ and $8 = 3 + 2 + 2 + 1$, where the irreps on the left are those of $SU(3)$ and those on the right are those of $SU(2)$. You can deduce them from the weight diagrams of the $SU(3)$ irreps simply by moving only in one of the $SU(2)$ subgroup directions on the diagrams (e.g., only moving in the X_3-direction). More generally, there is a way to decompose any irrep of a group as a sum of irreps of its subgroups. We do not elaborate on this any further.

Finally, we mention that for $U(1)$, we use the charge to define the irrep. The singlet is the $q = 0$ state, but there is no notion of the fundamental or the adjoint. Conjugate states have opposite charges. All the irreps of $U(1)$ have 1 DoF, and only the singlet is real.

A.9 Combining Representations

When we study spin, we learn how to combine $SU(2)$ representations. The canonical example is that of combining two spin-1/2s to generate a singlet (spin-0) and a triplet (spin-1). We often write it as $1/2 \times 1/2 = 0 + 1$. There is a similar method to combine representations for any Lie group. The basic idea is, just as in $SU(2)$, that we need to find all the possible ways to combine the indexes and then assign them to the various irreps. That way, we know what irreps are in the product representation and the corresponding Clebsch-Gordan (CG) coefficients.

Here, however, we do not explain how to construct the product representation. The reason for this is that, often, all we want to know is what irreps appear in the product representation, without the need to get all the CG coefficients. (In particular, we often only care about generating the singlet.) There is a simple way to do just this for a general $SU(N)$. This method is called *Young tableaux* or *Young diagrams*. The details of the method are well explained in several places, such as in the Particle Data Group (PDG). In question A.8, you are asked to learn how to use it.

Here, we give a few examples of combining irreps in some simple cases. For $U(1)$, when combining several irreps, the result is one irrep with a charge that is the sum of all the charges. For example, combining two states, one with charge a and one with charge b, the result is a state with charge $a + b$.

For $SU(2)$, combining two irreps, (a) and (b) with $a \geq b$, written in naming scheme 2, we get

$$a \times b = \sum_{k=1}^{b} a - b - 1 + 2k. \qquad \text{(A.41)}$$

The following results are generic to any non-Abelian group:

- When combining two nonsinglet irreps, the result is a direct sum of the irreps. For $SU(2)$, combining (a) and (b) with $a \geq b$, there are b such irreps.
- Combining a singlet to any irrep does not change it (i.e., $(a) \times (0) = (a)$).
- The product of the dimensions of the irreps on the left-hand side of equation (A.41) is equal to the sum of the dimensions on the right-hand side.

For $SU(3)$, using naming scheme 2, we have

$$3 \times \bar{3} = 1 + 8, \qquad 3 \times 3 = \bar{3} + 6, \qquad 3 \times 6 = 10 + 8. \tag{A.42}$$

From this, we can conclude that

$$3 \times 3 \times 3 = 10 + 8 + 8 + 1. \tag{A.43}$$

Note that the number of DoF on both sides are equal, as they should be. Of particular interest to us is that $3 \times \bar{3}$ and $3 \times 3 \times 3$ both contain the singlet irrep.

When we combine irreps, the results have well-defined symmetry properties under exchange of two of them (i.e., they are even or odd under an exchange). For example, when combining two spin doublets, the singlet is odd, while the triplet is even under the exchange. For $SU(2)$, the rule is simple: the highest irrep is symmetric, and then going down, it alternates. For other groups, it is more complicated, and we do not discuss this point here. When it is relevant, a subscript is usually added: s for a symmetric representation and a for an antisymmetric one. For example, in $SU(2)$, we have

$$2 \times 2 = 1_a + 3_s, \qquad 3 \times 3 = 5_s + 3_a + 1_s, \tag{A.44}$$

and in $SU(3)$, we have

$$3 \times 3 = \bar{3}_a + 6_s, \qquad 3 \times 3 \times 3 = 10_s + 8_m + 8_m + 1_a, \tag{A.45}$$

where 8_m refers to a mixed symmetry. (A mixed symmetry state is one where some of the exchanges are symmetric and some are antisymmetric. Such states can appear when we combine three or more irreps.)

The symmetry properties are important when the two irreps that we combine are identical. For example, the antisymmetric combination of two vectors in real, three-dimensional space is the cross product. Being antisymmetric, it vanishes identically for two identical vectors, $\vec{a} \times \vec{a} = 0$. This result is general, and it applies to all fully antisymmetric combinations in any group.

Combining representations plays a very important role in physics. In particular, we would like to find representations that can be combined into the singlet one, as such a combination is invariant under rotation in the relevant space. For that purpose, we mention the following properties:

Statement: The combination of an irrep with its conjugate contains the singlet, $1 \in R \times \overline{R}$.

Statement: The combination of two irreps that are not the conjugate of each other (i.e., $R_1 \neq \overline{R}_2$) does not contains the singlet, $1 \notin R_1 \times R_2$.

Finally, we discuss combining the irreps of nonsimple Lie groups. In such a case, we combine the irreps for each simple group separately and then have all the combinations. An $SU(3) \times SU(2)$ example should make this statement clear:

$$(3,2) \times (\bar{3},2) = (8,3) + (8,1) + (1,3) + (1,1). \tag{A.46}$$

Our notation is such that $(3,2)$ refers to a triplet under $SU(3)$ and a doublet under $SU(2)$. Note that the number of DoF is the same on both sides: $6 \times 6 = 24 + 8 + 3 + 1$. (More examples are given in question A.9.)

For Further Reading

For reviews of Lie groups that are very useful for particle physics, see Georgi [60], Zee [61], Slansky [62], Ramond [63], and chapter 4 of Goldberg [11].

Problems

Question A.1: S_3

In this question, we study the group S_3, which is the smallest finite non-Abelian group. One can think about it as all possible permutations of three elements. The group has six elements. In the language of permutations, we get the following representation of the group:

$$() = \begin{pmatrix} 1 & 0 & 0 \\ 0 & 1 & 0 \\ 0 & 0 & 1 \end{pmatrix} \qquad (12) = \begin{pmatrix} 0 & 1 & 0 \\ 1 & 0 & 0 \\ 0 & 0 & 1 \end{pmatrix}$$

$$(13) = \begin{pmatrix} 0 & 0 & 1 \\ 0 & 1 & 0 \\ 1 & 0 & 0 \end{pmatrix} \qquad (23) = \begin{pmatrix} 1 & 0 & 0 \\ 0 & 0 & 1 \\ 0 & 1 & 0 \end{pmatrix}$$

$$(123) = \begin{pmatrix} 0 & 1 & 0 \\ 0 & 0 & 1 \\ 1 & 0 & 0 \end{pmatrix} \qquad (321) = \begin{pmatrix} 0 & 0 & 1 \\ 1 & 0 & 0 \\ 0 & 1 & 0 \end{pmatrix} \tag{A.47}$$

The names are informative. For example, (12) represents exchanging the first and second elements and (123) and (321) are cyclic permutations to the right or left.

1. Write explicitly the 6×6 multiplication table for the group.
2. Show that the group is non-Abelian. Note that to prove this, it is enough to find one example.
3. Z_3 is a subgroup of S_3. Find three generators that correspond to Z_3.

4. Equation (A.7) states the following theorem for finite groups:

$$\sum_i [\dim(R_i)]^2 = N, \tag{A.48}$$

where N is the number of elements in the group and R_i runs over all the irreps. Based on this, prove that the representation in equation (A.47) is reducible. Then, write it explicitly in a $(1+2)$ block diagonal representation. (Hint: Find a vector that is an eigenvector of all these matrices.)

5. In the previous item, you found a two-dimensional and a one-dimensional irrep of S_3. Based on equation (A.48), there is only one more irrep, and it is one-dimensional. Find it.

Question A.2: Lie algebras

Consider two general elements of a Lie group:

$$A \equiv \exp(i\lambda X_a), \qquad B \equiv \exp(i\lambda X_b), \tag{A.49}$$

where X_i is a generator and we take λ to be a small parameter. Then, consider a third element in the group:

$$C = BAB^{-1}A^{-1} \equiv \exp(i\beta_c X_c). \tag{A.50}$$

Expand C in powers of λ and show that at the lowest order, you get the Lie algebra (see equation (A.15)),

$$[X_a, X_b] = if_{abc}X_c, \qquad f_{abc} \equiv \frac{\beta_c}{\lambda^2}. \tag{A.51}$$

Question A.3: The adjoint irrep

The structure constants constitute an irrep of their group. In this question, you are asked to show it explicitly.

1. All Lie algebras satisfy the Jacobi identity, equation (A.16). Using equation (A.15), $[X_a, X_b] = if_{abc}X_c$, show that the Jacobi identity leads to the following identity in terms of the structure constants:

$$f_{ade}f_{bcd} + f_{bde}f_{cad} + f_{cde}f_{abd} = 0. \tag{A.52}$$

2. We define the generators of the adjoint group as

$$(X_c)_{ab} \equiv -if_{abc}. \tag{A.53}$$

Show that

$$[X_a, X_b]_{ed} = if_{abc}(X_c)_{ed}; \tag{A.54}$$

that is, the generators satisfy the algebra of the group.

3. There are two equivalent sets of the generators of the adjoint of $SU(2)$ that are commonly used: one that is usually used in QM,

$$L_1 = \frac{1}{\sqrt{2}}\begin{pmatrix} 0 & 1 & 0 \\ 1 & 0 & 1 \\ 0 & 1 & 0 \end{pmatrix} \quad L_2 = \frac{1}{\sqrt{2}}\begin{pmatrix} 0 & -i & 0 \\ i & 0 & -i \\ 0 & i & 0 \end{pmatrix} \quad L_3 = \begin{pmatrix} 1 & 0 & 0 \\ 0 & 0 & 0 \\ 0 & 0 & -1 \end{pmatrix}, \tag{A.55}$$

and one that corresponds to the structure constants,

$$(L_k)_{ij} = -i\varepsilon_{ijk}. \tag{A.56}$$

Show that the two sets are equivalent by finding the matrix U that rotates between the bases.

Question A.4: More on the adjoint representation of SU(2)

Here, we consider a vector in real space, $\vec{a} = (a_x, a_y, a_z)$. It transforms as a vector under $SU(2)$:

$$a \to a' = U_V a, \qquad U_V = (e^{i\alpha_k X_k}), \tag{A.57}$$

where X_k refers to the generators of $SU(2)$ in the adjoint representation and α_k refers to the three rotation angles. Instead of the three-vector a, we can equivalently use a 2×2 traceless matrix A. (Note that both the vector and the traceless matrix have 3 DoF.) To prove this, we define

$$A \equiv \sigma \cdot a, \tag{A.58}$$

where σ are the Pauli matrices. Explicitly, we write

$$A_{ij} \equiv (\sigma_k)_{ij} a_k. \tag{A.59}$$

Next, you are asked to show that there is a mapping between a and A that holds under rotations.

1. Write A explicitly as a 2×2 matrix, and check that it is traceless.
2. Find the inverse operation (i.e., show how to get the vector a, given the matrix A).
3. A transforms as

$$A \to A' = U_M^\dagger A U_M, \qquad U_M = e^{i\alpha_k \sigma_k/2}. \tag{A.60}$$

Consider a rotation around the z-axis by an angle θ. Write explicitly the corresponding rotation matrices: U_V for a in equation (A.57) and U_M for A in equation (A.60).

4. Consider a unit vector in the x-direction:

$$a = (1, 0, 0), \quad A = \begin{pmatrix} 0 & 1 \\ 1 & 0 \end{pmatrix}. \tag{A.61}$$

Write explicitly a' of equation (A.57) and A' of equation (A.60) for rotation around the z-axis by angle θ. Using equation (A.58), show that A' corresponds to a'. This result demonstrates that the mapping between a and A is valid also under rotation.

Question A.5: SU(3)

1. Draw the weight diagram of the fundamental irrep of $SU(3)$.
2. Draw the root diagram of $SU(3)$.
3. The three matrices, $a\lambda_1$, $a\lambda_2$, and $a\lambda_3$, where λ_i are the Gell-Mann matrices and a is a constant, satisfy an $SU(2)$ algebra, also known as *Isospin*. What is a?
4. Explain why this fact does not imply that $SU(3)$ is not a simple Lie group.
5. There are other independent combinations of Gell-Mann matrices that satisfy $SU(2)$ algebras. Here, we discuss two of them, called *U-spin* and *V-spin*. Their raising operators are given by

$$V_+ = \frac{1}{2}(\lambda_4 + i\lambda_5), \qquad U_+ = \frac{1}{2}(\lambda_6 + i\lambda_7). \tag{A.62}$$

   Find these algebras. Hint: Look at the root diagram.

Question A.6: Dynkin diagrams

1. Argue that $SO(N)$ has $N(N-1)/2$ generators. This can be seen by the fact that $SO(N)$ corresponds to rotations in real, N-dimensional spaces.
2. Draw the Dynkin diagram of $SO(10)$. Note that you do not need to derive it—just use the general results of equation (A.33).
3. What is the rank of $SO(10)$?
4. Based on the Dynkin diagram, show that $SO(10)$ has the following subalgebras:

$$SO(8), \qquad SU(5), \qquad SU(4) \times SU(2), \qquad SU(3) \times SU(2) \times SU(2). \tag{A.63}$$

   In each case, show which simple root you can remove from the $SO(10)$ Dynkin diagram.

Question A.7: Representations

Here, we practice finding the number of DoF in a given irrep.

1. Find the dimension of the $(1, 1, 0, 0)$ irrep of $SU(5)$.
2. Find the number of DoF in the following $SU(3)$ irreps:

$$(3, 0), \qquad (2, 2). \tag{A.64}$$

3. Consider the $(3, 0)$ irrep of $SU(3)$. Draw its weight diagram and use the diagram to decompose the irrep into its $SU(2)$ irreps.

Question A.8: Combining irreps

Here, we practice the use of Young tableaux. The details of this method can be found in the PDG. Study the algorithm and perform the following calculations. Make sure to check that the number of DoF is the same on both sides. Write your answer in terms of both naming schemes 2 and 3. For example, for $SU(3)$, you should write

$$(1,0) \times (0,1) = (0,0) + (1,1), \qquad 3 \times \bar{3} = 1 + 8. \tag{A.65}$$

1. For $SU(3)$, calculate 3×3, 3×8, and $\overline{10} \times 8$.
2. For $SU(5)$, calculate $\bar{5} \times 10$ and 10×10.

Question A.9: Combining irreps in nonsimple groups

Consider $SU(3) \times SU(2) \times U(1)$. We denote a representation as $(R_3, R_2)_Q$, such that R_N is the representation under $SU(N)$ in naming scheme 2, and Q is the charge under $U(1)$.

1. Calculate

$$(3,2)_{+1} \times (3,2)_{+1}, \qquad (3,2)_{+1} \times (\bar{3},2)_{-1}, \qquad (3,2)_{+1} \times (\bar{3},1)_{+2} \times (1,2)_{-3}. \tag{A.66}$$

   Which cases contain the singlet?
2. In each case, verify that the number of DoF is the same on both sides (i.e., write the number of DoF for each term).

Question A.10: Looking for singlets

Consider an irrep, R. Where relevant, we use naming scheme 2.

1. Using the statements made above equation (A.46), show that $1 \in R \times R$ if and only if R is a real irrep.
2. All the irreps of $SU(2)$ are real. Show that indeed for any $SU(2)$ irrep R, $1 \in R \times R$.
3. Consider R to be an adjoint of $SU(N)$. Is $1 \in R \times R$?

Question A.11: Singlets in SU(3)

As discussed in chapter 5, the strong interaction is described by an $SU(3)$ gauge group. The particles observed in nature are bound states that are singlets under that $SU(3)$. We assign the quark (q), antiquark ($\bar{q}$), and gluon (g) fields to the triplet, anti-triplet, and octet irreps, respectively:

$$q(3), \qquad \bar{q}(\bar{3}), \qquad g(8). \tag{A.67}$$

Which of the following could be an observable bound state?

$$q\bar{q}, \qquad qq, \qquad qg, \qquad gg, \qquad q\bar{q}g, \qquad qqq. \tag{A.68}$$

Note that you do not need to calculate much—all you need is to use the properties that you proved in question A.10 and the results of equation (A.42).

BIBLIOGRAPHY

[1] Particle Data Group, R. L. Workman, V. D. Burkert, V. Crede, E. Klempt, U. Thoma, L. Tiator, et al. "Review of Particle Physics," *PTEP* 2022, no. 8 (2022): 083C01. https://doi.org/10.1093/ptep/ptac097.

[2] Peskin, M. E., and D. V. Schroeder. *An Introduction to Quantum Field Theory*. Addison-Wesley, 1995.

[3] Georgi, H. *Weak Interactions and Modern Particle Theory*. Dover Publications, 1984.

[4] Quigg, C. *Gauge Theories of the Strong, Weak, and Electromagnetic Interactions: Second Edition*. Princeton University Press, 2013.

[5] Peskin, M. E. *Concepts of Elementary Particle Physics*. Oxford University Press, 2019.

[6] Burgess, C. P., and G. D. Moore. *The Standard Model: A Primer*. Cambridge University Press, 2006.

[7] Langacker, P. *The Standard Model and Beyond*. CRC Press, 2009.

[8] Ramond, P. *Journeys beyond the Standard Model*. CRC Press, 1999.

[9] Cottingham, W. N., and D. A. Greenwood. *An Introduction to the Standard Model of Particle Physics*. Cambridge University Press, 2007.

[10] Donoghue, J. F., E. Golowich, and B. R. Holstein. *Dynamics of the Standard Model*. Cambridge University Press, 1994.

[11] Goldberg, D. *The Standard Model in a Nutshell*. Princeton University Press, 2017.

[12] Buras, A. J. *Gauge Theory of Weak Decays: The Standard Model and the Expedition to New Physics Summits*. Cambridge University Press, 2020.

[13] Zee, A. *Quantum Field Theory in a Nutshell*. Princeton University Press, 2003.

[14] Srednicki, M. *Quantum Field Theory*. Cambridge University Press, 2007.

[15] Schwartz, M. D. *Quantum Field Theory and the Standard Model*. Cambridge University Press, 2014.

[16] Ramond, P. *Field Theory: A Modern Primer*. Westview Press, 2001.

[17] Dine, M. *Supersymmetry and String Theory: Beyond the Standard Model*. Cambridge University Press, 2016.

[18] Nagashima, Y. *Elementary Particle Physics*. Vol. 1, *Quantum Field Theory and Particles*. Wiley-VCH, 2010.

[19] Nagashima, Y. *Elementary Particle Physics*. Vol. 2, *Foundations of the Standard Model*. Wiley-VCH, 2010.

[20] Petrov, A. A., and A. E. Blechman. *Effective Field Theories*. World Scientific Publishing Co., 2016

[21] Peskin, M. E. "Spin, Mass, and Symmetry." In *21st Annual SLAC Summer Institute on Particle Physics: Spin Structure in High-Energy Processes (SSI93)*, 1994. hep-ph/9405255: https://arxiv.org/pdf/hep-ph/9405255.pdf.

[22] Streater, R. and A. Wightman. *PCT, Spin and Statistics, and All That*. Princeton University Press, 2000.

[23] Eboli, O. J. P., and M. C. Gonzalez-Garcia. "Probing Trilinear Gauge Boson Interactions via Single Electroweak Gauge Boson Production at the CERN LHC." *Phys. Rev. D* 70 (2004): 074011. hep-ph/0405269: https://arxiv.org/pdf/hep-ph/0405269.pdf.

[24] Erler, J., and S. Su. "The Weak Neutral Current." *Prog. Part. Nucl. Phys.* 71 (2013): 119–149. arXiv:1303.5522 [hep-ph]: https://arxiv.org/pdf/1303.5522.pdf.

[25] Marciano, W. J., and Z. Parsa. "Neutrino Electron Scattering Theory." *J. Phys. G* 29 (2003): 2629. hep-ph/0403168: https://arxiv.org/pdf/hep-ph/0403168.pdf.

[26] Cahn, R., and G. Goldhaber. *The Experimental Foundations of Particle Physics.* Cambridge University Press, 2009.

[27] Hook, A. "TASI Lectures on the Strong CP Problem and Axions." *PoS* TASI2018, 004 (2019). arXiv:1812.02669 [hep-ph]: https://arxiv.org/pdf/1812.02669.pdf.

[28] CKMfitter Group, J. Charles, A. Höcker, H. Lacker, S. Laplace, F. R. Le Diberder, J. Malclès, et al. "CP Violation and the CKM Matrix: Assessing the Impact of the Asymmetric *B* Factories." *Eur. Phys. J.* C 41, no.1, 1–131 (2005). arXiv:hep-ph/0406184: https://arxiv.org/pdf/hep-ph/0406184.pdf. Updated results and plots available at: http://ckmfitter.in2p3.fr.

[29] Branco, G. C., L. Lavoura, and J. P. Silva. *CP Violation.* Oxford University Press, 2014.

[30] Nir, Y. "Flavour Physics and CP Violation." Proceedings of CLASHEP2009. arXiv:1010.2666 [hep-ph]: https://arxiv.org/pdf/1010.2666.pdf.

[31] Grossman, Y., and P. Tanedo. "Just a Taste: Lectures on Flavor Physics." Contribution to TASI 2016. arXiv:1711.03624 [hep-ph]: https://arxiv.org/pdf/1711.03624.pdf.

[32] Shifman, M. A. "Quark Hadron Duality." *PoS Heavy Flavor* 8 (1999). arXiv:hep-ph/0009131: https://arxiv.org/pdf/hep-ph/0009131.pdf.

[33] Manohar, A. V., and M. B. Wise. *Heavy Quark Physics.* Cambridge University Press, 2009.

[34] Grinstein, B. "An Introduction to the Theory of Heavy Mesons and Baryons." Contribution to TASI 1994. arXiv:hep-ph/9411275 [hep-ph]: https://arxiv.org/pdf/hep-ph/9411275.pdf.

[35] Neubert, M. "Heavy Quark Symmetry." *Phys. Rept.* 245 (1994): 259–396. arXiv:hep-ph/9306320 [hep-ph]: https://arxiv.org/pdf/hep-ph/9306320.pdf.

[36] Sterman, G. F. "QCD and Jets." Contribution to TASI 2004. arXiv:hep-ph/0412013 [hep-ph]: https://arxiv.org/pdf/hep-ph/0412013.pdf.

[37] Csaki, C., S. Lombardo, and O. Telem. "TASI Lectures on Non-supersymmetric BSM Models." Contribution to TASI 2016. arXiv:1811.04279 [hep-ph]: https://arxiv.org/pdf/1811.04279.pdf.

[38] Terning, J. *Modern Supersymmetry: Dynamics and Duality.* Oxford University Press, 2009.

[39] Haller, J., A. Hoecker, R. Kogler, K. Mönig, T. Peiffer, M. Schott, and J. Stelzer (The Gfitter Group). "A Generic Fitter Project for HEP Model Testing." The Gfitter Group. http://project-gfitter.web.cern.ch/project-gfitter/.

[40] Skiba, W. "TASI Lectures on Effective Field Theory and Precision Electroweak Measurements." Contribution to TASI 2009. arXiv:1006.2142 [hep-ph]: https://arxiv.org/pdf/1006.2142.pdf.

[41] Willenbrock, S. "Symmetries of the Standard Model." Contribution to TASI 2004. arXiv:hep-ph/0410370: https://arxiv.org/pdf/hep-ph/0410370.pdf.

[42] Logan, H. E. "TASI 2013 Lectures on Higgs Physics within and beyond the Standard Model." Contribution to TASI 2013. arXiv:1406.1786 [hep-ph]: https://arxiv.org/pdf/1406.1786.pdf.

[43] Gunion J. F., H. E. Haber, G. L. Kane and S. Dawson. *The Higgs Hunter's Guide.* Addison-Wesley, 2000.

[44] Isidori, G., Y. Nir, and G. Perez. "Flavor Physics Constraints for Physics beyond the Standard Model." *Ann. Rev. Nucl. Part. Sci.* 60 (2010): 355. arXiv:1002.0900 [hep-ph]: https://arxiv.org/pdf/1002.0900.pdf.

[45] Gonzalez-Garcia, M., and Y. Nir. "Neutrino Masses and Mixing: Evidence and Implications." *Rev. Mod. Phys.* 75 (2003): 345–402. arXiv:hep-ph/0202058 [hep-ph]: https://arxiv.org/pdf/hep-ph/0202058.pdf.

[46] Grossman, Y. "TASI 2002 Lectures on Neutrinos." Contribution to TASI 2002. arXiv:hep-ph/0305245 [hep-ph]: https://arxiv.org/pdf/hep-ph/0305245.pdf.

[47] de Gouvea, A. "TASI Lectures on Neutrino Physics." Contribution to TASI 2004. arXiv:hep-ph/0411274 [hep-ph]: https://arxiv.org/pdf/hep-ph/0411274.pdf.

[48] Vogel, P. "Neutrinoless Double Beta Decay." Contribution to TASI 2006. arXiv:hep-ph/0611243 [hep-ph]: https://arxiv.org/pdf/hep-ph/0611243.pdf.

[49] Conrad, J. M. "Neutrino Experiments." Contribution to TASI 2006. arXiv:0708.2446 [hep-ex]: https://arxiv.org/pdf/0708.2446.pdf.

[50] Beacom, J. F. "TASI Lectures on Astrophysical Aspects of Neutrinos." Contribution to TASI 2006. arXiv:0706.1824 [astro-ph]: https://arxiv.org/pdf/0706.1824.pdf.

[51] Kolb, E. W., and M. S. Turner. *The Early Universe*. CRC Press, 1994.

[52] Carroll, S. M. "TASI Lectures: Cosmology for String Theorists." Contribution to TASI 1999. arXiv:hep-th/0011110 [hep-th]: https://arxiv.org/pdf/hep-th/0011110.pdf.

[53] Trodden, M., and S. M. Carroll. "TASI Lectures: Introduction to Cosmology." Contribution to TASI 2003. arXiv:astro-ph/0401547: https://arxiv.org/pdf/astro-ph/0401547.pdf.

[54] Rubakov, V. A., and D. S. Gorbunov. *Introduction to the Theory of the Early Universe: Hot Big Bang Theory*. World Scientific, 2017.

[55] Gorbunov, D. S., and V. A. Rubakov. *Introduction to the Theory of the Early Universe: Cosmological Perturbations and Inflationary Theory*. World Scientific, 2011.

[56] Piattella, O. F. *Lecture Notes in Cosmology*. arXiv:1803.00070 [astro-ph.CO]: https://arxiv.org/pdf/1803.00070.pdf.

[57] Lisanti, M. "Lectures on Dark Matter Physics." Contribution to TASI 2015. arXiv:1603.03797 [hep-ph]: https://arxiv.org/pdf/1603.03797.pdf.

[58] Dolgov, A. "Neutrinos in Cosmology." *Phys. Rept.* 370 (2002): 333–535. arXiv:hep-ph/0202122 [hep-ph]: https://arxiv.org/pdf/hep-ph/0202122.pdf.

[59] Davidson, S., E. Nardi, and Y. Nir. "Leptogenesis." *Phys. Rept.* 466 (2008): 105–177. arXiv:0802.2962 [hep-ph]: https://arxiv.org/pdf/0802.2962.pdf.

[60] Georgi, H. *Lie Algebras in Particle Physics*. CRC Press, 1999.

[61] Zee, A. *Group Theory in a Nutshell for Physicists*. Princeton University Press, 2016.

[62] Slansky, R., "Group Theory for Unified Model Building." *Phys. Rept.* 79 (1981): 1.

[63] Ramond, P. *Group Theory: A Physicist's Survey*. Cambridge University Press, 2010.

INDEX

Printed in the USA
CPSIA information can be obtained
at www.ICGtesting.com
JSHW051945010124
54621JS00007B/37